Mark Fordek
-1985-

Alomir Favero
2017

VISCOPLASTICITY

Mechanics of plastic solids

Editor-in-chief: G.Æ. Oravas
Editor: J. Schroeder

1. Foundations of plasticity
   A. Sawczuk (ed.)

2. Problems of plasticity
   A. Sawczuk (ed.)

3. Introduction to the mechanics of plastic forming of metals
   W. Szczepiński

4. Limit analysis of structures at thermal cycling
   D.A. Gokhfeld and O.F. Cherniavsky

5. Viscoplasticity
   N. Cristescu and I. Suliciu

Viscoplasticity

N. CRISTESCU
University
of Bucharest

I. SULICIU
National Institute
for Scientific and Technical Creation,
Bucharest

1982

Martinus Nijhoff Publishers
The Hague / Boston / London

Distributors:

for the United States and Canada

Kluwer Boston, Inc.
190 Old Derby Street
Hingham, MA 02043
USA

for all other countries

Kluwer Academic Publishers Group
Distribution Center
P.O. Box 322
3300 AH Dordrecht
The Netherlands

For the Socialist Countries:

Hungary, Albania, Bulgaria, China, Cuba, Czechoslovakia, German Democratic Republic, Democratic People's Republic of Korea, Mongolia, Poland, Roumania, Soviet Union, Democratic Republic of Vietnam, and Yugoslavia

Editura Technica
Piata Scinteii 1
7000 Bucharest, Romania

Library of Congress Cataloging in Publication Data CIP

Cristescu, N.
Viscoplasticity.

(Mechanics of plastic solids ; 5)
Bibliography: p.
Includes index.
1. Viscoplasticity. I. Suliciu, I. (Ion)
II. Title. III. Series.
TA418.2.C743 1982 620.1'1233 82-18950
ISBN 90-247-2777-4

ISBN 90-247-2777-4 (this volume)
ISBN 90-247-2740-5(series)

This is a copublication by
Martinus Nijhoff Publishers
P.O. Box 566
2501 CN The Hague, The Netherlands

and

Editura Technica
Piata Scinteii 1
7000 Bucharest, Romania

Revised and reorganized English translation of Viscoplasticitate: Ciclul „ Mecanica teoretica si aplicata" (Romanian, published in 1976 by Editura Technica)

PRINTED IN ROMANIA

PREFACE

Interest in viscoplasticity, i.e. in mathematical models called viscoplastic, goes back to the 1930s when the mechanics of materials such as pastes, paint oils and dough were first investigated. Some research is even older and aims to describe creep of metals. Later elastic-viscoplastic models were applied to various materials such as metals, soils, polymers, rocks, mainly for quasistatic problems.

We would like to remind that the classical time-independent theory of plasticity is generally attributed to Barré de Saint-Venant, though in his early paper published in Journal de Mathématique Sér. II, vol. 16 (1871) he mentions the possibility to consider two additive components for strains, a plastic one and a viscous one. However, a significant effort to include time effects in constitutive equations describing plasticity was made in the last decades only.

The present volume is a revised and reorganized English version of a book first published in 1976 in Romanian. For instance, in the new version the chapters "Thermodynamics of rate-type constitutive equations" and "Some applications to metal working: drawing theory" are no longer included, mainly since both subjects are now rapidly developing so that in the next future each would need a separate treatment. The two new chapters added instead show the way in which mathematical models based on rate-type constitutive equations can be formulated to describe the corona losses in electric-transmission lines as well as the creep behaviour of rocks.

Most parts of the book are based on the research work by the authors and their collaborators.

Maybe a more accurate title for the book would have been: "A short introduction to the phenomenological modelling of solid deforming bodies with special references to dynamic problems". However, the present much shorter title "Viscoplasticity" was chosen not only for convenience but also in order to suggest the intention of the authors to present some mathematical models in order to describe the property of bodies exhibiting elastic, viscous and plastic deformation at the same time, as shown by experimental evidence.

A major part of this book constitutes an attempt to determine effective mathematical models by using experimental, computational (i.e. fitting of a constitutive equation to a set of experimental data as well as computational simulation) and theoretical procedures — the three main approaches nowadays in continuum physics research.

Chapter I contains general notions of continuum mechanics which allow the reader to follow the content of the book more easily. Chapter II presents several rate-independent and rate-dependent constitutive equations. In this chapter we do not discuss constitutive equations which depend directly on plastic work; the interested reader may find infoımation about this elsewhere (HILL [1950], OLSZAK, PERZYNA and SAWCZUK [1970], NOWACKI [1978], OWEN [1968], MROZ, SHRIVASTAVA and DUBEY [1976]). Initial- and boundary-value problems for the one-dimensional theory are formulated. One of the purposes of this chapter is also to prepare the reader for a better understanding of the experimental results discussed in Chapter III. Some of the results presented in Chapter III are used in Chapter IV for the formulation of the rate-type constitutive equation for the one-dimensional case. Some of the experimental results described are then qualitatively used in Chapter V in order to determine general restrictions on the form of constitutive equations. Chapters II to V deal with one-dimensional problems in the sense that only one spatial coordinate as well as only one component of the stress and strain tensors are involved. Chapters VI and VII discuss constitutive equations with internal state variables. A thermomechanical theory is presented here which allows finite speed of propagation for both mechanical and thermal waves. The relationship between these constitutive equations and other rheological models is considered. In Chapter VIII we present constitutive equations used in viscoplasticity. Some of these constitutive equations form a subclass of those discussed in Chapters VI and VII. The other constitutive equations, i.e. those which do not possess elastic-instantaneous response (eq. (8.3.1)), do not belong to this subclass. These are discussed in Chapter VII section 8 and in GURTIN and SULICIU [1976]. Chapter IX is devoted to the formulation of a rate type-model based on experimental data, which describes creep of rocks for both the change of shape and the compression of the volume. As mentioned above, Chapter X deals with mathematical models of electric-transmission lines, obtained by using rate-type constitutive equations. An effective model based on experimental data is also obtained in this case. For convenience, methods used in the book are given in the Appendix for the integration of non-linear hyperbolic equations typical for the theory of plasticity, where the major difficulty is the determination of the loading/unloading boundary.

The book will be of interest to students, engineers and research workers. Parts of the book have been used as text book for graduate courses at various universities (in Romania and in the United States).

The book has been prepared for printing as follows: Chapters III, IV, VIII, IX and the Appendix were written by N. Cristescu, Chapters I, II, V, VI, VII and X by I. Suliciu. The content of the whole book, however was largely discussed and worked out by both authors.

The authors express thanks to Dr. Cornelia Cristescu for help in preparing the manuscript and for performing most of the numerical computations and also to Dr. Mihaela Mihăilescu-Suliciu for help in preparing the manuscript as well as for many useful comments on the content of the book.

CONTENTS

Chapter I

PRELIMINARIES

1. THE ONE-DIMENSIONAL CASE

In some cases it is possible and useful to describe the motion of a material body $\mathscr{B}$ by means of only one spatial coordinate.

The general balance laws for the one-dimensional motion with coupled thermo-mechanical effects are represented by the integral equations $(1.1.5)_{1-2}$ or their differential form (1.1.6). Inequality $(1.1.5)_3$ or $(1.1.6)_3$ is a mathematical expression of the second law of thermodynamics.

During the body motion certain discontinuities may appear; there may be different causes to generate them, for instance the initial or boundary conditions, the form of the constitutive equations, etc. The formulas (1.1.10) and (1.1.11) represent the conditions that have to be satisfied by the thermo-mechanical quantities when crossing a shock wave; relations (1.1.12), (1.1.14) and (1.1.15) are the conditions to be satisfied when crossing an acceleration wave. To these two groups of jump conditions one must add the constitutive equations of the body under consideration.

1. A finite or infinite interval on the real line R will be called a *one-dimensional reference configuration* $\mathscr{R}$ of a body $\mathscr{B}$. A *one-dimensional motion* of the body $\mathscr{B}$ will be a mapping $\chi:\mathscr{R}\times I \to R$, where I is an interval on the real line, that is

$$x = \chi(X, t), \quad X \in \mathscr{R}, \quad t \in I \subset R, \tag{1.1.1}$$

with the property that for any fixed $t \in I$, $\chi(X, t)$ is injective and bicontinuous with respect to X. One will often assume that χ is of class C^2 with respect to (X, t). X is called the *initial* (or Lagrangean) *coordinate*, x is the *actual* (or Eulerian) *coordinate* and t denotes time. When χ is of class

C^1, one may introduce the functions

$$v = \dot{x} = \frac{\partial \chi}{\partial t},\qquad (1.1.2)$$

$$\varepsilon = \frac{\partial \chi}{\partial X} - 1,\qquad (1.1.3)$$

called the *particle velocity* and *strain*, respectively. The quantities

$$\sigma, b, e, \theta, \eta, \psi = e - \eta\theta, q, g = \frac{\partial \theta}{\partial X}, r\qquad (1.1.4)$$

are frequently used in the thermodynamics of a continuous body (see for instance COLEMAN and GURTIN [1965], III, TRUESDELL [1969] and also Chapter VI of the present book). In the context of this work they have the following meaning:

σ — is called the *stress* and it is the force per unit area in the reference configuration; it is oriented toward the exterior normal;

θ — is the *absolute temperature* and is always positive;

e — is the *internal energy* per unit volume in the reference configuration; *per unit mass ??*

η — is the *entropy* per unit volume in the reference configuration;

ψ — is the *free energy* per unit volume in the reference configuration;

q — is the *heat flux* across a unit area in the reference configuration and it is oriented toward the interior normal;

g — is the *temperature gradient* in the reference configuration;

b — is the *body force*;

r — is the *heat supply* received from the exterior of the body.

The *balance of momentum and energy* as well as the Clausius-Duhem nequality may be written as (see for instance COLEMAN and GURTIN [1965], III)

Momentum

$$\frac{d}{dt}\int_{X_1}^{X_2} \dot{x}\rho_0 dX = \int_{X_1}^{X_2} b\rho_0 dX + \sigma(X_2, t) - \sigma(X_1, t),$$

Energy

$$\frac{d}{dt}\int_{X_1}^{X_2}\rho_0\left(\frac{\dot{x}^2}{2} + e\right)dX = \int_{X_1}^{X_2}(\dot{x}b + r)\rho_0 dX + \sigma(X_2, t)\,\dot{x}\,(X_2, t) -$$

$$\qquad (1.1.5)$$

$$-\sigma(X_1, t)\,\dot{x}(X_1, t) - q(X_2, t) + q(X_1, t),$$

(C-D)

$$\frac{d}{dt}\int_{X_1}^{X_2}\eta dX \geqslant \int_{X_1}^{X_2}\frac{r}{\theta}\,\rho_0 dX + \frac{q(X_1, t)}{\theta(X_1, t)} - \frac{q(X_2, t)}{\theta(X_2, t)},$$

where ρ_0 is the *mass density* in the reference configuration.

14

Rate of Entropy Increase   *Rate of Entropy Input*

When all functions present in (1.1.5) are sufficiently smooth, these equations can be written in a differential form

Momentum
$$\rho_0 \ddot{x} - \rho_0 b = \frac{\partial \sigma}{\partial X},$$

Energy
$$\rho_0 \frac{\partial}{\partial t}\left(\frac{\dot{x}^2}{2} + e\right) = \rho_0(\dot{x}b + r) + \frac{\partial}{\partial X}(\sigma \dot{x}) - \frac{\partial q}{\partial X}, \qquad (1.1.6)$$

C-D
$$\dot{\eta} \geq \rho_0 \frac{r}{\theta} - \frac{\partial}{\partial X}\left(\frac{q}{\theta}\right) = \rho_0 \frac{r}{\theta} + \frac{q}{\theta^2}\frac{\partial \theta}{\partial X} - \frac{1}{\theta}\frac{\partial q}{\partial X}$$

In the following it will be convenient to write equations (1.1.6) by using the quantities v and ε. Thus, from (1.1.2), (1.1.3) and (1.1.6)$_{1-2}$, one has

Momentum
$$\rho_0 \frac{\partial v}{\partial t} - \frac{\partial \sigma}{\partial X} = \rho_0 b,$$

from 1.1.3 $\frac{\partial^2}{\partial x \partial t} = \frac{\partial^2}{\partial t \partial x}$

Kinematic
$$\frac{\partial \varepsilon}{\partial t} - \frac{\partial v}{\partial X} = 0, \qquad (1.1.7)$$

Energy
$$\rho_0 \frac{\partial e}{\partial t} - \sigma \frac{\partial \varepsilon}{\partial t} + \frac{\partial q}{\partial X} = \rho_0 r.$$

The system (1.1.7), as a system of partial differential equations with the unknown functions $v, \varepsilon, \sigma, e$ and q, is not complete. This system consisting of balance laws and compatibility equations only, has to be completed with the so called *constitutive equations*; this will be considered in the next chapters.

A substantial part of this book will be devoted to the study of purely mechanical phenomena, that is those phenomena in which thermal effects may be neglected. In this case we shall add to (1.1.7)$_{1-2}$ a constitutive equation relating σ and ε which will give us a complete system of three partial differential equations for the three unknown functions v, ε and σ.

2. A *wave* through the body $\mathscr{B}$ will be a smooth curve in the (X, t)-plane

$$\mathscr{C} = (Y(t), t), \qquad U(t) = \frac{dY(t)}{dt} = \dot{Y}(t), \qquad t \in I, \qquad (1.1.8)$$

with the property that certain quantities among $v, \varepsilon, \sigma, e, \eta, \psi, \theta, q, g$ or their derivatives have *jump discontinuities* across this curve and are continuous with respect to (X, t) elsewhere (a function $f(X, t)$ is said to have a jump discontinuity at (X, t) if, for t fixed, $\lim\limits_{\substack{Y \to X \\ Y < X}} f(Y, t)$ and $\lim\limits_{\substack{Y \to X \\ Y > X}} f(Y, t)$ exist, are finite and differ from each other).

A smooth curve $\mathscr{C}$ will be called a *shock wave* if the motion $\chi(X, t)$ is continuous across it but some of the quantities $v, \varepsilon, \sigma, e, \eta, \psi, \theta, q, g$ have jump discontinuities across $\mathscr{C}$, being smooth functions of (X, t) everywhere else. That is, if $f(X, t)$ denotes one of the above quantities, the limits

$$f^-(t) = \lim_{\substack{X \to Y(t) \\ X < Y(t)}} f(X, t), \quad f^+(t) = \lim_{\substack{X \to Y(t) \\ X > Y(t)}} f(X, t) \tag{1.1.9}$$

exist and

$$[f(t)] = f^+(t) - f^-(t) \neq 0 \tag{1.1.10}$$

for at least one of these quantities.

Let us apply the balance laws (1.1.5) in case of a shock wave $\mathscr{C}$ that propagates across a body $\mathscr{B}$, under the assumption that b and r are continuous across $\mathscr{C}$. We get the following relations between jumps (see for instance CHEN and GURTIN [1972]).

$$\begin{gathered} \rho_0 U[v] + [\sigma] = 0, \\ \rho_0 U\left[\frac{v^2}{2} + e\right] + [\sigma v] - [q] = 0, \\ U[\eta] \geqslant -\left[\frac{q}{\theta}\right], \end{gathered} \tag{1.1.11}$$

while (1.1.2) and (1.1.3) imply

$$U[\varepsilon] + [v] = 0. \tag{1.1.12}$$

Relations (1.1.11) are called *dynamic compatibility conditions* and (1.1.12) is called the *kinematic compatibility condition.*

Relations (1.1.11) are deduced from (1.1.5). Let us show for instance. how one establishes $(1.1.11)_1$. Let X_1 and X_2 be such that $X_1 < Y(t) < X_2$. Then $(1.1.5)_1$ can be written as

$$\frac{\mathrm{d}}{\mathrm{d}t}\int_{X_1}^{Y(t)} \rho_0 v(X, t)\mathrm{d}X + \frac{\mathrm{d}}{\mathrm{d}t}\int_{X_1}^{X_2} \rho_0 v(X, t)\mathrm{d}X = \int_{X_1}^{X_2} \rho_0 b(X, t)\mathrm{d}X + \\ + \sigma(X_2, t) - \sigma(X_1, t).$$

Differentiating the integrals in the standard manner and using the notation (1.1.9), one obtains

$$\rho_0 v^-(Y(t), t)\dot{Y}(t) - \rho_0 v^+(Y(t),t)\,\dot{Y}(t) + \int_{X_1}^{Y(t)} \rho_0 \frac{\partial v(X, t)}{\partial t} \mathrm{d}X +$$

$$+ \int_{Y(t)}^{X_2} \rho_0 \frac{\partial v(X, t)}{\partial t} \mathrm{d}X = \int_{X_1}^{X_2} \rho_0 b(X, t)\mathrm{d}X + \sigma(X_2, t) - \sigma(X_1, t)^{\cdot}$$

Thus, according to $(1.1.8)_2$ and (1.1.10), for $X_1 \to Y(t)$, $X_2 \to Y(t)$ one gets

$$-\rho_0 U[v] = [\sigma].$$

3. A smooth curve $\mathscr{C}$ given by an equation $Y = \varphi(t)$ will be called an *acceleration wave* if the motion $\chi(X, t)$ is of class C^1 everywhere while the velocity $v(X, t)$ and the strain $\varepsilon(X, t)$, as well as all quantities $\theta, \sigma, \psi, \eta$ and q are continuous functions but their derivatives may have jump discontinuities across $\mathscr{C}$, being continuous everywhere else but on $\mathscr{C}$.

As v and ε are continuous functions, we can write the *kinematic compatibility conditions*, to be satisfied when crossing the curve $\mathscr{C}$, in the form

$$\left[\frac{\partial v}{\partial t}\right] + c\left[\frac{\partial v}{\partial X}\right] = 0,$$

$$\left[\frac{\partial \varepsilon}{\partial t}\right] + c\left[\frac{\partial \varepsilon}{\partial X}\right] = 0, \tag{1.1.13}$$

where c denotes the slope of the curve $\mathscr{C}$ at $(\varphi(t), t)$

$$c = \frac{\mathrm{d}\varphi(t)}{\mathrm{d}t}, \tag{1.1.14}$$

and the square brackets denote, as they did above (see (1.1.10)), the jump of the respective quantity across $\mathscr{C}$. Since v and ε are the partial derivatives of the function χ, one obtains another compatibility condition

$$\left[\frac{\partial \varepsilon}{\partial t}\right] = \left[\frac{\partial v}{\partial X}\right]. \tag{1.1.15}$$

The *dynamic compatibility conditions* are obtained from (1.1.6) as follows: we consider relations (1.1.6) on both sides of the curve $\mathscr{C}$, we write their limit values at a point of $\mathscr{C}$ and calculate the difference of these two limits. The result is

$$\rho_0\left[\frac{\partial v}{\partial t}\right]=\left[\frac{\partial \sigma}{\partial X}\right],$$

$$\rho_0 v\left[\frac{\partial v}{\partial t}\right]+\rho_0\left[\frac{\partial e}{\partial t}\right]=v\left[\frac{\partial \sigma}{\partial X}\right]+\sigma\left[\frac{\partial v}{\partial X}\right]-\left[\frac{\partial q}{\partial X}\right], \qquad (1.1.16)$$

$$\left[\frac{\partial \eta}{\partial t}\right]+\left[\frac{\partial}{\partial X}\left(\frac{q}{\theta}\right)\right]\geqslant 0,$$

provided b and r are assumed to be continuous functions across the curve $\mathscr{C}$.

The jump conditions (1.1.13), (1.1.15) and $(1.1.16)_{1-2}$ form a system of homogeneous linear algebraic equations for the unknowns $\left[\frac{\partial v}{\partial t}\right], \left[\frac{\partial v}{\partial X}\right]$, etc. This system is underdetermined and it is independent of the constitutive equations of the body. In order to discuss the existence conditions of acceleration waves, the above relations have to be completed with the constitutive equations, which relate the quantities ε, θ, σ, e, η and q. This problem will be partly discussed in the following chapters. In case of perfect fluids, a systematic treatment of this problem can be found in HADAMARD's book [1903] as well as in COURANT and FRIEDRICHS [1948]. For elastic-plastic bodies one can find this problem considered in RAKHMATULIN and DEMIANOV [1961] and CRISTESCU [1967]. For a study of the properties of shock and acceleration waves in non-linear elastic and visco-elastic materials, see also CHEN [1973, 1976], NUNZIATO, WALSH, SCHULER and BARKER [1974] and THURSTON [1974].

There exists a close relation between characteristics and acceleration waves: The acceleration waves represent a subset of the family of all characteristics (see for instance COURANT and FRIEDRICHS [1948]). For the determination of the set of all acceleration waves as a subset of the set of all characteristics, see SULICIU [1973, 1974c].

2. THE GENERAL BALANCE LAWS AND ELEMENTS OF WAVE PROPAGATION THEORY

1. A *reference configuration* of a body $\mathscr{B}$ will be a domain $\mathscr{R}$ in R^3 with the property there exists a one-to-one correspondence between the points of $\mathscr{R}$ and those of $\mathscr{B}$. A *three-dimensional motion* of the body $\mathscr{B}$ is a mapping $\chi:\mathscr{R}\times I\to R^3$, where I is an interval on the real line, that is

$$\mathbf{x}=\chi(\mathbf{X}, t), \quad \mathbf{X}\in\mathscr{R}, \quad t\in I\subset R, \qquad (1.2.1)$$

with the property that for any fixed $t \in I, \chi(.,t)$ is injective and bicontinuous with respect to $\mathbf{X}$. For simplicity, we shall use Cartesian coordinates to describe the points $\mathbf{X}$ of $\mathscr{R}$ as well as the points $\mathbf{x}$ of the actual configuration $\mathscr{D}_t = \chi(\mathscr{R}, t)$. As for the one-dimensional case, χ will be assumed of class C^p, $p \geqslant 2$, with respect to (X, t) unless another smoothness assumption is explicitely adopted.

The quantities

$$\mathbf{v} = \dot{\mathbf{x}} = \frac{\partial \chi}{\partial t}(\mathbf{X}, z), \quad \ddot{\mathbf{x}} = \dot{\mathbf{v}} = \frac{\partial^2 \chi}{\partial t^2}(\mathbf{X}, t) \tag{1.2.2}$$

are called the *velocity* and *acceleration* of the material particle, respectively. Since χ in (1.2.1) is invertible, for any fixed t, the particle velocity and acceleration may be also expressed as functions of the actual coordinates $\mathbf{x}$ of the particle, i.e.

$$\mathbf{v}(\mathbf{x}, t) = \dot{\mathbf{x}}(\chi^{-1}(\mathbf{x}, t), t) \tag{1.2.3}$$

$$\mathbf{a}(\mathbf{x}, t) = \ddot{\mathbf{x}}(\chi^{-1}(\mathbf{x}, t), t) = \frac{d\mathbf{v}}{dt}(\mathbf{x}, t) = \frac{\partial \mathbf{v}}{\partial t} + (\operatorname{grad} \mathbf{v})\mathbf{v}.$$

The quantity

$$\mathbf{L}(\mathbf{x}, t) = \operatorname{grad} \mathbf{v}(\mathbf{x}, t) = \left(\frac{\partial v_i}{\partial x_j}(\mathbf{x}, t)\right) = (L_{ij}(\mathbf{x}, t)) \tag{1.2.4}$$

is called the *velocity gradient* (or the velocity of deformation). The derivative

$$\frac{d}{dt} = \frac{\partial}{\partial t} + (\operatorname{grad})\mathbf{v} = \frac{\partial}{\partial t} + v_i \frac{\partial}{\partial x_i} \tag{1.2.5}$$

is called the *material derivative*.

Another measure of the motion χ of a continuous medium is the *deformation gradient*, defined as

$$\mathbf{F}(\mathbf{X}, t) = \operatorname{Grad} \chi(\mathbf{X}, t) = \left(\frac{\partial \chi_i}{\partial X_j}(\mathbf{X}, t)\right) = (F_{ij}(\mathbf{X}, t)). \tag{1.2.6}$$

The deformation gradient and the velocity gradient are related by

$$\mathbf{L} = \dot{\mathbf{F}}\mathbf{F}^{-1}, \tag{1.2.7}$$

where $\mathbf{F}^{-1}$ is the inverse of $\mathbf{F}$ and $\dot{\mathbf{F}}\mathbf{F}^{-1}$ denotes the usual product of two matrices.

2. Relative to the forces acting upon the body, one considers here the non-polar case (see for instance TRUESDELL and TOUPIN [1960], Section 200, SOLOMON [1969], Chapter II, JACOB [1959]), that is the case when the body is subjected only to the body forces $\mathbf{b}$, where $\mathbf{b}$ is a function of

$(\mathbf{X}, t)$, and to contact forces that can be characterized by a tensor $\mathbf{T} = \mathbf{T}(\mathbf{X}, t)$ called the *Cauchy stress tensor* (see for instance TRUESDELL and TOUPIN [1960], Section 203, MISICU [1967], Subsections 1.2, 2.1—2.3). If $\mathbf{n}$ is a unit vector, then $\mathbf{t_n} = \mathbf{Tn}$ is the force acting on the unit area with normal $\mathbf{n}$ in the actual configuration.

3. The balance of the moment of momentum leads to

$$\mathbf{T} = \mathbf{T}^T, \ (T_{ij}) = (T_{ji}), \tag{1.2.8}$$

where $(\ )^T$ denotes the transpose of the considered second order tensor.

The balance of momentum and energy is written in the actual coordinates as the system of differential equations

$$\rho\dot{\mathbf{v}} - \operatorname{div}\mathbf{T} = \rho\mathbf{b}, \qquad \rho\dot{v}_i - \frac{\partial T_{ij}}{\partial x_j} = \rho b_i, \tag{1.2.9}$$

$$\rho\dot{e} - \mathbf{T}\cdot\mathbf{L} + \operatorname{div}\mathbf{q} = \rho r, \qquad \rho\dot{e} - T_{ij}L_{ij} + \frac{\partial q_i}{\partial x_i} = \rho r, \tag{1.2.10}$$

where ρ is the actual mass density, e is the internal energy per unit mass, $\mathbf{q}$ is the heat flux (the heat quantity that enters through the actual unit area during a unit time interval), and $r = r(\mathbf{x}, t)$ is the heat supply per unit mass in a unit time interval (absorbed by the particle $\mathbf{X}$ of $\mathscr{B}$ and supplied by the external world by radiation).

The balance of mass can be written in several ways; we give here two forms:

$$\rho_0 = J\rho, \qquad \frac{\partial\rho}{\partial t} + \operatorname{div}(\rho\mathbf{v}) = 0. \tag{1.2.11}$$

$(1.2.11)_1$ expresses the balance of mass in initial coordinates, where ρ_0 is the mass density in the initial configuration and J is the determinant of $\mathbf{F}$, $J = \det\mathbf{F}$. $(1.2.11)_2$ represents the balance of mass if actual coordinates are used.

Let us introduce the *Piola-Kirchhoff stress tensor*

$$\mathbf{S} = \rho^{-1}\mathbf{T}(\mathbf{F}^T)^{-1} = \frac{1}{\rho_0}\widetilde{\mathbf{S}} \tag{1.2.12}$$

(see for instance COLEMAN and GURTIN [1967], TRUESDELL and TOUPIN [1960], Section 210, MISICU [1967], Subsections 1.6, 2.1—2.3). Then (1.2.10) can be written in the form

$$\rho\dot{e} - \rho\mathbf{S}\cdot\dot{\mathbf{F}} + \operatorname{div}\mathbf{q} = \rho r. \tag{1.2.13}$$

The *entropy rate* γ is defined as

$$\rho\gamma = \rho\dot{\eta} - \left(\frac{\rho r}{\theta} - \operatorname{div}\left(\frac{q}{\theta}\right)\right), \tag{1.2.14}$$

where $\eta = \eta(\mathbf{X}, t)$ is the entropy per unit mass while $\theta = \theta(\mathbf{X}, t) > 0$ is the absolute temperature.

The second law of thermodynamics, or the Clausius-Duhem inequality, states that (see for instance TRUESDELL and TOUPIN [1960], Section 254)

$$\gamma \geqslant 0. \tag{1.2.15}$$

From (1.2.13), (1.2.14) and (1.2.15) one gets

$$\gamma = \dot{\eta} - \frac{\dot{e}}{\theta} + \theta^{-1}\mathbf{S}\cdot\dot{\mathbf{F}} - \frac{1}{\rho\theta^2}\mathbf{q}\cdot\mathbf{g} \geqslant 0, \tag{1.2.16}$$

where

$$\mathbf{g} = \operatorname{grad}\theta, \qquad \mathbf{q}\cdot\mathbf{g} = q_i\frac{\partial\theta}{\partial x_i}\,. \tag{1.2.17}$$

If one introduces the free energy

$$\psi = e - \theta\eta, \tag{1.2.18}$$

then (1.2.16) can also be expressed as

$$\theta\gamma = -\dot{\psi} - \eta\dot{\theta} + \mathbf{S}\cdot\dot{\mathbf{F}} - \frac{1}{\rho\theta}\mathbf{q}\cdot\mathbf{g} \geqslant 0. \tag{1.2.19}$$

Using the tensor $\tilde{\mathbf{S}}$ given by (1.2.12), we can write the balance equations (1.2.9) and (1.2.10) in initial coordinates,

$$\begin{aligned} &\rho_0\frac{\partial\mathbf{v}}{\partial t} - \operatorname{Div}\tilde{\mathbf{S}} = \rho_0\mathbf{b}, \\ &\rho_0\frac{\partial e}{\partial t} - \tilde{\mathbf{S}}\cdot\frac{\partial\mathbf{F}}{\partial t} + \operatorname{Div}\tilde{\mathbf{q}} = \rho_0 r, \end{aligned} \tag{1.2.20}$$

with $\tilde{\mathbf{q}} = J\mathbf{F}^{-1}\mathbf{q}$. (For the general form of the balance equations see also CARLSON [1972], GREEN and RIVLIN [1964 a, b]).

4. A regular surface Σ with equation $\varphi(\mathbf{X}, t) = 0$, where $\mathbf{X} \in \mathscr{R}$ and $t \in I$, is called an *acceleration wave through the body* $\mathscr{B}$ if φ is of class C^1, $\mathbf{v}$, $\mathbf{F}$, θ, $\tilde{\mathbf{S}}$, ψ, η and $\mathbf{q}$ are continuous and their derivatives may have jump discontinuities across Σ, while being continuous at all $(\mathbf{X}, t) \in \mathscr{R}\times I - \Sigma$.

The quantities

$$U = -\frac{\dfrac{\partial\varphi}{\partial t}}{|\operatorname{Grad}\varphi|}, \qquad n_i = \frac{\dfrac{\partial\varphi}{\partial X_i}}{|\operatorname{Grad}\varphi|}, \tag{1.2.21}$$

are called the *propagation speed* and the *direction of propagation* of the acceleration wave, respectively.

The jumps of the derivatives of $\mathbf{v}, \mathbf{F}, \theta, \widetilde{\mathbf{S}}, \psi, \eta$ and $\mathbf{q}$ cannot be independent; they have to satisfy three types of conditions. The *geometric and kinematic compatibility conditions* give those relations between the jumps of $\frac{\partial \mathbf{v}}{\partial t}, \frac{\partial \mathbf{v}}{\partial \mathbf{X}}$, etc., that are due to the continuity of $\mathbf{v}$ or follow from the fact that $\mathbf{v}$ and $\mathbf{F}$ are the partial derivatives of the same vector function $\boldsymbol{\chi}$. These compatibility conditions can be written as

$$\left[\frac{\partial v_k}{\partial t}\right] = U^2 a_k, \quad \left[\frac{\partial F_{kl}}{\partial X_j}\right] = a_k n_l n_j, \quad \left[\frac{\partial v_k}{\partial X_l}\right] = \left[\frac{\partial F_{kl}}{\partial t}\right] = -\,U a_k n_l, \tag{1.2.22}$$

$$\left[\frac{\partial \theta}{\partial t}\right] = -U\nu, \quad \left[\frac{\partial \theta}{\partial X_j}\right] = \nu n_j, \tag{1.2.23}$$

$$\left[\frac{\partial \widetilde{S}_{ij}}{\partial t}\right] = -U s_{ij}, \quad \left[\frac{\partial \widetilde{S}_{ij}}{\partial X_k}\right] = s_{ij} n_k, \text{ etc.} \tag{1.2.24}$$

Here $[f] = f^+ - f^-$, f^+ being the limit value of f at a point on the surface Σ, reached from the positive side of the normal $(\mathbf{n}, -U)$ to Σ at that point, and f^- being the limit value of f at the same point but reached from the other side of the surface. The vector $\mathbf{a} = (a_1, a_2, a_3)$ is called the *mechanical amplitude of the wave*, the scalar ν is called the *thermal amplitude of the wave* and s_{ij} may be called the *stress amplitude*, etc.

The second group of restrictions imposed on the jumps comes from the balance equations. They are called *dynamic compatibility equations* and they are obtained from equations (1.2.20) as follows: one considers (1.2.20) on each side of the surface Σ and then, after taking limit values at a point on Σ, one calculates the difference between the two obtained relations. Since $\mathbf{b}$ and r are assumed continuous, one obtains the following homogeneous dynamic compatibility conditions

$$\rho_0\left[\frac{\partial v_i}{\partial t}\right] - \left[\frac{\partial \widetilde{S}_{ij}}{\partial X_j}\right] = 0, \tag{1.2.25}$$

$$\rho_0\left[\frac{\partial e}{\partial t}\right] - \widetilde{S}_{ij}\left[\frac{\partial F_{ij}}{\partial t}\right] + \left[\frac{\partial \tilde{q}_j}{\partial X_j}\right] = 0.$$

The third group of restrictions is imposed by the *constitutive equations*, i.e. those relations that must exist between $\mathbf{F}$, θ, $\mathrm{Grad}\theta$, $\tilde{\mathbf{S}}$, η, ψ and $\mathbf{q}$. These restrictions will be discussed in other chapters.

In case of an isolated discontinuity surface, the geometric-kinematic and dynamic compatibility conditions have been presented for the first time by HADAMARD [1903] for the general three-dimensional case (for the deduction of the compatibility conditions across acceleration and shock waves as well as for further references, see TRUESDELL and TOUPIN [1960]). Similar compatibility conditions are also obtained in a more general case when the discontinuities of the motion are not necessarily located on an isolated discontinuity surface (see SULICIU [1973]).

Chapter II

ONE-DIMENSIONAL CONSTITUTIVE EQUATIONS IN PLASTICITY AND VISCOPLASTICITY. INITIAL AND BOUNDARY VALUE PROBLEMS

In order to complete the system of balance and compatibility equations $(1.1.7)_1$ and $(1.1.7)_2$, we need a relation between stresses and strains. Such a relation is called a *constitutive equation*. Constitutive equations are subject to two fundamental requirements. On one hand, they have to reflect a certain number of experimental facts that are considered as essential for a chosen framework; on the other hand, their mathematical form has to be simple enough to enable the prediction of certain behaviour that can be confirmed by experimental observations. Because of these requirements, the constitutive equations used in plasticity have serious limitations.

Throughout this book constitutive equations will be formulated locally with respect to the material particle, that is for a fixed material particle or for material bodies where stresses and strains are constant in space.

Let us introduce now several preliminary notions. A plane set $\mathscr{D}$ will be called the *set of all states* while its points, denoted by (ε, σ), will be called *states*. A smooth function (i.e. a function with continuous derivative) $\varepsilon(t)$ (or $\sigma(t)$) with $t \in [0, T]$, $T > 0$, is called a *strain history* (or a *stress history)* of a particle X of the body. The time derivatives $\dot{\varepsilon} = \dot{\varepsilon}(t)$ and $\dot{\sigma} = \dot{\sigma}(t)$ are called the *strain rate* and *stress rate* respectively at the state $(\varepsilon = \varepsilon(t),\ \sigma = \sigma(t))$. Whenever a strain history $\varepsilon(t)$ or a stress history $\sigma(t)$ will be mentioned, it will be understood they are of class C^1, if not specified otherwise.

Throughout this chapter, by a constitutive equation is meant a rule which, at a given state $(\varepsilon, \sigma) \in \mathscr{D}$ determines the stress rate $\dot{\sigma}$ (or the strain rate) when the strain rate $\dot{\varepsilon}$ (or the stress rate) is known.

1 Dimensional Constitutive Equation

$\varepsilon, \sigma, \dot{\sigma} \longmapsto \dot{\varepsilon}$ or $\varepsilon, \sigma, \dot{\varepsilon} \longmapsto \dot{\sigma}$

1. SOME CONSTITUTIVE EQUATIONS OF ONE-DIMENSIONAL CLASSICAL PLASTICITY

1.1. **Elastic constitutive equation at a given state.** Let $(\varepsilon_0, \sigma_0) \in \mathscr{D}$, let $I = (a, b)$, $a < b$, be an interval on the real line and let $f: I \to \mathbf{R}$ be an increasing continuous function. One says that a material has an elastic behaviour with respect to the state $(\varepsilon_0, \sigma_0) \in \mathscr{D}$ if there exist $I \neq \emptyset$, $\varepsilon_0 \in I$ and a function f with the following properties: $(I, f(I)) \subset \mathscr{D}$ and, for any strain history $\varepsilon(t) \in I$, $\varepsilon(0) = \varepsilon_0$, the stress history $\sigma(t)$ is given by

$$\sigma(t) = f(\varepsilon(t)), \tag{2.1.1}$$

while f is a one-to-one map. Both I and f depend on the state $(\varepsilon_0, \sigma_0)$.

If f is a linear function on I, that is if

$$\sigma = f(\varepsilon) = \sigma_0 + E(\varepsilon - \varepsilon_0), \tag{2.1.2}$$

one says the material is *linearly elastic* with respect to the state $(\varepsilon_0, \sigma_0) \in \mathscr{D}$. $E > 0$ is a constant called the *elastic modulus.* For strain histories $\varepsilon(t) \in I$ one can also write (2.1.2) in the form

$$\dot{\sigma} = E\dot{\varepsilon} \tag{2.1.2'}$$

which does not explicitely depend on $(\varepsilon_0, \sigma_0)$.

1.2. **Constitutive equations of one-dimensional classical plasticity.** We shall give here a general formulation of these constitutive equations. At any state $(\varepsilon, \sigma) \in \mathscr{D}$, we suppose that the stress rate is proportional to the strain rate. The proportionality coefficient is a function of the state $(\varepsilon, \sigma) \in \mathscr{D}$ and of the strain rate and it is a positive homogeneous function of degree zero in the strain rate argument. This means there exists a function $\bar{\mathscr{E}} : \mathscr{D} \times R \to R$ such that the stress rate is given by

$$\dot{\sigma} = \bar{\mathscr{E}}(\varepsilon, \sigma, \dot{\varepsilon})\dot{\varepsilon}. \tag{2.1.3}$$

Quite often in classical plasticity one can find the strain rate expressed as a function of the state (ε, σ) and the stress rate. One therefore assumes there exists a function $\bar{\mathscr{F}} : \mathscr{D} \times R \to R$ such that

$$\dot{\varepsilon} = \bar{\mathscr{F}}(\varepsilon, \sigma, \dot{\sigma})\dot{\sigma}. \tag{2.1.4}$$

One of the most important aspects of the classical plasticity theories is represented by the fact that both proportionality coefficients $\bar{\mathscr{E}}$ and $\bar{\mathscr{F}}$ are positively homogeneous functions in their last argument. Indeed, if equation (2.1.3) (or (2.1.4)) is satisfied along a curve $(\varepsilon(t), \sigma(t)) \in \mathscr{D}$, $t \in [0,T]$, $T > 0$, then $(\varepsilon(\tau), \sigma(\tau)) \in \mathscr{D}$ with $\tau = h(t)$, $h \in C^1$ and $h'(t) > 0$ will also satisfy this equation. This is the reason why any plasticity theory based on a constitutive equation of the form (2.1.3) (or (2.1.4)) is called a *time-independent theory.* *

It is easy to show that both functions $\bar{\mathscr{E}}$ and $\bar{\mathscr{F}}$ depend on their last argument only by depending on its sign, i.e.

$$\bar{\mathscr{E}}(\varepsilon, \sigma, \dot{\varepsilon}) = \mathscr{E}(\varepsilon, \sigma, \operatorname{sign}\dot{\varepsilon}), \quad \bar{\mathscr{F}}(\varepsilon, \sigma, \dot{\sigma}) = \mathscr{F}(\varepsilon, \sigma, \operatorname{sign}\dot{\sigma}). \tag{2.1.5}$$

Positive Homogenaity of Degree k:

* $f(tx_1, tx_2, \ldots, tx_n) = t^k f(x_1, x_2, \ldots, x_n)$

So $\bar{\mathscr{E}}(\varepsilon, \sigma, k\dot{\varepsilon}) = k\,\bar{\mathscr{E}}(\varepsilon, \sigma, \dot{\varepsilon})$

Indeed, let $\lambda = |\dot{\varepsilon}| > 0$; then

$$\bar{\mathscr{E}}(\varepsilon, \sigma, \dot{\varepsilon}) = \bar{\mathscr{E}}\left(\varepsilon, \sigma, \lambda \frac{\dot{\varepsilon}}{|\dot{\varepsilon}|}\right) = \bar{\mathscr{E}}\left(\varepsilon, \sigma, \frac{\dot{\varepsilon}}{|\dot{\varepsilon}|}\right)$$

and (2.1.5) follows.

By means of (2.1.5) the constitutive equations (2.1.3) and (2.1.4) can be written as

$$\dot{\sigma} = \mathscr{E}(\varepsilon, \sigma, \operatorname{sign}\dot{\varepsilon})\dot{\varepsilon}, \tag{2.1.3'}$$

$$\dot{\varepsilon} = \mathscr{F}(\varepsilon, \sigma, \operatorname{sign}\dot{\sigma})\dot{\sigma}. \tag{2.1.4'}$$

Plasticity theories based on equation (2.1.3′) or (2.1.4′) are *rate independent* theories in the sense given below. Let $(\varepsilon_0, \sigma_0) \in \mathscr{D}$ be an arbitrary state. Let $\varepsilon(t)$ with $\varepsilon(0) = \varepsilon_0$ and $\dot{\varepsilon}(t)>0$ or $\dot{\varepsilon}(t) < 0$; the stress is then determined as the solution of the equation

$$\frac{d\sigma}{d\varepsilon} = \mathscr{E}(\varepsilon, \sigma, 1), \qquad \sigma(\varepsilon_0) = \sigma_0, \qquad \text{for} \quad \varepsilon \geqslant \varepsilon_0$$

and

$$\frac{d\sigma}{d\varepsilon} = \mathscr{E}(\varepsilon, \sigma, -1), \qquad \sigma(\varepsilon_0) = \sigma_0, \quad \text{for } \varepsilon \leqslant \varepsilon_0.$$

If the functions $\mathscr{E}(\varepsilon, \sigma, \pm 1)$ have the required properties that ensure the existence of unique piecewise smooth solutions for the above equations (and $\mathscr{E}$ is assumed to have them), then these solutions

$$\sigma = \begin{cases} \hat{\sigma}^{+}(\varepsilon, \varepsilon_0, \sigma_0), & \varepsilon \geqslant \varepsilon_0, \\ \hat{\sigma}^{-}(\varepsilon, \varepsilon_0, \sigma_0), & \varepsilon \leqslant \varepsilon_0, \end{cases} \tag{2.1.3''}$$

are independent of $\dot{\varepsilon}$. The existence of a unique curve (2.1.3″) for a fixed state $(\varepsilon_0, \sigma_0) \in \mathscr{D}$ justifies to some extent why the constitutive equations of classical plasticity are called "finite laws" by certain authors (especially when referring to the state $(\varepsilon_0 = 0,\ \sigma_0 = 0)$) (see for instance CRISTESCU [1967]).

Now, if at the state $(\varepsilon_0, \sigma_0)$ a strain $\bar{\varepsilon} > \varepsilon_0$ (or $\bar{\varepsilon} < \varepsilon_0$) is suddenly applied, the corresponding stress $\bar{\sigma}$ is obtained by means of the same relation (2.1.3″), where $\varepsilon = \bar{\varepsilon}$. Therefore, this relation can be also called the *instantaneous response curve*. All constitutive equations of classical plasticity that are discussed in § 1.4—1.5 have the important property of leading to constant propagation velocity for any shock wave; this follows from the local solution of the Riemann problem (see SULICIU [1973]).

Another important feature of the constitutive equations of classical plasticity lies in their property of having a domain of (linear) elastic behaviour, that is a domain where the constitutive equation (2.1.2′) is satisfied.

A relatively general example of such a constitutive equation is given below. Let $f, g: I \to R$ be two functions of class C^2 with the following properties:

$$f(\varepsilon) > g(\varepsilon), \qquad f(0) > 0 > g(0),$$
$$E > f'(\varepsilon) \geqslant 0, \qquad E > g'(\varepsilon) \geqslant 0, \tag{2.1.6}$$
$$f''(\varepsilon) \leqslant 0, \qquad g''(\varepsilon) \geqslant 0.$$

The set $\mathscr{D}$ of all states is defined as the set of all points (ε, σ) in the plane, with $\varepsilon \in I$, for which $g(\varepsilon) \leqslant \sigma$ and $\sigma \leqslant f(\varepsilon)$. The function $\mathscr{E}$ of (2.1.3′) is given by

$$\mathscr{E}(\varepsilon, \sigma, \operatorname{sign} \dot{\varepsilon}) = \begin{cases} E & \text{if } g(\varepsilon) < \sigma < f(\varepsilon) \\ & \text{or} \\ & \quad \sigma = f(\varepsilon) \text{ and } \dot{\varepsilon} < 0 \\ & \text{or} \\ & \quad \sigma = g(\varepsilon) \text{ and } \dot{\varepsilon} > 0, \\ \hline f'(\varepsilon) & \text{if } \sigma = f(\varepsilon) \text{ and } \dot{\varepsilon} \geqslant 0, \\ \hline g'(\varepsilon) & \text{if } \sigma = g(\varepsilon) \text{ and } \dot{\varepsilon} \leqslant 0. \end{cases} \tag{2.1.7}$$

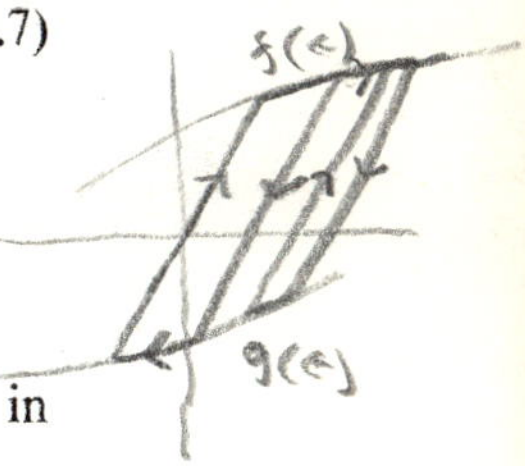

Here, as everywhere in this chapter, E is the Young modulus.

One can see that the set $\{(\varepsilon, \sigma) \in \mathscr{D},\ g(\varepsilon) < \sigma < f(\varepsilon)\}$ is a domain in which (2.1.2′) or (2.1.2) are satisfied.

The point where the straightline $\sigma = E\varepsilon$ intersects the curve $\sigma = f(\varepsilon)$ is called the *yield point in tension*, with respect to the state $(0, 0)$, and $\sigma_Y = E\varepsilon_Y = f(\varepsilon_Y)$ is called the *yield stress in tension*. The *yield point in compression* with respect to the state $(0, 0)$, is then the point where $\sigma = E\varepsilon$ intersects $\sigma = g(\varepsilon)$. $\bar{\sigma}_Y = E\bar{\varepsilon}_Y = g(\bar{\varepsilon}_Y)$ is the *yield stress in compression*. For many metals one finds $\bar{\sigma}_Y = -\sigma_Y$ and $g(\varepsilon) = -f(-\varepsilon)$ for $\varepsilon < \bar{\varepsilon}_Y = -\varepsilon_Y = -\dfrac{\sigma_Y}{E}$.

We present below several examples of constitutive equations that are considered in the literature.

1.3. **Elastic perfectly-plastic body and elastic body with the locking property.** Let $\sigma_Y > 0$ be a constant called the *yield stress*. The functions in (2.1.6) are taken as $f(\varepsilon) = \sigma_Y$ and $g(\varepsilon) = -\sigma_Y$. Then (2.1.7) will be

$$\mathscr{E}(\varepsilon, \sigma, \operatorname{sign} \dot{\varepsilon}) = \begin{cases} E & \text{if } |\sigma| < \sigma_Y \\ & \text{or} \\ & \quad |\sigma| = \sigma_Y \text{ and } J < 0, \\ \hline 0 & \text{if } |\sigma| = \sigma_Y \text{ and } J \geqslant 0, \end{cases} \tag{2.1.8}$$

J = sign σ sign ε̇

J < 0 ⇒ either σ or ε̇ is negitive but not both

J ≥ 0 ⇒ both σ and ε̇ are pos. or both neg. or one or both is zero where

$$J = \text{sign}\,\sigma\;\text{sign}\,\dot{\varepsilon}. \tag{2.1.9}$$

If a smooth strain history $\varepsilon(t)$, $t \in [0, T]$, is given and if

$$\sigma(0) = \sigma_0, \quad |\sigma_0| \leqslant \sigma_Y, \tag{2.1.10}$$

then the stress history $\sigma(t)$, $t \in [0, T]$ is the solution of equation (2.1.3′) with the initial condition (2.1.10), where $\mathscr{E}$ is given by (2.1.8). In this case is not possible to use the form (2.1.4), as the strain rate $\dot{\varepsilon}$ is not uniquely determined when the state (ε, σ) and the stress rate $\dot{\sigma}$ are known.

Permanent strain. Let $(\varepsilon^0, \sigma^0) \in \mathscr{D}$ be an arbitrary state. Let $\varepsilon(t)$, $t \in [0, T]$ be a smooth strain history such that $\varepsilon(0) = \varepsilon^0$, $\dot{\varepsilon}(t) \neq 0$ for any $t \in [0, T]$ and $J(0) = \text{sign}\,\sigma^0\;\text{sign}\,\dot{\varepsilon}(0) < 0$ if $\sigma^0 \neq 0$. Then

$$\sigma(t) = \sigma^0 + E(\varepsilon(t) - \varepsilon^0)$$

on an interval $[0, T_1]$, $0 < T_1 \leqslant T$. One can easily see from (2.1.8) that $T_1 \geqslant \tilde{T}$, where $\tilde{T}$ is the first value of t at which $J(t)$ vanishes. Then $\sigma(\tilde{T}) = 0$ and the quantity

$$\varepsilon_0^r = \varepsilon(\tilde{T}) = \varepsilon^0 - \frac{\sigma^0}{E} \quad \text{and} \quad \varepsilon_0^r = \varepsilon^0 \;\text{ for }\; \sigma^0 = 0, \tag{2.1.11}$$

is called the *permanent strain attached to the state* $(\varepsilon^0, \sigma^0)$.

Elastic and plastic strain. For any state $(\varepsilon, \sigma) \in \mathscr{D}$, we call

$$\varepsilon^e = \frac{\sigma}{E} \tag{2.1.12}$$

the *elastic strain* at the state (ε, σ) and

$$\varepsilon^p = \varepsilon - \varepsilon^e \tag{2.1.13}$$

the *plastic strain* at the state (ε, σ).

Here the plastic and permanent strains coincide. Thus, for this constitutive equation, the plastic strain can be interpreted as the permanent strain which one obtains when the stress vanishes.

An elastic body with the *locking property* has complementary properties to those of an elastic perfectly-plastic body. For such a body, the domain of elastic behaviour is the set of all points (ε, σ) with $|\varepsilon| < \varepsilon_L$, where $\varepsilon_L > 0$ is called the *locking strain*. At a given state (ε, σ), with $|\varepsilon| \leqslant \varepsilon_L$ and a given stress rate $\dot{\sigma}$, the strain rate $\dot{\varepsilon}$ is determined by a rela-ion (2.1.4′) where $\mathscr{F}(\varepsilon, \sigma, \operatorname{sign}\dot{\sigma})$ has the form

$$\mathscr{F}(\varepsilon, \sigma, \operatorname{sign}\dot{\sigma}) = \begin{cases} \dfrac{1}{E} & \text{if } |\varepsilon| < \varepsilon_L \\ & \text{or} \\ & |\varepsilon| = \varepsilon_L \text{ and } J' < 0, \\ 0 & \text{if } |\varepsilon| = \varepsilon_L \text{ and } J' \geqslant 0. \end{cases} \tag{2.1.8′}$$

with

$$J' = \operatorname{sign}\varepsilon \operatorname{sign}\dot{\sigma}. \tag{2.1.9′}$$

Just as for an elastic perfectly-plastic body, a constitutive equation (2.1.4′) with $\mathscr{F}$ given by (2.1.8′) cannot be written in the form (2.1.3′); indeed, for any state (ε, σ) with $|\varepsilon| = \varepsilon_L$, the knowledge of the strain rate $\dot{\varepsilon}$ does not suffice for determining a unique stress rate $\dot{\sigma}$.

Typical examples of real bodies whose behaviour can be approximately described by such a constitutive equation are furnished by certain kinds of soils.

1.4. **Elastic-plastic body with linear work-hardening and idealized Bauschinger effect.** The constitutive equation of such a body is a constitutive equation of the form (2.1.7) with $f(\varepsilon) = E_1\varepsilon + \sigma_Y^0$ and $g(\varepsilon) = E_1\varepsilon - \sigma_Y^0$, where $0 < E_1 < E$, E_1 is a constant called the *work-hardening modulus* and $\sigma_Y^0 > 0$ is also constant. The function $\mathscr{E}$ is given by

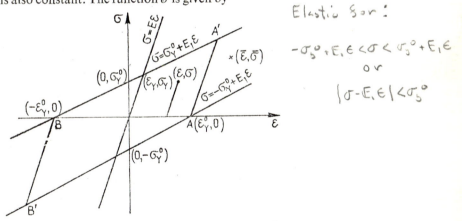

Elastic for:

$-\sigma_y^0 + E_1\epsilon < \sigma < \sigma_y^0 + E_1\epsilon$

or

$|\sigma - E_1\epsilon| < \sigma_y^0$

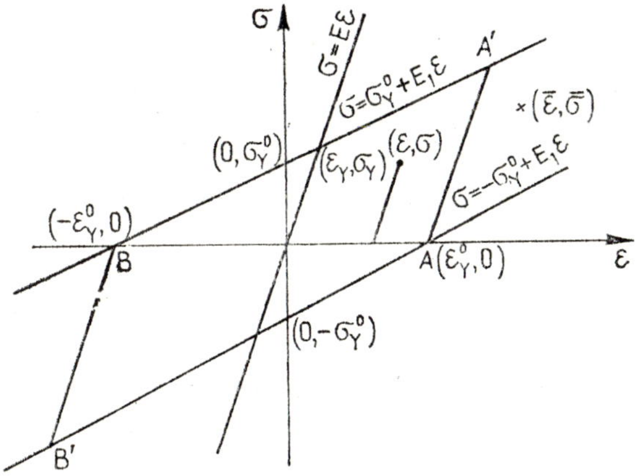

Fig. 2.1.1 Elastic-plastic body with linear work-hardening and idealized Bauschinger effect

$$\mathscr{E}(\varepsilon, \sigma, \operatorname{sign}\dot{\varepsilon}) = \begin{cases} E & \text{if } |\sigma - E_1\varepsilon| < \sigma_Y^0 \\ & \text{or} \\ & |\sigma - E_1\varepsilon| = \sigma_Y^0 \text{ and } J_1 < 0, \\ E_1 & \text{if } |\sigma - E_1\varepsilon| = \sigma_Y^0 \text{ and } J_1 \geqslant 0 \end{cases} \tag{2.1.14}$$

$J_1 = \operatorname{sign}(\sigma - E_1\epsilon) \operatorname{sign}\dot{\epsilon}$

with

$$J_1(\varepsilon, \sigma, \dot{\varepsilon}) = \text{sign}\,(\sigma - E_1\varepsilon)\ \text{sign}\,\dot{\varepsilon}. \tag{2.1.15}$$

In this case the constitutive equation can also be written in the form (2.1.4′) with $\mathscr{F}: \mathscr{D} \times R \to (0, \infty)$ given by

$$\mathscr{F}(\varepsilon, \sigma, \text{sign}\,\dot{\sigma}) = \begin{cases} \dfrac{1}{E} & \text{if } |\sigma - E_1\varepsilon| < \sigma_Y^0 \\ & \text{or} \\ & |\sigma - E_1\varepsilon| = \sigma_Y^0 \text{ and } J_1' < 0, \\ \dfrac{1}{E_1} & \text{if } |\sigma - E_1\varepsilon| = \sigma_Y^0 \text{ and } J_1' \geqslant 0, \end{cases} \tag{2.1.16}$$

where

$$J_1'(\varepsilon, \sigma, \text{sign}\,\dot{\sigma}) = \text{sign}\,(\sigma - E_1\varepsilon)\ \text{sign}\,\dot{\sigma}. \tag{2.1.17}$$

The permanent strain has a similar definition to that given in the previous case (see (2.1.11)). For any state $(\varepsilon, \sigma) \in AA'BB'$ (see fig. 2.1.1.) one has

$$\varepsilon^r = \varepsilon - \frac{\sigma}{E}, \tag{2.1.18}$$

whereas for any state $(\bar{\varepsilon}, \bar{\sigma}) \in \mathscr{D}$ and $(\bar{\varepsilon}, \bar{\sigma}) \notin AA'BB'$ one has

$$\varepsilon^r = \begin{cases} \varepsilon_Y^0 & \text{if } \quad \bar{\sigma} - E\bar{\varepsilon} > E\varepsilon_Y^0, \\ -\varepsilon_Y^0 & \text{if } \quad \bar{\sigma} - E\bar{\varepsilon} < -E\varepsilon_Y^0, \end{cases} \tag{2.1.19}$$

where $\varepsilon_Y^0 = \dfrac{1}{E_1}\sigma_Y^0$.

Elastic and plastic strains can be defined as in (2.1.12) and (2.1.13) respectively. Let us observe that for these bodies, plastic and permanent strains no longer coincide for states $(\varepsilon, \sigma) \in \mathscr{D}$ with $|\sigma - E\varepsilon| > E\varepsilon_Y^0$. One can avoid this kind of behaviour if $\mathscr{D}$ is chosen to contain all points $(\varepsilon, 0)$, with any $\varepsilon \in R$.

1.5. **Elastic-plastic body with non-linear work-hardening.** For most materials a simple tension test (let us say $\sigma = at$, $a > 0$) starting from the state (0, 0), will not corroborate the constitutive equations discussed in Subsections 1.3 and 1.4. In fact, if one prescribes the applied stress (starting from state (0, 0)) as an increasing function of t and one measures the corresponding strain, then for most materials one obtains a curve $\sigma = f(\varepsilon)$ in the (ε, σ)-plane, having the same properties as $f(\varepsilon)$ in (2.1.6) but satisfying strict inequalities for the derivatives if $\sigma > \sigma_Y > 0$, where σ_Y is the yield stress.

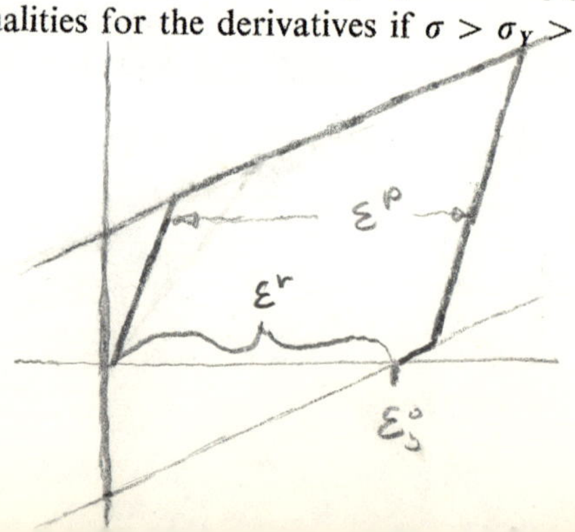

A constitutive equation with non-linear work-hardening is formally a constitutive equation (2.1.3′) (or (2.1.4′)) where $\mathscr{E}$ is defined by (2.1.7) and the functions f and g satisfy (2.1.6) with strict inequalities, i.e. *

$$\begin{gathered} f(\varepsilon) > g(\varepsilon), \qquad f(0) > 0 > g(0), \\ E > f'(\varepsilon) > 0, \qquad E > g'(\varepsilon) > 0, \\ f''(\varepsilon) \begin{cases} < 0 & \text{if} \quad \varepsilon \in (\varepsilon_Y, \infty), \\ \leqslant 0 & \text{if} \quad \varepsilon \in (-\infty, \varepsilon_Y], \end{cases} \\ g''(\varepsilon) \begin{cases} > 0 & \text{if} \quad \varepsilon \in (-\infty, \bar{\varepsilon}_Y), \\ \geqslant 0 & \text{if} \quad \varepsilon \in (\bar{\varepsilon}_Y, \infty), \end{cases} \end{gathered} \tag{2.1.20}$$

Bars Denote Compression

where $\bar{\varepsilon}_Y < 0$ and $\varepsilon_Y > 0$ are defined in Subsection 1.3 following formula (2.1.7).

In applications some authors have taken $f'(\varepsilon_Y) = E$ and/or $g'(\bar{\varepsilon}_Y) = E$. Also the function $\mathscr{F}$ in (2.1.4′) can be written explicitely in this case

Matching Slope at Yield

$$\mathscr{F}(\varepsilon, \sigma, \operatorname{sign}\dot{\sigma}) = \begin{cases} \dfrac{1}{E} & \text{if} \quad g(\varepsilon) < \sigma < f(\varepsilon) \\ & \text{or} \\ & \quad \sigma = f(\varepsilon) \text{ and } \dot{\sigma} < 0, \\ & \text{or} \\ & \quad \sigma = g(\varepsilon) \text{ and } \dot{\sigma} > 0, \\ \dfrac{1}{f'(\varepsilon)} & \text{if} \quad \sigma = f(\varepsilon) \text{ and } \dot{\sigma} \geqslant 0, \\ \dfrac{1}{g'(\varepsilon)} & \text{if} \quad \sigma = g(\varepsilon) \text{ and } \dot{\sigma} \leqslant 0. \end{cases} \tag{2.1.21}$$

A constitutive equation with $\mathscr{E}$ given by (2.1.7) or $\mathscr{F}$ given by (2.1.21) and f, g satisfying (2.1.20) allows the formulation of a consistent mathematical theory; in the framework of such a theory one may find unique solutions for initial and boundary value problems.

However, from the experimental point of view, both functions f and g are known only for $\varepsilon > \varepsilon_Y = \dfrac{\sigma_Y}{E} > 0$ and $\varepsilon < \bar{\varepsilon}_Y = \dfrac{\bar{\sigma}_Y}{E} < 0$ respectively. In fact, one generally does not know yet if for $\varepsilon > \bar{\varepsilon}_Y$ (respectively $\varepsilon < \varepsilon_Y$) the set of all states $\mathscr{D}$ has a boundary that can be written as $\sigma = g(\varepsilon)$ (respectively $\sigma = f(\varepsilon)$), at least for certain stress histories (or strain histories).

In reality σ may not be bound by $f(\varepsilon)$ & $g(\varepsilon)$

Therefore wave propagation problems solved in the framework of a theory based on (2.1.21) have been limited, in most cases, to problems

* The following are excluded

etc

with zero initial values and boundary values leading only to states (ε, σ) with $\varepsilon \geqslant 0$ and $\sigma \geqslant 0$ or $\varepsilon \leqslant 0$ and $\sigma \leqslant 0$.

Let us quote some of the general facts that are experimentally observed but cannot be described by means of a constitutive equation (2.1.3′) or (2.1.4′): creep and relaxation phenomena (Chapter III, Section 1.2), rate dependent effects (Chapter III, Subsection 1.3), peak stress phenomenon (Chapter III, Section 2) and the phenomenon observed in dynamic reloading processes (Chapter III, Section 3).

2. SEMILINEAR AND QUASILINEAR RATE-TYPE CONSTITUTIVE EQUATIONS

2.1. The simplest constitutive equations that may describe the experimental facts just mentioned at the end of Section 1 which also preserve the main features of the classical constitutive equations are the quasilinear rate-type constitutive equations. Let us present the general form of such a constitutive equation.

A plane domain $\mathscr{D}$, containing the origin, will be called a *state domain*. The functions $\varphi, \psi: \mathscr{D} \to R$ will be called *material functions* and they are assumed to be sufficiently smooth on $\mathscr{D}$ if not otherwise specified. The domain $\mathscr{D}$ together with the functions φ and ψ characterize a given material. For any state $(\tilde{\varepsilon}, \tilde{\sigma}) \in \mathscr{D}$ and any history $\varepsilon(t)$ (or $\sigma(t)$), $t \in [0, T)$, $T > 0$, with $\varepsilon(0) = \tilde{\varepsilon}$ (or $\sigma(0) = \tilde{\sigma}$), we shall call $\dot{\varepsilon} = \dot{\varepsilon}(0)$ (or $\dot{\sigma} = \dot{\sigma}(0)$) the *strain rate* (or *stress rate)* at the state $(\tilde{\varepsilon}, \tilde{\sigma})$.

The material behaviour in $\mathscr{D}$ is said to be described by a *quasilinear rate-type* constitutive equation if, for any state $(\varepsilon, \sigma) \in \mathscr{D}$ and for any strain rate $\dot{\varepsilon}$ at this state, the stress rate at the state (ε, σ) is given by

$$\dot{\sigma} = \varphi(\varepsilon, \sigma)\dot{\varepsilon} + \psi(\varepsilon, \sigma). \tag{2.2.1}$$

Equation (2.2.1) together with the equations $(1.1.7)_1$ and $(1.1.7)_2$ form a system of partial differential equations. The necessary and sufficient condition for this system to be hyperbolic and have real acceleration waves (see Section 4) is

$$\varphi(\varepsilon, \sigma) > 0. \tag{2.2.2}$$

We shall assume in the following that (2.2.2) holds on $\mathscr{D}$. Thus, in defining a rate type constitutive equation, the role of $\dot{\varepsilon}$ can be interchanged with that of $\dot{\sigma}$.

Let $(\varepsilon_0, \sigma_0) \in \mathscr{D}$ be a state with $\psi(\varepsilon_0, \sigma_0) \neq 0$; if a strain increment $\Delta\varepsilon$ is applied at $(\varepsilon_0, \sigma_0)$, then the corresponding stress increment $\Delta\sigma$ will be known if and only if the time interval Δt required to apply $\Delta\varepsilon$ is also known. In fact, if $\varepsilon(t) = \varepsilon_0$ for $t \in [0, T)$, $T > 0$, then $\sigma(t)$ is determined as the solution of the initial value problem (2.2.3)

$$\dot{\sigma} = \psi(\varepsilon_0, \sigma), \quad \sigma(0) = \sigma_0;$$

therefore equation (2.2.1) has the relaxation property at the state $(\varepsilon_0, \sigma_0)$ (see Chapter III, Subsection 1.2). If $\sigma(t) = \sigma_0$ for $t \in [0, T)$, $T > 0$, then $\varepsilon(t)$ is the solution of the initial value problem

$$\dot{\varepsilon} = -\frac{1}{\varphi(\varepsilon, \sigma_0)}\psi(\varepsilon, \sigma_0), \qquad \varepsilon(0) = \varepsilon_0, \tag{2.2.4}$$

and this implies that, at the state $(\varepsilon_0, \sigma_0)$, equation (2.2.1) has the creep property (see Chapter III, Subsection 1.2). These properties indicate an essential difference between the constitutive equation (2.2.1) and the constitutive equation (2.1.3′) (or (2.1.4′)) and they describe, by means of the term ψ, the viscoplastic properties of the material.

All points $(\varepsilon_0, \sigma_0) \in \mathscr{D}$ with

$$\psi(\varepsilon_0, \sigma_0) = 0 \tag{2.2.5}$$

$\psi = -\lambda\sigma$

will be called *equilibrium points or points of elastic behaviour*.

The first constitutive equation (2.2.1) (with $\varphi = E$ and $\psi = -\sigma$) has * been proposed by MAXWELL in 1867 and this explains why certain authors are calling Maxwellian materials all materials that can be described by a constitutive equation (2.2.1). However, almost a century later it became clear that a constitutive equation (2.2.1) may describe also the viscoplastic properties of a material.

2.2. A semilinear rate-type constitutive equation closely related to the classical constitutive equation discussed in Section 1.1.3, has been proposed by SOKOLOVSKIĬ [1948 a, b]. The domain $\mathscr{D}$ is the whole ε, σ plane; the functions φ and ψ of (2.2.1) are given by

$$\varphi(\varepsilon, \sigma) = E, \tag{2.2.6}$$

$$\psi(\varepsilon, \sigma) = \begin{cases} 0 & \text{if } |\sigma| < \sigma_Y, \\ (-\operatorname{sign}\sigma)F(|\sigma - \sigma_Y|) & \text{if } |\sigma| \geqslant \sigma_Y, \end{cases} \tag{2.2.7}$$

where $\sigma_Y > 0$ is the yield limit and F is a smooth function with $F(r) > 0$ and $F'(r) > 0$ for $r > 0$ and $F(0) = 0$. The simplest example of such a function considered in the literature is $F(r) = kEr$, where $k > 0$ is a viscosity constant. For other examples of functions F see CRISTESCU [1958], PERZYNA [1963 a, b], TING and SYMONDS [1964], etc.

* For $\varphi = E$ & $\psi = -\lambda\sigma$ then:

Sol $\varepsilon(t) = \varepsilon_0$, $\sigma(0) = \sigma_0$

$\sigma = \sigma_0 e^{-\lambda t}$

Sol $\sigma(t) = \sigma_0$, $\varepsilon(0) = \varepsilon_0$

$\varepsilon = \frac{\sigma_0 \lambda}{E} t + \varepsilon_0$

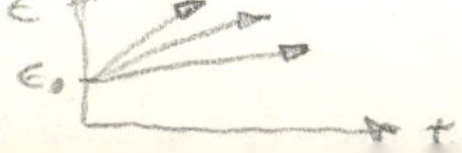

For any state (ε, σ) with $|\sigma| < \sigma_Y$, the constitutive equation described in Subsection 1.3 coincides with SOKOLOVSKIĬ's constitutive equation, that is their domains of elastic behaviour coincide, i.e. equation (2.1.2′) is satisfied in the same domain in both cases.

Let us consider SOKOLOVSKIĬ's case (2.2.6) and (2.2.7). For a given state $(\tilde{\varepsilon}, \tilde{\sigma})$ one can easily study the curves $(\varepsilon(t), \sigma(t))$ for different strain histories $\varepsilon(t)$, $\varepsilon(0) = \tilde{\varepsilon}$, where $\sigma(t)$ is determined as the solution of (2.2.1) with $\sigma(0) = \tilde{\sigma}$. In this way one finds that SOKOLOVSKIĬ's constitutive equation gives , in principle, a much better description of many experimentally observed general facts than a constitutive equation of type (2.1.3′) (see Chapter III). From any state (ε, σ) with $|\sigma| > \sigma_Y$, the material always relaxes to the state (ε, σ_Y), if $\sigma > \sigma_Y$ or to the state $(\varepsilon, -\sigma_Y)$, if $\sigma < -\sigma_Y$. * Creep from any state (ε, σ), $|\sigma| > \sigma_Y$, always takes place with a constant strain rate. However, in general, these two conclusions are quantitatively not in agreement with the experimental results.

2.3. MALVERN [1951 a, b] has generalized SOKOLOVSKIĬ's ideas by assuming the function ψ dependent on both state variables ε and σ. His constitutive equation is closely related to that presented in Subsection 1.5. The domain $\mathscr{D}$ on which ψ is defined is either $\{(\varepsilon, \sigma);\ \varepsilon \geqslant 0,\ \sigma \geqslant 0\}$ or $\{(\varepsilon, \sigma);\ \varepsilon \leqslant 0,\ \sigma \leqslant 0\}$; the function φ is the same as in (2.2.6) while ψ is given by

$$\psi(\varepsilon, \sigma) = \begin{cases} 0 & \text{if } 0 \leqslant \sigma \leqslant f(\varepsilon), \quad \varepsilon \geqslant 0, \\ -kF(\sigma - f(\varepsilon)) & \text{if } \sigma > f(\varepsilon), \quad \varepsilon \geqslant 0, \end{cases} \tag{2.2.9}$$

where F has the same properties as the function F in (2.2.7) and $k > 0$ is a viscosity constant. The continuous curve

$$\sigma = f(\varepsilon), \qquad \varepsilon \in [0, \infty), \tag{2.2.10}$$

called the *quasistatic loading curve* from the state $(0, 0)$, relates the stress and strain in tension ($\dot{\sigma} > 0$) for very small strain rates. In most cases, on the interval $[0, \varepsilon_Y = \sigma_Y/E]$ (where σ_Y is the yield stress) $f(\varepsilon)$ may be identified with $E\varepsilon$ if the deformation processes do not leave the plane quadrant $\sigma \geqslant 0$, $\varepsilon \geqslant 0$. Here the function f plays a similar role to that of the function f of (2.1.21) (see the properties of f in (2.1.20)) in the following sense: The material behaves elastically for $\sigma < f(\varepsilon)$ and any loading process with relatively small strain rates leads to a $(\varepsilon(t), \sigma(t))$-curve that is close to $(\varepsilon, \sigma = f(\varepsilon))$. However, in this case no requirement such as $f'(\varepsilon) > 0$, $\varepsilon \in [0, \infty)$ or $f''(\varepsilon) < 0$ for $\varepsilon \in [\varepsilon_Y, \infty)$ is necessary to ensure real acceleration waves (see Section 4).

* For $\varepsilon(t) = \varepsilon_0$, $\sigma(0) = \sigma_0 > \sigma_y$

$\sigma = (\sigma_0 - \sigma_y) e^{-kEt} + \sigma_y$

For $\sigma(t) = \sigma_0 > \sigma_y$, $\varepsilon(0) = \varepsilon_0$

$\varepsilon = k(\sigma_0 - \sigma_y)t + \varepsilon_0$

Here are some examples of functions F considered in the literature (and therefore of functions Ψ, according to (2.2.9)):

$$F(r) = r, \qquad F(r) = e^{\lambda r} - 1, \quad r \geqslant 0, \tag{2.2.11}$$

where $\lambda > 0$ is a constant, are two cases considered by MALVERN [1951];

$$F(r) = \left(\frac{r}{a}\right)^n, \qquad r \geqslant 0, \tag{2.2.12}$$

where $a > 0$ and $n > 0$ are constants, has been considered by KUKUDJANOV [1967]. This last example has interesting properties for $0 < n < 1$ (see SULICIU [1974 a] and Chapter V, Section 1).

A study of the $(\varepsilon(t), \sigma(t))$-curves for different strain histories will lead to almost the same remarks as in SOKOLOVSKIĬ's case; the main difference consists in the fact that both relaxation and creep from a state (ε, σ), $\sigma > f(\varepsilon)$, can take place to a fixed state on the curve $\sigma = f(\varepsilon)$. This property could be in a better agreement with the experimental results if the function f is suitably chosen (see Chapter III, Subsection 1.2).

2.4. A relatively simple semilinear rate-type constitutive equation with the same domain of elastic behaviour as the constitutive equation (2.1.3') + (2.1.7), can be defined by using the function

$$\psi(\varepsilon, \sigma) = \begin{cases} -k_1 F_1(\sigma - f(\varepsilon)) & \text{if } \sigma \geqslant f(\varepsilon), \\ 0 & \text{if } g(\varepsilon) < \sigma < f(\varepsilon), \\ k_2 F_2(g(\varepsilon) - \sigma) & \text{if } \sigma \leqslant g(\varepsilon), \end{cases} \tag{2.2.13}$$

where $k_1 > 0$, $k_2 > 0$ are two viscosity constants and the functions F_1 and F_2 have the properties

$$\begin{aligned} &F_1(r) > 0, \qquad F_2(r) > 0, \\ & \qquad\qquad\qquad\qquad\qquad \text{for } r > 0 \\ &F_1'(r) > 0, \qquad F_2'(r) > 0, \\ &F_1(0) = F_2(0) = 0, \end{aligned} \tag{2.2.14}$$

while f and g satisfy (2.1.6).

By a suitable choice of both functions f and g one can obviously change the domain of elastic behaviour of the semilinear constitutive equation with ψ given by (2.2.13) so as to make it coincide with those discussed in Subsections 1.3—1.5. Let us notice that the present case includes SOKOLOVSKIĬ's constitutive equation as well as MALVERN's. Concerning the experimental determination of the functions f and g, all the remarks at the end of Subsection 1.5 hold in this case too.

2.5. One can also find in the literature several examples of functions ψ that do not depend only on the overstress $\sigma - f(\varepsilon)$. Thus, CRISTESCU [1972*b*] used a function $\psi(\varepsilon, \sigma)$ of the form

$$\psi(\varepsilon, \sigma) = \begin{cases} k(\varepsilon)(\sigma - f(\varepsilon)) & \text{if } \sigma \geqslant f(\varepsilon), \\ 0 & \text{if } \sigma < f(\varepsilon), \end{cases} \tag{2.2.15}$$

defined for $\varepsilon \geqslant 0$ and $\sigma \geqslant 0$, where

$$k(\varepsilon) = k_0 \left(1 - \exp\left(-\frac{\varepsilon}{\hat{\varepsilon}}\right)\right), \tag{2.2.16}$$

with $k_0 > 0$ and $\hat{\varepsilon} > 0$ two constants, or

$$k(\varepsilon) = \begin{cases} 0 & \text{if } 0 \leqslant \varepsilon < \varepsilon_0, \\ k_1 & \text{if } \varepsilon_0 \leqslant \varepsilon \leqslant \varepsilon_1, \\ \dfrac{k_1 - k_2}{\varepsilon_1 - \varepsilon_2}(\varepsilon - \varepsilon_2) + k_2 & \text{if } \varepsilon_1 < \varepsilon < \varepsilon_2, \\ k_2 & \text{if } \varepsilon_2 \leqslant \varepsilon, \end{cases} \tag{2.2.17}$$

with $0 < \varepsilon_0 < \varepsilon_1 < \varepsilon_2$, $0 \leqslant k_1 < k_2$ all constant. In particular, if $\varepsilon_0 > \varepsilon_Y$ and $k_1 > 0$, the corresponding constitutive equation incorporates a so called "delayed plasticity" effect (see CRISTESCU [1972 b]).

Other forms for the function ψ have been proposed within the framework of dislocation theory (see JOHNSTON and GILMAN [1959]). One interesting form is given by (see KURIYAMA and KAWATA [1973])

$$\psi(\varepsilon, \sigma) = \begin{cases} a\left[N_0 + \dfrac{3}{4} M\left(\varepsilon - \dfrac{\sigma}{E}\right)\right] \exp\left(-\dfrac{2D}{\sigma - 2H\left[\dfrac{3}{4}\left(\varepsilon - \dfrac{\sigma}{E}\right)\right]^n}\right) \\ \qquad\qquad \text{if } 0 \leqslant \dfrac{\sigma}{E} \leqslant \varepsilon < \dfrac{\sigma}{E} + \dfrac{4}{3}\left(\dfrac{\sigma}{2H}\right)^{\frac{1}{n}}, \\ 0 \qquad\qquad \text{if } \dfrac{\sigma}{E} + \dfrac{4}{3}\left(\dfrac{\sigma}{2H}\right)^{\frac{1}{n}} \leqslant \varepsilon, \ \sigma \geqslant 0, \end{cases} \tag{2.2.18}$$

where a, D, H, M, N_0 and n are positive constants. The partial derivatives of any order of the function ψ given by (2.2.18) are continuous across the quasistatic curve

$$\varepsilon = \frac{\sigma}{E} + \frac{4}{3}\left(\frac{\sigma}{2H}\right)^{\frac{1}{n}}$$

and therefore ψ is of class C^∞ in the domain $0 < \sigma < E\varepsilon$.

2.6. The general form of a semilinear constitutive equation is

$$\dot{\sigma} = E\dot{\varepsilon} + \psi(\varepsilon, \sigma), \tag{2.2.19}$$

where $(\varepsilon, \sigma) \in \mathscr{D}$ and $\dot{\varepsilon}$ and $\dot{\sigma}$ are the strain rate and the stress rate respectively at the state (ε, σ).

Let $(\tilde{\varepsilon}, \tilde{\sigma})$ be an arbitrary state in $\mathscr{D}$. If a smooth strain history $\varepsilon(t)$, $t \in [0, T]$, $T > 0$, with $\varepsilon(0) = \tilde{\varepsilon}$ is given, then the corresponding stress history $\sigma(t)$, $t \in [0, T_1)$, $T_1 \in (0, T]$, can be determined as the solution of equation (2.2.19) with the initial condition $\sigma(0) = \tilde{\sigma}$ (for existence, uniqueness and behaviour of the solutions of equation (2.2.1) for $\varepsilon = \varepsilon(t)$ in different classes of functions, see SULICIU [1973] and Chapter V, Section 3).

If one gives a smooth strain history $\varepsilon(t)$ with $\varepsilon(0) = \bar{\varepsilon} \neq \tilde{\varepsilon}$ (i.e., the strain jumps in time at the state $(\tilde{\varepsilon}, \tilde{\sigma})$ from $\tilde{\varepsilon}$ to $\bar{\varepsilon}$, see Chapter V, Section 3), then the initial condition $\sigma(0+) = \bar{\sigma}$ to be attached to equation (2.2.19) will also have the property $\bar{\sigma} \neq \tilde{\sigma}$. All such states $(\bar{\varepsilon}, \bar{\sigma}) \in \mathscr{D}$ will be called *states of instantaneous response relative to the state* $(\tilde{\varepsilon}, \tilde{\sigma}) \in \mathscr{D}$ and they satisfy the equation

$$\mathrm{d}\bar{\sigma} = E\,\mathrm{d}\bar{\varepsilon}. \tag{2.2.20}$$

Equation (2.2.20) has a unique solution passing through $(\tilde{\varepsilon}, \tilde{\sigma})$; this is the solution given by

$$\bar{\sigma} = \tilde{\sigma} + E(\bar{\varepsilon} - \tilde{\varepsilon}). \tag{2.2.21}$$

The straight line (2.2.21) is called the *instantaneous response straight line at the state* $(\tilde{\varepsilon}, \tilde{\sigma}) \in \mathscr{D}$. The instantaneous response for a semilinear rate type constitutive equation is therefore always linearly elastic. Thus, if the strain history $\varepsilon(t)$ from a state $(\tilde{\varepsilon}, \tilde{\sigma}) \in \mathscr{D}$ is such that $\varepsilon(0) = \bar{\varepsilon} \neq \tilde{\varepsilon}$, the initial condition in stress is $\sigma(0+) = \bar{\sigma} = \tilde{\sigma} + E(\bar{\varepsilon} - \tilde{\varepsilon})$ (concerning the notion of instantaneous response for equation (2.2.1) see SULICIU [1973] and Chapter V, Section 3).

Let $(\varepsilon, \sigma) \in \mathscr{D}$ be an arbitrary state; the elastic strain ε^e and the anelastic strain ε^n (or viscoplastic strain ε^{vp}) are defined as

$$\varepsilon^e = \frac{\sigma}{E}, \qquad \varepsilon^n = \varepsilon - \varepsilon^e. \tag{2.2.22}$$

Combining (2.2.22) with (2.2.19) we get

$$\dot{\varepsilon}^n = -\frac{1}{E}\psi\left(\frac{\sigma}{E} + \varepsilon^n, \sigma\right). \tag{2.2.23}$$

Hence the anelastic strain rate does not depend on the strain rate and the stress rate and vanishes at any point of elastic behaviour (of equilibrium).

Let us discuss now the notion of permanent strain attached to a state $(\tilde{\varepsilon}, \tilde{\sigma}) \in \mathscr{D}$ with $\psi(\tilde{\varepsilon}, \tilde{\sigma}) \neq 0$. Let ψ in (2.2.19) have one of its particular forms, say (2.2.9) with $(2.2.11)_1$. Let $\sigma(t)$ be a strictly decreasing smooth stress history with $\sigma(0) = \tilde{\sigma} > f(\tilde{\varepsilon})$. The anelastic strain is determined as the solution of equation (2.2.23) with the initial condition $\varepsilon^n(0) = \tilde{\varepsilon} - \dfrac{\tilde{\sigma}}{E}$; it increases as long as $\sigma(t) > f\left(\dfrac{\sigma(t)}{E} + \varepsilon^n(t)\right)$ and remains constant for $t > t_0$, where t_0 is the time for which $\sigma(t_0) = f\left(\dfrac{\sigma(t_0)}{E} + \varepsilon^n(t_0)\right)$. $\varepsilon^n(t_0)$ could be called the permanent strain attached to the state $(\tilde{\varepsilon}, \tilde{\sigma})$ if, in contrast to the similar notion given in Subsections 1.3—1.4, it did not depend on the form of the history $\sigma(t)$ used to determine it by means of the equation (2.2.19). Therefore one cannot define within this theory such a notion, attached to a given state. However, one may introduce the notion of *permanent strain for instantaneous unloading* ($\bar{\sigma} = 0$), attached to a state $(\tilde{\varepsilon}, \tilde{\sigma})$. Indeed, if at the state $(\tilde{\varepsilon}, \tilde{\sigma}) \in \mathscr{D}$ a stress history $\sigma(t) = 0$, $t \in [0, T)$ is given then, by taking $\bar{\sigma} = 0$ in (2.2.21) one gets, as $\bar{\sigma} = 0 < f(\bar{\varepsilon})$,

$$\varepsilon^R = \bar{\varepsilon} = \tilde{\varepsilon} - \frac{\tilde{\sigma}}{E} = \varepsilon^n(t) = \text{const.} \tag{2.2.24}$$

Hence the permanent strain corresponding to an instantaneous unloading from a given state $(\tilde{\varepsilon}, \tilde{\sigma})$ is identical with the anelastic strain (2.2.22) attached to that state.

2.7. The general form (2.2.1) of a quasilinear rate type constitutive equation has been proposed by CRISTESCU [1963, 1964] and used afterwards by many authors (see LUBLINER [1964], CRISTESCU [1967], etc.). Consideration of a non-constant function $\varphi(\varepsilon, \sigma)$ is a necessity that is experimentally justified. Thus, for many materials, during impact experiments of two bars one finds that the elastic modulus E increases with the impact velocity (compare the elastic bar velocity obtained by EFRON and MALVERN [1969], BELL [1960] and FILBEY [1961] for the same type of alluminium). Therefore one cannot assume that $\varphi(\varepsilon, \sigma) = E = \text{const.}$ for all $(\varepsilon, \sigma) \in \mathscr{D}$, where $\mathscr{D}$ includes all strain and stress states of interest in studying the behaviour of a given material (see also CAMPBELL [1973]). A more detailed approach to the properties of the constitutive equation (2.2.1) will be given in Chapter V.

3. RATE-TYPE CONSTITUTIVE EQUATIONS WITH INSTANTANEOUS PLASTIC STRAIN

3.1. Rate-type constitutive equations with instantaneous plastic strain may be formulated in several ways (see CRISTESCU [1963], LUBLINER [1964] and also CRISTESCU [1967], Chapter III, Sections 1—4). We shall give here a relative general formulation including the constitutive equations used for the numerical computations in Chapter IV.

The main idea consists in bringing together in the same constitutive equation the properties of a semilinear rate-type constitutive equation as well as those of a classical one (time-independent). One assumes that a state domain $\mathscr{D}$ is given and, for any state $(\varepsilon, \sigma) \in \mathscr{D}$ and any stress rate $\dot{\sigma}$ at that state, there are also given two functions $\mathscr{F}(\varepsilon, \sigma, \text{sign}\, \dot{\sigma})$ and $\psi(\varepsilon, \sigma)$. Then the strain rate $\dot{\varepsilon}$ at the state $(\varepsilon, \sigma) \in \mathscr{D}$ is determined by $\dot{\sigma}$ according to the relation

$$\dot{\varepsilon} = \left(\frac{1}{E} + \mathscr{F}(\varepsilon, \sigma, \text{ sign } \dot{\sigma})\right) \dot{\sigma} + \psi(\varepsilon, \sigma). \tag{2.3.1}$$

Equation (2.3.1) may be interpreted as follows: the strain rate $\dot{\varepsilon}$ is the sum of two strain rates; one of them is given by a classical constitutive equation (2.1.4′) while the other one is given by a semilinear rate-type constitutive equation (2.2.19).

3.2. Let us consider in the following the case $\varepsilon \geqslant 0$, $E\varepsilon \geqslant \sigma \geqslant 0$ only. In Chapter IV the function $\mathscr{F}(\varepsilon, \sigma, \text{sign}\, \dot{\sigma})$ will be chosen in the form (see CRISTESCU [1972*b*])

$$\mathscr{F}(\varepsilon, \sigma, \text{sign}\, \dot{\sigma}) = \begin{cases} \Phi(\varepsilon, \sigma) & \text{if } \dot{\sigma} \geqslant 0, \\ 0 & \text{if } \dot{\sigma} < 0, \end{cases} \tag{2.3.2}$$

where $\Phi(\varepsilon, \sigma)$ is a non-negative function of class C^1 for $0 \leqslant \varepsilon$, $0 \leqslant \sigma \leqslant E\varepsilon$.

The elastic strain ε^e and anelastic strain ε^n can be defined the same way as in the previous sections, that is

$$\varepsilon^e = \frac{\sigma}{E}, \qquad \varepsilon^n = \varepsilon - \varepsilon^e.$$

Then (2.3.1) can be written in the form

$$\dot{\varepsilon}^n = \mathscr{F}\left(\varepsilon^n + \frac{\sigma}{E}, \sigma, \text{sign}\, \dot{\sigma}\right)\dot{\sigma} + \psi\left(\varepsilon^n + \frac{\sigma}{E}, \sigma\right). \tag{2.3.3}$$

The solution $\varepsilon = f_I(\sigma)$, $\sigma > \sigma_0$, of equation

$$\frac{d\varepsilon}{d\sigma} = \frac{1}{E} + \mathscr{F}(\varepsilon, \sigma, \operatorname{sign} \dot{\sigma} \geqslant 0) = \frac{1}{E} + \Phi(\varepsilon, \sigma) \tag{2.3.4}$$

through the point ($\varepsilon = \varepsilon_0$, $\sigma = \sigma_0$) will be called the *instantaneous response curve relative to the state* $(\varepsilon_0, \sigma_0) \in \mathscr{D}$. This curve has the same meaning as the curve (2.2.21); it contains all states (ε, σ), $\sigma > \sigma_0$, which can be obtained from $(\varepsilon_0, \sigma_0)$ by a jump in stress (or strain).

Let us consider a fixed state on the instantaneous response curve, say $(f_I(\tilde{\sigma}), \tilde{\sigma})$, $\tilde{\sigma} > \sigma_0$; according to (2.3.2), the instantaneous curve through this state is for $0 \leqslant \sigma \leqslant \tilde{\sigma}$ a solution of the equation

$$\frac{d\varepsilon}{d\sigma} = \frac{1}{E} + \mathscr{F}(\varepsilon, \sigma, -1) = \frac{1}{E}, \tag{2.3.5}$$

and therefore it is the straight line

$$\varepsilon = f_I(\tilde{\sigma}) + \frac{1}{E}(\sigma - \tilde{\sigma}). \tag{2.3.6}$$

If Φ does not vanish on $\varepsilon = \varepsilon_0 + \frac{1}{E}(\sigma - \sigma_0)$, $\sigma > \sigma_0$, i.e. if $\Phi\left(\varepsilon, \varepsilon_0 + \frac{1}{E}(\sigma - \sigma_0)\right) \neq 0$, then $f_I(\tilde{\sigma}) - \frac{\tilde{\sigma}}{E} > 0$.

Let us suppose now that at the state $(\varepsilon_0, \sigma_0)$ a stress $\sigma = \tilde{\sigma} > \sigma_0$ is suddenly applied, followed by a sudden application of a stress $\sigma = 0$, which is kept equal to zero for $t > 0$; the result is a state $(f_I(\tilde{\sigma}), \tilde{\sigma})$, followed by a state $\left(\varepsilon = f_I(\tilde{\sigma}) - \frac{\tilde{\sigma}}{E}, 0\right)$. If $\psi(\varepsilon, 0) = 0$, then

$$\varepsilon^n(t) = f_I(\tilde{\sigma}) - \frac{\tilde{\sigma}}{E} = \text{const.} > 0, \quad t \geqslant 0 \tag{2.3.7}$$

is a solution of equation (2.3.3). Therefore a sudden loading followed by a sudden unloading to zero stress leads to a non-zero permanent strain. This behaviour is the reason why one calls equation (2.3.1) a *rate-type constitutive equation with instantaneous plasticity*.

In some applications in Chapter IV we shall assume that

$$\Phi(\varepsilon, \sigma) \geqslant 0, \quad \psi(\varepsilon, \sigma) \geqslant 0 \quad \text{for} \quad (\varepsilon, \sigma) \in \mathscr{D}, \tag{2.3.8}$$

and further, that there exists a curve $(\varepsilon, \sigma = f(\varepsilon)) \in \mathscr{D}$ with the properties

$$\psi(\varepsilon, \sigma) > 0, \qquad f(\varepsilon) < \sigma,$$

$$\psi(\varepsilon, \sigma) = 0, \qquad f(\varepsilon) \geqslant \sigma; \qquad\qquad (2.3.9)$$

which will be called the *relaxation boundary*.

By means of the above mentioned curves one can indicate various significant domains (see fig. 2.3.1). If in equation (2.3.4) one has

$$\Phi\left(\frac{\sigma}{E}, \sigma\right) > 0 \text{ and}$$

$$\left(\frac{1}{E} + \Phi(\varepsilon, \sigma)\right) \frac{\partial \Phi(\varepsilon, \sigma)}{\partial \varepsilon} + \frac{\partial \Phi(\varepsilon, \sigma)}{\partial \sigma} > 0 \text{ for } \sigma > \sigma_Y, \ f(\varepsilon) < \sigma < E\varepsilon,$$

D_2 & D_1 Are Possible States From the State $(0,0)$

D_1 & D_3 Are Not

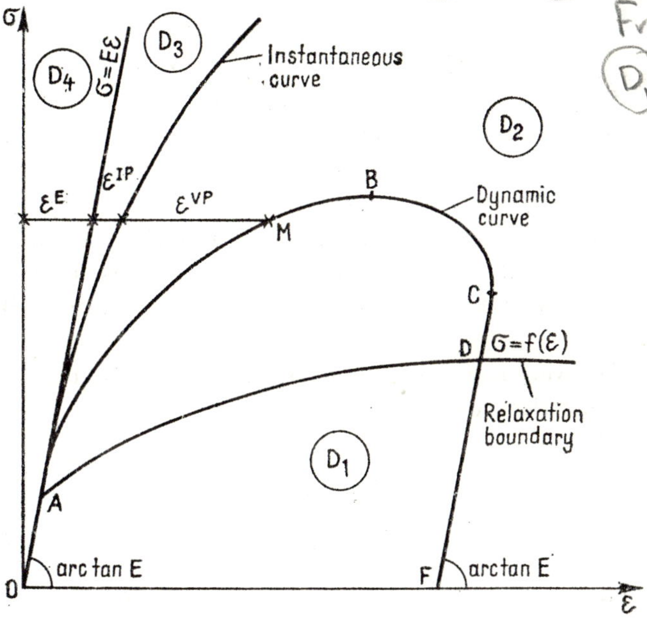

Fig. 2.3.1 The position of the stress-strain curve at a given section of a bar in an impact experiment (the dynamic curve) and different possible decompositions of the total strain

and one takes $\varepsilon_0 = 0$, $\sigma_0 = 0$, then the obtained instantaneous curve relative to the state $(0, 0)$ is the same as in fig. 2.3.1. All points in D_1 and D_2 are possible states of stress and strain. The domains D_3 and D_4 have no physical meaning for a loading or unloading process from $\varepsilon \geqslant 0$, $\sigma \geqslant 0$, i.e. no state $(\varepsilon, \sigma) \in D_3 \cup D_4$ can be reached from $(0, 0)$ when $\sigma(t) \geqslant 0$, $t \geqslant 0$ is prescribed and $\varepsilon(t)$ is determined as the solution of equation (2.3.1) with $\varepsilon(0) = 0$. A possible solution of (2.3.1), (2.3.2) through $(0, 0)$, for a given smooth $\sigma(t)$, has been represented in fig. 2.3.1 and called a *dynamic curve*. If one considers an arbitrary point M on this curve, the strain at this point can be written as the sum of three components called the *elastic*, *instantaneous plastic* and *viscoplastic* strains respectively, that is

$$\varepsilon = \varepsilon^e + \varepsilon^{Ip} + \varepsilon^{vp}. \tag{2.3.10}$$

One obviously has

$$\varepsilon^n = \varepsilon^{Ip} + \varepsilon^{vp}, \tag{2.3.11}$$

and ε^n is determined as the solution of equation (2.3.3) when $\sigma(t)$ is known. The relation between the instantaneous strain ε^I, the instantaneous plastic strain ε^{Ip} and the elastic strain ε^e for a given stress can be written by means of the function f_I as follows

$$\varepsilon^I = \varepsilon^e + \varepsilon^{Ip} = \frac{\sigma}{E} + \varepsilon^{Ip} = f_I(\sigma). \tag{2.3.12}$$

3.3. The constitutive equation (2.3.1) obviously reduces to the semi-linear rate-type equation (2.2.19) when $\mathscr{F}(\varepsilon, \sigma, \operatorname{sign} \dot{\sigma}) \equiv 0$. In this case the domain D_3 in fig. 2.3.1 disappears. It also disappears when the instantaneous curve has a greater slope than the elastic slope E (and this may happen during dynamic experiments in compression with large loading rates), but in this case Φ can no longer stay non-negative on the whole domain $\mathscr{D}$.

The following is an example of a function Φ (Cristescu [1972 b]) used in the numerical computations in Chapter IV:

$$\Phi(\varepsilon, \sigma) = \begin{cases} 0 & \text{if } 0 \leqslant \varepsilon \leqslant \check{\varepsilon},\ E(\varepsilon - \check{\varepsilon}) \leqslant \sigma \leqslant E\varepsilon, \\[2ex] \dfrac{3\left[\varepsilon - \dfrac{\sigma}{E} - \check{\varepsilon} + \left(\dfrac{a(\varepsilon)}{3E}\right)^{3/2}\right]^{2/3}}{a(\varepsilon)} - \dfrac{1}{E} & \text{if } f(\varepsilon) < \sigma < E(\varepsilon - \check{\varepsilon}), \\[2ex] 0 & \text{if } 0 \leqslant \sigma \leqslant f(\varepsilon), \end{cases} \tag{2.3.13}$$

where

$$a(\varepsilon) = m + n\sqrt{\varepsilon} \tag{2.3.14}$$

and $\check{\varepsilon} \geqslant 0$, $m > 0$ and $n > 0$ are constants.

For $\Phi(\varepsilon, \sigma)$ given by (2.3.13) the instantaneous curve (2.3.4) relative to the state $(0, 0)$ reduces to the elastic line $\sigma = E\varepsilon$, while the permanent strain (2.3.7) vanishes. The domain D_3 in fig. 2.3.1 disappears.

4. CHARACTERISTIC FIELDS, INITIAL AND BOUNDARY VALUE PROBLEMS

4.1. **Characteristic fields. Relations along characteristics.** The system of partial differential equations that describes the one-dimensional motion of an *elastic-plastic* or *elastic-visco-plastic body* can be written in the form

$$\rho_0 \frac{\partial v}{\partial t} - \frac{\partial \sigma}{\partial x} = 0,$$

$$\frac{\partial \varepsilon}{\partial t} - \frac{\partial v}{\partial x} = 0, \tag{2.4.1}$$

$$\frac{\partial \sigma}{\partial t} - \left(\varphi(\varepsilon, \sigma) + \mathscr{E}\left(\varepsilon, \sigma, \operatorname{sign} \frac{\partial \varepsilon}{\partial t}\right)\right) \frac{\partial \varepsilon}{\partial t} = \psi(\varepsilon, \sigma)$$

where $(2.4.1)_1$ is the *balance of momentum*, $(2.4.1)_2$ is the *compatibility equation* and $(2.4.1)_3$ is the *constitutive equation*. One can easily see that all constitutive equations (2.1.3′), (2.2.1), (2.2.19) and (2.3.1) can be obtained from (2.4.1) by a suitable choice of the functions φ, $\mathscr{E}$ and ψ.

Substituting $(2.4.1)_2$ into $(2.4.1)_3$ one gets

$$\frac{\partial \sigma}{\partial t} - \left(\varphi(\varepsilon, \sigma) + \mathscr{E}\left(\varepsilon, \sigma, \operatorname{sign} \frac{\partial v}{\partial x}\right)\right) \frac{\partial v}{\partial x} = \psi(\varepsilon, \sigma). \tag{2.4.2}$$

The system $(2.4.1)_{1-3}$ or $(2.4.1)_{1-2}$ and (2.4.2) is not a quasilinear system since $\mathscr{E}$ depends on $\partial\varepsilon/\partial t$, depending on sign $(\partial\varepsilon/\partial t)$ (or on $\partial v/\partial x$ respectively). However, it is an "almost quasilinear" system of partial differential equations. The notions of characteristic directions and characteristic vectors may be introduced in the same way as for quasilinear systems, yet they depend not only on the unknown functions but also on the sign of the partial derivative $\partial\varepsilon/\partial t$ (or $\partial v/\partial x$ respectively) at the considered point (x, t).

Let us introduce the following notations

$$V_1 = v, \; V_2 = \varepsilon, \; V_3 = \sigma,$$

$$A_{11} = A_{12} = A_{22} = A_{23} = A_{32} = A_{33} = 0, \tag{2.4.3}$$

$$A_{21} = -1, \; A_{13} = -1/\rho_0, \; A_{31} = -\left[\varphi(\varepsilon, \sigma) + \mathscr{E}\left(\varepsilon, \sigma, \operatorname{sign} \frac{\partial v}{\partial x}\right)\right] = -a,$$

$$b_1 = b_2 = 0, \quad b_3 = \psi(\varepsilon, \sigma).$$

Using these, we can write the system $(2.4.1)_{1-2}$ and (2.4.2) in the "normal" form

$$\frac{\partial V_i'}{\partial t} + A_{ij} \frac{\partial V_j}{\partial x} = b_i, \qquad i = 1, 2, 3. \tag{2.4.4}$$

Let $P_0=(x_0, t_0)$ be a point in the plane and assume that a smooth solution of the system (2.4.4) is given in a neighbourhood D of P_0. Let $x = \hat{x}(t)$ be a smooth curve $\mathscr{C}$ through P_0. Along this curve, one defines the smooth functions $\hat{V}_i(t) = V_i(\hat{x}(t), t)$, $i = 1, 2, 3$; therefore along the curve

$$\frac{\partial V_i}{\partial t} + \frac{\partial V_i}{\partial x}\lambda = \frac{\mathrm{d}\hat{V}_i}{\mathrm{d}t}, \qquad i = 1, 2, 3, \tag{2.4.5}$$

holds, where

$$\lambda = \lambda(t) = \frac{\mathrm{d}\hat{x}(t)}{\mathrm{d}t}. \tag{2.4.6}$$

If one substitutes $\partial V_i/\partial t$ from (2.4.5) into (2.4.4), one finds

$$(A_{ij} - \lambda\delta_{ij})\frac{\partial V_j}{\partial x} = b_i - \frac{\mathrm{d}\hat{V}_i}{\mathrm{d}t}. \tag{2.4.7}$$

The eigenvalues of $\mathbf{A}$ in P_0, i.e. the roots of the algebraic equation in λ

$$\det(A_{ij} - \lambda\delta_{ij}) = 0, \tag{2.4.8}$$

will be called *the eigenvalues of* $\mathbf{A}$ *at the point* $P_0 \in D$.

The three-dimensional vector $\mathbf{r}$ that verifies the relation

$$\mathbf{A}\mathbf{r} = \lambda\mathbf{r}, \tag{2.4.9}$$

will be called the *right eigenvector at* P_0, *corresponding to the eigenvalue* λ.

The system (2.4.4) is said to be *hyperbolic* in D if, at any $P_0 \in D$: 1) all three roots $\lambda_1, \lambda_2, \lambda_3$ of equation (2.4.8) are real and 2) the three eigenvectors $\mathbf{r}_1, \mathbf{r}_2, \mathbf{r}_3$ corresponding to the eigenvalues $\lambda_1, \lambda_2, \lambda_3$ respectively, are linearly independent.

According to (2.4.3), equation (2.4.8) may be written as

$$\lambda\left(\lambda^2 - \frac{a}{\rho_0}\right) = 0; \tag{2.4.10}$$

hence, the first hyperbolicity condition is satisfied if and only if $a \geqslant 0$. If $a = 0$ at a certain point, then $\lambda = 0$ is a triple root of (2.4.10) and all eigenvectors defined by (2.4.9) are proportional to the vector (0, 1, 0). If

$$a\left(\varepsilon, \sigma, \operatorname{sign}\frac{\partial v}{\partial x}\right) = a\left(\varepsilon, \sigma, \operatorname{sign}\frac{\partial \varepsilon}{\partial t}\right) = \varphi(\varepsilon, \sigma) + \mathscr{E}\left(\varepsilon, \sigma, \operatorname{sign}\frac{\partial \varepsilon}{\partial t}\right) > 0, \tag{2.4.11}$$

the three eigenvectors will be linearly independent. Therefore, condition (2.4.11) represents the necessary and sufficient hyperbolicity condition for the system (2.4.1) or (2.4.4).

The characteristic curves through an arbitrary point $P_0 \in D$ are the solutions of the equations

$$\lambda = \frac{dx}{dt} = 0,$$

$$\lambda^2 = \left(\frac{dx}{dt}\right)^2 = \frac{1}{\rho_0}\left[\varphi(\varepsilon(x,t),\sigma(x,t)) + \right. \tag{2.4.12}$$

$$\left. + \mathscr{E}\left(\varepsilon(x,t),\sigma(x,t),\operatorname{sign}\frac{\partial\varepsilon(x,t)}{\partial t}\right)\right] = \frac{a}{\rho_0}$$

through P_0. Equation $(2.4.12)_2$ cannot have smooth solutions in D, even if φ and $\mathscr{E}$ are continuous functions of ε and σ, unless $\frac{\partial\varepsilon}{\partial t}$ does not change its sign in D.

The relations along characteristics are obtained from the conditions imposed on (2.4.7), considered as an algebraic system for the unknowns $\frac{\partial V_i}{\partial x}$, which ensure that there are solutions for those values of λ that verify equation (2.4.8). Under conditions (2.4.11) the relations along the characteristics are

$$\begin{aligned}
&\frac{d\hat{\sigma}}{dt} - a\frac{d\hat{\varepsilon}}{dt} - \psi = 0 && \text{for } \lambda = \frac{dx}{dt} = 0,\\
&\frac{d\hat{\sigma}}{dt} - \rho_0\lambda\frac{d\hat{v}}{dt} - \psi = 0 && \text{for } \lambda = \frac{dx}{dt} = \sqrt{\frac{a}{\rho_0}},\\
&\frac{d\hat{\sigma}}{dt} + \rho_0\lambda\frac{d\hat{v}}{dt} - \psi = 0 && \text{for } \lambda = \frac{dx}{dt} = -\sqrt{\frac{a}{\rho_0}}
\end{aligned} \tag{2.4.13}$$

If the constitutive equation is given as

$$\frac{\partial\varepsilon}{\partial t} = \left(\Phi(\varepsilon,\sigma) + \mathscr{F}\left(\varepsilon,\sigma,\operatorname{sign}\frac{\partial\sigma}{\partial t}\right)\right)\frac{\partial\sigma}{\partial t} + \Psi(\varepsilon,\sigma), \tag{2.4.2'}$$

the system $(2.4.1)_{1-2}$,(2.4.2′) can no longer be written in the form (2.4.4). However, one may consider the relations $(2.4.1)_{1-2}$, (2.4.2′) and (2.4.5) as an algebraic system consisting of six equations for the unknowns $\partial V_i/\partial t$, $\partial V_i/\partial x$, $i = 1, 2, 3$. By equating the determinant of this system to zero, one obtains the characteristic directions $(\lambda, 1)$ through a point $P_0 \in D$. The relations along the characteristics are determined as in the previous case. Thus the characteristic directions are

$$\lambda = \frac{dx}{dt} = 0,$$

$$\lambda^2 = \left(\frac{dx}{dt}\right)^2 = \frac{1}{\rho_0\left(\bar{\Phi}(\varepsilon, \sigma) + \mathscr{F}\left(\varepsilon, \sigma, \operatorname{sign}\frac{\partial \sigma}{\partial t}\right)\right)}; \tag{2.4.10'}$$

and the necessary and sufficient condition for the system to be hyperbolic is

$$\bar{\Phi}(\varepsilon, \sigma) + \mathscr{F}\left(\varepsilon, \sigma, \operatorname{sign}\frac{\partial \sigma}{\partial t}\right) > 0. \tag{2.4.11'}$$

The relations along the characteristics become in this case,

$$\frac{d\hat{\varepsilon}}{dt} - \left(\bar{\Phi}(\varepsilon, \sigma) + \mathscr{F}\left(\varepsilon, \sigma, \operatorname{sign}\frac{\partial \sigma}{\partial t}\right)\right)\frac{d\hat{\sigma}}{dt} - \Psi(\varepsilon, \sigma) = 0 \quad \text{for } \lambda = \frac{dx}{dt} = 0,$$

$$\frac{d\hat{\sigma}}{dt} \mp \rho_0 \lambda \frac{d\hat{v}}{dt} + \rho_0 \lambda^2 \Psi = 0 \quad \text{for } \lambda = \frac{dx}{dt} = \pm \frac{1}{\sqrt{\rho_0\left(\bar{\Phi}(\varepsilon, \sigma) + \mathscr{F}\left(\varepsilon, \sigma, \operatorname{sign}\frac{\partial \sigma}{\partial t}\right)\right)}} \tag{2.4.13'}$$

with the signs in $(2.4.13')_2$ corresponding to each other.

4.2. Let us study presently the characteristic directions, the characteristic curves and the relations satisfied along them for the different types of constitutive equations defined in Sections 1—3.

The constitutive equations of classical plasticity presented in Section 1 are obtained from $(2.4.1)_3$ by taking

$$\varphi(\varepsilon, \sigma) \equiv 0, \quad \psi(\varepsilon, \sigma) \equiv 0, \quad (\varepsilon, \sigma) \in \mathscr{D}. \tag{2.4.14}$$

The system (2.4.1), with $(2.4.1)_3$ replaced by the constitutive equation (2.1.3'), is hyperbolic if and only if

$$\mathscr{E}\left(\varepsilon, \sigma, \operatorname{sign}\frac{\partial \varepsilon}{\partial t}\right) > 0 \tag{2.4.15}$$

for any $(\varepsilon, \sigma) \in \mathscr{D}$ and any strain rate $\frac{\partial \varepsilon}{\partial t}$. Thus, for $\mathscr{E}$ defined by (2.1.7) the system (2.4.1) is hyperbolic for all states $(\varepsilon, \sigma) \in \mathscr{D}$ if and only if the functions f and g satisfy the conditions

$$f'(\varepsilon) > 0, \quad g'(\varepsilon) > 0. \tag{2.4.16}$$

For the constitutive equation defined in Subsection 1.3, the conditions (2.4.16) are not satisfied since $f(\varepsilon) = \sigma_Y$ and $g(\varepsilon) = -\sigma_Y$, $\varepsilon \in R$. Thus one may have the following situation in the domain D in the (x, t)-plane where the system (2.4.1) has a smooth solution: The characteristics (2.4.12) are

$$\lambda = \frac{\mathrm{d}x}{\mathrm{d}t} = 0, \qquad \lambda^2 = \left(\frac{\mathrm{d}x}{\mathrm{d}t}\right)^2 = c_0^2 = \frac{E}{\rho_0}, \tag{2.4.17}$$

at any point $(x, t) \in D$ with $|\sigma(x, t)| < \sigma_Y$ or $|\sigma(x, t)| = \sigma_Y$ and $\operatorname{sign} \sigma(x, t) \operatorname{sign} \frac{\partial \varepsilon}{\partial t}(x, t) < 0$, and

$$\lambda^3 = \left(\frac{\mathrm{d}x}{\mathrm{d}t}\right)^3 = 0 \tag{2.4.18}$$

at any point $(x, t) \in D$ where

$$|\sigma(x, t)| = \sigma_Y \text{ and } \operatorname{sign} \sigma(x, t) \operatorname{sign} \frac{\partial \varepsilon}{\partial t}(x, t) \geqslant 0. \tag{2.4.19}$$

Therefore, if D contains points where (2.4.19) is satisfied, the system (2.4.1) no longer will be hyperbolic in D. One can easily verify that the system $(2.4.1)_{1-2}$, (2.4.2′) with $\Phi = 0$, $\Psi = 0$ and $\mathscr{F}$ given by (2.1.8′) is not a hyperbolic system on any open set D containing points (x, t) satisfying $|\varepsilon(x, t)| = \varepsilon_l$ and $\operatorname{sign} \varepsilon(x, t) \times \operatorname{sign} \dot{\sigma}(x, t) \geqslant 0$.

One of the simplest constitutive equations of classical plasticity is that presented in Subsection 1.4. In this case the characteristics of the system (2.4.1) are those given by (2.4.17) at any point $(x, t) \in D$ where

$$|\sigma(x, t) - E_1\varepsilon(x, t)| < \sigma_Y^0 \tag{2.4.20}$$

or

$$|\sigma(x, t) - E_1\varepsilon(x, t)| = \sigma_Y^0 \text{ and } \operatorname{sign}(\sigma(x, t) - E_1\varepsilon(x, t)) \operatorname{sign} \frac{\partial \varepsilon}{\partial t}(x, t) < 0 \tag{2.4.21}$$

and by

$$\lambda = \frac{\mathrm{d}x}{\mathrm{d}t} = 0, \qquad \lambda^2 = \left(\frac{\mathrm{d}x}{\mathrm{d}t}\right)^2 = c_1^2 = \frac{E_1}{\rho_0} \tag{2.4.22}$$

at any point $(x, t) \in D$ where

$$|\sigma(x, t) - E_1\varepsilon(x, t)| = \sigma_Y^0 \text{ and}$$
$$\operatorname{sign}(\sigma(x, t) - E_1\varepsilon(x, t)) \operatorname{sign} \frac{\partial \varepsilon}{\partial t}(x, t) \geqslant 0. \tag{2.4.23}$$

All characteristics are straight lines with slopes $\pm c_0$ or $\pm c_1$ but D may contain certain points where the slope of a characteristic has jumps.

For any point $(x, t) \in D$ where (2.4.20) or (2.4.21) are satisfied, the relations along the characteristics (2.4.17) are

$$\frac{d\hat{\sigma}}{dt} - E\frac{d\hat{\varepsilon}}{dt} = 0 \qquad \text{for } \lambda = \frac{dx}{dt} = 0,$$
$$\frac{d\hat{\sigma}}{dt} \mp \rho_0 c_0 \frac{d\hat{v}}{dt} = 0 \qquad \text{for } \lambda = \frac{dx}{dt} = \pm\sqrt{\frac{E}{\rho_0}} = \pm c_0, \tag{2.4.24}$$

while along the characteristics (2.4.22) one has

$$\frac{d\hat{\sigma}}{dt} - E_1\frac{d\hat{\varepsilon}}{dt} = 0 \qquad \text{for } \lambda = \frac{dx}{dt} = 0,$$
$$\frac{d\hat{\sigma}}{dt} \mp \rho_0 c_1 \frac{d\hat{v}}{dt} = 0 \qquad \text{for } \lambda = \frac{dx}{dt} = \pm\sqrt{\frac{E_1}{\rho_0}} = \pm c_1. \tag{2.4.25}$$

In both relations $(2.4.24)_2$ and $(2.4.25)_2$ the lower and upper signs correspond to each other.

The characteristic equations and the relations along characteristics for the constitutive equation described in Subsection 1.5, formula (2.1.21), can be written in a similar way. The essential difference consists in the fact that there exist subdomains of D where $\sigma = f(\varepsilon)$ and $\frac{\partial \sigma}{\partial t} > 0$ or $\sigma = g(\varepsilon)$ and $\frac{\partial \sigma}{\partial t} < 0$, with the property that the characteristic slopes are not constant but depend on the value of ε at the considered point in D.

When $\mathscr{E}\left(\varepsilon, \sigma, \frac{\partial \varepsilon}{\partial t}\right) \equiv 0$ in $(2.4.1)_3$, the system (2.4.1) becomes a quasilinear system and it is hyperbolic if and only if condition (2.2.2) is satisfied. The characteristic curves through a point $P_0 \in D$ are solutions of the system (2.4.12) and they are of class C^2, at least. Thus, the constitutive equations considered in Section 2 are simpler than those considered in Sections 1 and 3. If $\varphi(\varepsilon, \sigma) = E$, i.e. case (2.2.19) holds, all the characteristic curves of the system (2.4.1) are straight lines of slope $\pm c_0 = \pm \sqrt{\frac{E}{\rho_0}}$ or the straight lines $x = \text{const.}$

The constitutive equation (2.3.1) is obtained from (2.4.2') for $\bar{\Phi}(\varepsilon, \sigma) = 1/E$. The characteristic equations (2.4.10') and the relations (2.4.13') hold in this case without further simplifications. Let us also notice that the characteristic curves of the system $(2.4.1)_{1-2}$, (2.4.2') in a domain D where this system has smooth solutions, may have a discontinuous slope.

4.3. **Initial and boundary value problems** (see COURANT [1962], COURANT and FRIEDRICHS [1948], ROZHDESTVENSKIĬ and YANENKO [1968] JEFFREY [1976]). We shall assume here that the hyperbolicity condition (2.4.11) (or (2.4.11')) for the system (2.4.1) (or $(2.4.1)_{1-2}$,(2.4.2')), is satisfied.

Let us consider an interval (α, β) (identified with the initial configuration $\mathscr{R}$ of a one-dimensional material body, see Chapter I). We assume that there are given three functions $v_0(x)$, $\varepsilon_0(x)$, $\sigma_0(x)$, defined on (α, β), such that $(\varepsilon_0(x), \sigma_0(x)) \in \mathscr{D}$ for any $x \in (\alpha, \beta)$. To solve an initial value problem for the system (2.4.1) we have to find the solution $v(x, t)$, $\varepsilon(x, t)$, $\sigma(x, t)$, $t \geqslant 0$, $x \in (\alpha, \beta)$, of the system (2.4.1) for which $(\varepsilon(x, t), \sigma(x, t)) \in \mathscr{D}$ for all $x \in (\alpha, \beta)$ and $t > 0$, and

$$v(x, 0) = v_0(x),\ \varepsilon(x, 0) = \varepsilon_0(x),\ \sigma(x, 0) = \sigma_0(x), x \in (\alpha, \beta). \qquad (2.4.26)$$

If at least one of the numbers α and β is finite, the initial conditions (2.4.26) do not determine a unique solution of the initial value problem. In this case it is necessary to supply each finite end $x = \alpha$ or (and) $x = \beta$ with additional information in the form of functions of t. The number of independent conditions generally required for the finite ends is equal to the number of characteristics pointing to the interior of the domain which one can draw through the corresponding finite end. (For further information and a justification of this assertion as well as for initial and boundary value problems formulated on curves other than $t = 0$ and $x = \text{const.}$, see the references quoted above).

The most frequent boundary conditions are

$$v(\alpha, t) = v_\alpha(t) \quad \text{or} \quad \sigma(\alpha, t) = \sigma_\alpha(t), \quad t \geqslant 0 \qquad (2.4.27)$$

and

$$v(\beta, t) = v_\beta(t) \quad \text{or} \quad \sigma(\beta, t) = \sigma_\beta(t), \ t \geqslant 0, \qquad (2.4.28)$$

where $v_\alpha(t)$, $\sigma_\alpha(t)$, $v_\beta(t)$ and $\sigma_\beta(t)$ are given functions of t for $t \geqslant 0$.

Since $\frac{\mathrm{d}\hat{\varepsilon}}{\mathrm{d}t}$ is not present in the relations along the characteristics, for $\lambda \neq 0$ (see $(2.4.13)_2$) this function cannot be specified, in general, as a boundary condition. This remark is in agreement with the experimental fact that strain cannot be measured at the end of a bar.

One can obviously consider more general boundary conditions, as for instance

$$
\begin{aligned}
&l_\alpha\left(v(\alpha, t),\ \sigma(\alpha, t),\ \frac{\partial v}{\partial t}(\alpha, t)\right) = h_\alpha(t), \\
&\qquad\qquad\qquad\qquad\qquad\qquad t \geqslant 0, \qquad\qquad (2.4.29)\\
&l_\beta\left(v(\beta, t),\ \sigma(\beta, t),\ \frac{\partial v}{\partial t}(\beta, t)\right) = h_\beta(t),
\end{aligned}
$$

where h_α, h_β, l_α and l_β are given functions. However, the functions l_α and l_β cannot be arbitrarily given; l_α has to be independent of the second relation of $(2.4.13)_2$ (i.e. independent of the relation along the characteristics of negative slope) while l_β has to be independent of the first relation $(2.4.13)_2$. Let us consider an example. Let l_α be defined by

$$
l_\alpha\left(v, \sigma, \frac{\partial v}{\partial t}\right) = Av + B\sigma
$$

with A and B given constants. Consider the constitutive equation $(2.4.1)_3$ with $\varphi = \psi = 0$ and $\mathscr{E}$ defined by (2.1.14). Then, according to the second relation $(2.4.24)_2$ and the second relation $(2.4.25)_2$, A/B has to be different from $\rho_0 c_0$ and $\rho_0 c_1$.

An initial and boundary value problem for the system (2.4.1) requires one to find that solution $v(x, t), \varepsilon(x, t), \sigma(x, t), t > 0,\ x \in (a, b)$, of the system (2.4.1), with $(\varepsilon(x, t), \sigma(x, t)) \in \mathscr{D}$ for $x \in [a, b]$, $t \geqslant 0$, which satisfies (2.4.26), (2.4.27) and (2.4.28) or (2.4.26) and (2.4.29).

Initial and boundary value problems for the system $(2.4.1)_{1-2}$, (2.4.2′) may be formulated in the same way.

5. EXAMPLE

The first initial and boundary value problems for the system (2.4.1) with $\varphi = 0$ and $\psi = 0$, have been solved during the second world war and published by RAKHMATULIN [1945], TAYLOR [1946] and von KÁRMÁN and DUWEZ [1950] (for many other initial and boundary value problems see also RAKHMATULIN and DEMIANOV [1961] and CRISTESCU [1967]).

In order to understand better the results of the dynamic experiments presented in Chapter III, Sections 2—3, we shall give here the solution of an initial and boundary value problem (the case of a semiinfinite bar) for the system

$$\rho_0 \frac{\partial v}{\partial t} - \frac{\partial \sigma}{\partial X} = 0,$$

$$\frac{\partial \varepsilon}{\partial t} - \frac{\partial v}{\partial X} = 0, \tag{2.5.1}$$

$$\frac{\partial \sigma}{\partial t} - \mathscr{E}\left(\varepsilon, \sigma, \operatorname{sgn} \frac{\partial \varepsilon}{\partial t}\right) \frac{\partial \varepsilon}{\partial t} = 0,$$

obtained from (2.4.1) by taking $\varphi = 0$ and $\psi = 0$. The initial conditions are

$$\varepsilon(X, 0) = 0, \quad \sigma(X, 0) = 0, \quad v(X, 0) = 0, \quad X > 0, \tag{2.5.2}$$

and the boundary conditions are

$$v(0, t) = \begin{cases} v_1 = \text{const. if } 0 \leqslant t < t_1, \\ v_2 = \text{const. if } t_1 \leqslant t < \infty \end{cases} \tag{2.5.3}$$

with $0 < v_1 < v_2$.

The function $\mathscr{E}\left(\varepsilon, \sigma, \operatorname{sgn} \frac{\partial \varepsilon}{\partial t}\right)$ is defined in (2.1.7). The functions $f(\varepsilon)$ and $g(\varepsilon)$ given by (2.1.6) satisfy (2.4.16) as well as the additional condition

$$f''(\varepsilon) < 0, \quad g''(\varepsilon) > 0. \tag{2.5.4}$$

The instantaneous curve (see the comments that follow relation (2.1.3″)) at the state (0, 0), is

$$\sigma = \hat{\sigma}(\varepsilon) = \begin{cases} f(\varepsilon) & \text{if } \varepsilon > \varepsilon_Y > 0, \\ E\varepsilon & \text{if } -\varepsilon_Y \leqslant \varepsilon \leqslant \varepsilon_Y, \\ g(\varepsilon) & \text{if } \varepsilon < -\varepsilon_Y, \end{cases} \tag{2.5.5}$$

where ε_Y is defined at the end of Subsection 1.2.

The solution to the initial and boundary value problem (2.5.1)—(2.5.3) can be constructed according to the method developed by von Kármán and Duwez [1950] (or another method, as for instance in Suliciu [1973] and Mihăilescu and Suliciu [1975*a, b*]); this method assumes that the unknown functions v, ε, σ in (2.5.1) depend on X and t only through the ratio

$$\xi = \frac{X}{t}. \tag{2.5.6}$$

Due to the initial and boundary conditions (2.5.2)—(2.5.3), the constitutive equation $(2.5.1)_3$ can be replaced by (2.5.5) in a domain of the quadrant $X > 0$, $t > 0$, for $t \leqslant t_1$. Thus $(2.5.1)_{1-2}$ together with (2.5.6) lead to

$$\rho_0 \xi \frac{\mathrm{d}v}{\mathrm{d}\xi} + \frac{\mathrm{d}\sigma}{\mathrm{d}\xi} = 0, \qquad \xi \frac{\mathrm{d}\varepsilon}{\mathrm{d}\xi} + \frac{\mathrm{d}v}{\mathrm{d}\xi} = 0. \tag{2.5.7}$$

Since at any point where (2.5.5) is differentiable with respect to ε one has

$$\frac{\mathrm{d}\sigma}{\mathrm{d}\xi} = \frac{\mathrm{d}\hat{\sigma}(\varepsilon)}{\mathrm{d}\varepsilon} \frac{\mathrm{d}\varepsilon}{\mathrm{d}\xi},$$

the equations (2.5.7) give

$$\frac{\mathrm{d}\varepsilon}{\mathrm{d}\xi}\left(\rho_0 \xi^2 - \frac{\mathrm{d}\hat{\sigma}(\varepsilon(\xi))}{\mathrm{d}\varepsilon}\right) = 0; \tag{2.5.8}$$

relation (2.5.8) is satisfied if either

$$\varepsilon(\xi) = \text{const.} \tag{2.5.9}$$

or

$$\xi^2 = \frac{X^2}{t^2} = \frac{1}{\rho_0} \frac{\mathrm{d}\hat{\sigma}(\varepsilon)}{\mathrm{d}\varepsilon}. \tag{2.5.10}$$

Now, since we have zero initial conditions, we obtain from (2.5.5)

$$\frac{\mathrm{d}\hat{\sigma}}{\mathrm{d}\varepsilon} = E, \qquad \varepsilon \in (-\varepsilon_Y, \varepsilon_Y); \tag{2.5.11}$$

hence, at any point (X, t) lying in the angle between the OX-axis and the straight line,

$$X = c_0 t, \qquad c_0 = \sqrt{\frac{E}{\rho_0}}, \tag{2.5.12}$$

(2.5.9) is satisfied and thus also

$$\varepsilon(X, t) = 0, \quad \sigma(X, t) = 0, \quad v(X, t) = 0. \tag{2.5.13}$$

The straight line (2.5.12) is a shock wave that produces a jump discontinuity in strain, stress and velocity from the values $(0, 0, 0)$ to the values

$$\varepsilon = -\varepsilon_Y, \quad \sigma = -E\varepsilon_Y, \quad v = v_Y = c_0\varepsilon_Y, \tag{2.5.14}$$

if

$$v_1 > v_Y = c_0\varepsilon_Y. \tag{2.5.15}$$

The stress and strain have negative values as the boundary conditions produce a compression of the bar. In order to obtain (2.5.14) one has to use the jump conditions $(1.1.11)_1$ and (1.1.12).

From (2.5.5) one has

$$\frac{d\hat{\sigma}}{d\varepsilon} = g'(\varepsilon) = \rho_0 c^2(\varepsilon), \qquad \varepsilon < -\varepsilon_Y. \tag{2.5.16}$$

Let us assume in the following that (2.5.15) is satisfied. Then, at any point (X, t) lying in the angle between the straight lines

$$\begin{aligned} X &= c(-\varepsilon_Y)\,t, \qquad c(-\varepsilon_Y) = \sqrt{\frac{g'(-\varepsilon_Y)}{\rho_0}}, \\ X &= c(\varepsilon_1)t, \qquad c(\varepsilon_1) = \sqrt{\frac{g(\varepsilon_1)}{\rho_0}}, \qquad \varepsilon_1 < -\varepsilon_Y, \end{aligned} \tag{2.5.17}$$

(2.5.10) is satisfied while ε and σ are determined as follows: For a straight line of slope $\xi = X/t \geqslant c(-\varepsilon_Y)$, we determine $\varepsilon(\xi)$ from (2.5.10), that is from $\xi = g'(\varepsilon(\xi))$ (which, due to $(2.5.4)_2$, gives a unique solution for ε), and then $\sigma(\xi) = \hat{\sigma}(\varepsilon(\xi))$ from (2.5.5). The velocity $v(\xi)$ is determined by using $(2.5.7)_1$, (2.5.6), (2.5.10) and (2.5.16) and one obtains

$$v(\xi) = v_Y - \frac{1}{\rho_0}\int_{-\varepsilon_Y}^{\varepsilon(\xi)} \sqrt{\frac{1}{\xi}\frac{d\sigma}{d\xi}}\, d\xi = v_Y - \int_{-\varepsilon_Y}^{\varepsilon(\xi)} \sqrt{\frac{g'(\varepsilon)}{\rho_0}}\, d\varepsilon. \tag{2.5.18}$$

The boundary conditions $(2.5.3)_1$ determine ε_1; it is the solution of the equation

$$v_1 = v_Y - \int_{-\varepsilon_Y}^{\varepsilon_1} \sqrt{\frac{g'(\varepsilon)}{\rho_0}}\, d\varepsilon. \tag{2.5.19}$$

At any point (X, t) that lies between the straight lines $X = c_0 t$ and $X = c(-\varepsilon_Y)t$, one has the solution (2.5.14).

In the domain lying above the straight line $X = c(\varepsilon_1)t$, for $t < t_1 + X/c(\varepsilon_1)$, one has the constant solution

$$\varepsilon(X, t) = \varepsilon_1, \quad \sigma(X, t) = g(\varepsilon_1), \quad v(X, t) = v_1. \tag{2.5.20}$$

Now, starting from the point $(0, t_1)$ of the (X, t)-plane, one can construct the solution by following a similar way as above. The constitutive equation $(2.5.1)_3$ may be replaced by a constitutive equation that is similar to (2.5.5) and has the form

$$\sigma = \hat{\sigma}(\varepsilon) = g(\varepsilon) \quad \text{for} \quad \varepsilon < \varepsilon_1. \tag{2.5.21}$$

At any point lying between the straight lines

$$X = c(\varepsilon_1)(t - t_1)$$
$$X = c(\varepsilon_2)(t - t_1), \quad \varepsilon_2 < \varepsilon_1, \tag{2.5.22}$$

one gets a solution of the form (2.5.10), i.e.

$$\xi^2 = \frac{X^2}{(t - t_1)^2} = \frac{1}{\rho_0} \frac{d\hat{\sigma}(\varepsilon)}{d\varepsilon}. \tag{2.5.23}$$

ε and σ are determined the same way as above and

$$v(\xi) = v_1 - \int_{\varepsilon_1}^{\varepsilon(\xi)} \sqrt{\frac{g'(\varepsilon)}{\rho_0}}\, d\varepsilon, \qquad \xi = \frac{X}{t - t_1}, \tag{2.5.24}$$

while ε_2 is determined from the equation

$$v_2 = v_1 - \int_{\varepsilon_1}^{\varepsilon_2} \sqrt{\frac{g'(\varepsilon)}{\rho_0}}\, d\varepsilon. \tag{2.5.25}$$

At any point (X, t) of the $(X),t)$-plane, we have

$$\varepsilon(X, t) = \varepsilon_2, \quad \sigma(X, t) = g(\varepsilon_2), \quad v(X, t) = v_2 \tag{2.5.26}$$

for $t > t_1 + X/c(\varepsilon_2)$.

Nothing can be said about the uniqueness of the constructed solution unless a uniqueness criterium is adopted. Such a criterium may be that considered by LAX [1957] or that adopted by MIHĂILESCU and SULICIU 1975a, b]. The last authors have also used another method to construct solutions to certain problems of the same type as the problems discussed in this section.

Chapter III

EXPERIMENTAL BACKGROUND FOR THE DEVELOPMENT OF THE THEORY

Several kinds of experiments played a significant role in the development of the theory. In order to formulate a mathematical model describing the main mechanical properties of a certain material, usually several diagnostic tests are performed. It is expected from these tests to reveal those mechanical properties which under certain circumstances (initial and boundary conditions) are dominant. From the point of view followed in the present book, the various possible tests can be classified by convention according to the rate of loading applied. The tests producing rates of strain ranging between 10^{-5}sec^{-1} and 10^{-3}sec^{-1} are termed *static tests*. Tests in which the rates of strains range between 10^{-2}sec^{-1} and 10^{2}sec^{-1} can be termed *intermediate tests* (these are the rates of strains usually met in various metal working processes, for instance), while the tests producing rates of strains above 10^{2}sec^{-1} (and certainly of the order of magnitude of 10^{3}sec^{-1} and higher) can be called *dynamic* in the sense that as a rule in such tests the inertia forces, the propagation of waves etc. must be considered in the analysis. Sometimes the inertia forces are considered in the analysis when rates of strains of approximately 1 sec^{-1} are reached. The classification of various testing machines according to the kind of test performed, i.e. tension, compression, bending, combined loading etc., is not important for the approach followed here. Concerning details we may refer the interested reader to Ilyushin and Lenskiĭ [1959], Goldsmith [1960], Krafft [1961], Kolsky [1963], Voloshenko-Klimovitskiĭ [1965], Johnson [1972], Bell [1973], Fowles [1973], Polukhin et al. [1976], Vasin et al. [1975].

1. STATIC EXPERIMENTS

1.1. **Standard tests.** The simplest diagnostic test is the *uniaxial compression or tension test of a cylindrical specimen.* For the description of the technical aspects of this kind of experiments see for instance Ponomarev

Static => 10^{-5} to 10^{-3} sec^{-1}
Intermediate => 10^{-2} to 10^{2} sec^{-1}
Dynamic => 10^{3} sec^{-1} and Higher
Strain Rate Classification

et al. [1956], ILYUSHIN and LENSKIĬ [1959], DALLY and RILEY [1965] BELL [1973] etc. We shall give a general description of several diagnostic experiments and we shall point out what sort of information they may reveal which could be useful for the formulation of a simple mathematical model. Basically there are two kinds of testing machines. With the so-called "hard" machines the variation in time of the length of the specimen can be prescribed, while the response of the material is known by registering the force necessary to produce this deformation. Other kind of machines called "soft" can prescribe the stress applied to the specimen (by a weight attached directly or not to the specimen, as in most of the creep testing machines, for instance), while the response of the material is observed by measuring the elongation of the specimen. Both kinds of testing machines provide some stress-strain diagrams as the one shown for instance on fig. 3.1.1. For the approach taken here it is important that these experiments are run within an ordinary (standard) time interval. The exact specification of "ordinary" is difficult since it depends on the nature of the tested material and on the experimental conditions (temperature etc.). Generally, for example, for most metals at room temperature, the "ordinary" time interval is of the order of a few minutes and only rarely otherwise. Usually it is the time necessary to perform such kinds of experiments with a "standard" universal testing machine. Therefore in various diagnostic tests which will be described, the interval of time in which the

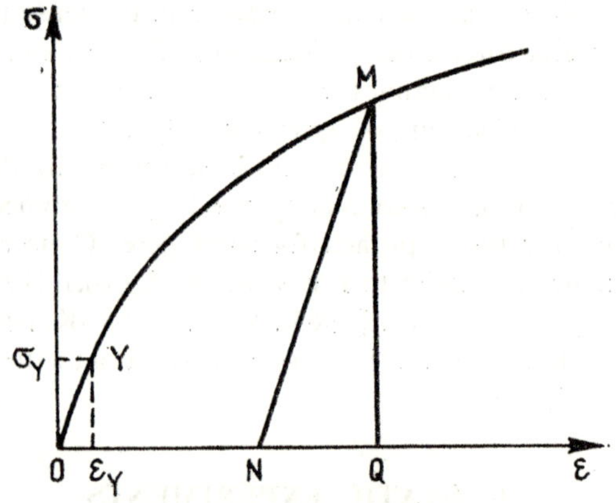

Fig. 3.1.1 Typical stress-strain curve

experiments are performed is a fundamental characteristic. We recall that in the "static" tests the rate of strain generally ranges between 10^{-5} sec^{-1} and 10^{-3} sec^{-1} both during loading and unloading.

A typical diagram obtained in tension or compression, or torsion etc. is shown on fig. 3.1.1. An ordinate represents the stress and the abscissa the strain. The definition used for either one of these two concepts is irrelevant for the qualitative discussion done here. Generally, in such experiments the initial state of the specimen is the state free of strain and of stress. Here we shall consider only loading in a single direction, i.e. the reversing of the stress sign will not occur. For convenience, stress and strain will be assumed to be positive throughout. For small stresses the diagram is linear. When the angle between the tangent to the stress-strain curve and the stress axis exceeds by a certain amount (established by convention) its initial value we say that the "proportionality" limit is reached. For stresses higher than this mentioned stress the stress-strain diagram is no longer linear.

For the purpose of establishing a mathematical model it is useful to perform unloading experiments which would follow the loading ones. The starting states are various magnitudes of the stress and strain reached in loading tests. If during unloading from a certain magnitude of strain (which has to be defined) the material response is non-linear and obeying the same non-linear law as during the loading process which led to that particular state, then the material is said to be *non-linear elastic*.

For most metals, if a loading first takes place up to a certain point M say, (fig. 3.1.1) and is then succeded by an unloading, then the unloading is following the path MN. If some other (sometimes less important) phenomena as hysteresis can be neglected, then with reasonable approximation the path MN can be replaced by a straight line

$$\sigma = \sigma_M + E(\varepsilon - \varepsilon_M) \text{ for } \varepsilon \in [\varepsilon_N, \varepsilon_M], \tag{3.1.1}$$

where E is the Young modulus. Then the portion ON of the strain is called the *plastic strain* while NQ is the *elastic reversible strain*. The straight line (3.1.1) is with good approximation parallel to the Hooke straight line

$$\sigma = E\varepsilon \tag{3.1.2}$$

which applies until the proportionality limit is surpassed, i.e. in the first loading for $\varepsilon \in (0, \varepsilon_Y]$.

Since after unloading we find that a part of the strain is reversible we write

$$\varepsilon = \varepsilon^p + \varepsilon^e \tag{3.1.3}$$

where ε^p is the plastic (irreversible) part of the strain ON and ε^e is the elastic (reversible) part NQ. The mathematical models where the strain can be decomposed according to (3.1.3) are called *elasto/plastic.* If in certain circumstances the elastic part of the strain can be neglected everywhere in the body as compared to the irreversible part, the model will be called *rigid/plastic.*

The *conventional elasticity limit* is defined as the maximum stress which corresponds to the magnitude of the plastic strain which is considered to be still negligible (0.05% say). For higher stresses the plastic component of the strain is generally no longer negligible. Point Y on fig. 3.1.1 marks the beginning of this portion. The stress corresponding to this point, called the *yield limit*, is denoted by σ_Y and the corresponding strain by ε_Y. We point out that in the literature two definitions for the yield limit are used. One of them associates the yield limit with a certain established by convention magnitude of the irreversible part of the strain as compared with the total strain (0.2% say), as would be obtained by an unloading which would follow immediately the loading. Making several loading-unloading tests and systematically increasing the maximum stress at each successive test, one finds that at a certain test the irreversible portion of the strain, called *plastic*, has reached a certain limit defined by convention; then we say that the *yield limit* is reached. The other definition associates the yield limit with the change of the slope of the tangent to the stress-strain curve in fig. 3.1.1, i.e. the yield stress is the smallest stress for which the specimen is deforming without a noticeable increase of the applied stress. For small stresses the curve is nearly linear; assuming that one can approximate the portion of the curve above σ_Y with another straight line with $\dot{\sigma} = 0$ or with $\dot{\sigma} > 0$, we can obtain σ_Y as the ordinate of the intersection point of these two straight lines found experimentally. The two definitions of the yield limit given above are certainly not equivalent. For our purposes here the "proportionality limit", the "elastic limit" and both "yield limits" practically coincide.

If the loading experiment is done with continuously increasing stress starting from $\sigma = \varepsilon = 0$, then the resulting stress-strain diagram is usually non-linear for $\sigma > \sigma_Y$. This behaviour of the material for a continuously increasing loading will be expressed by a function

$$\sigma = f(\varepsilon) \quad \text{for} \quad \sigma > \sigma_Y \tag{3.1.4}$$

whose graphic representation must run as close as possible to the experimental data. If $f' > 0$, that is when f is a strictly increasing function, we say that the material is *work-hardening.* If starting from the yield point for a certain strain interval $\varepsilon \in [\varepsilon_Y, \varepsilon_1)$ we get $\sigma = \sigma_Y =$ const., then the material is called *perfectly plastic.* The case $f' < 0$ on some interval of strain, i.e. the case of softening materials is only seldom met in practice.

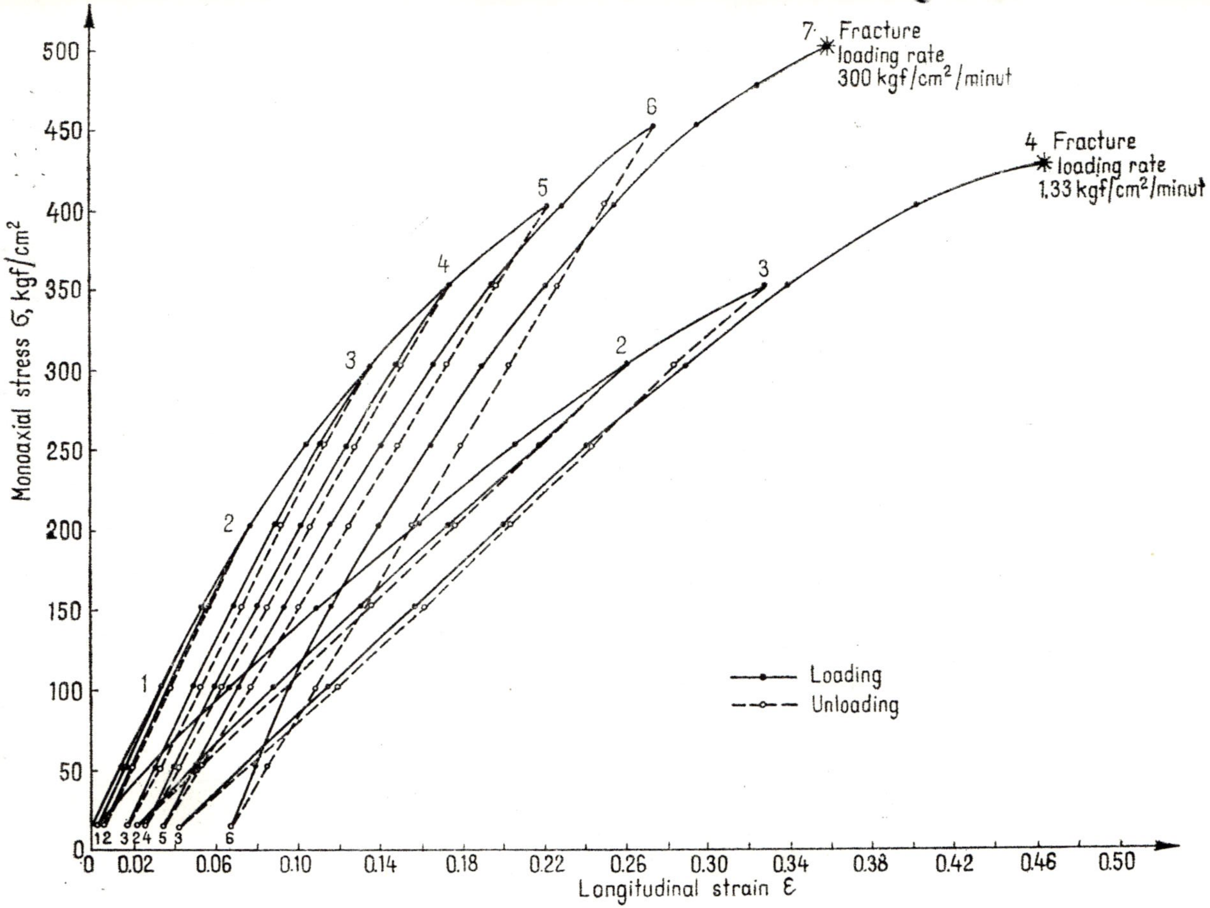

Fig. 3.1.2 Stress-strain curves for schist

Experiments carried out with greater precision have shown that for higher strains, even for metals, the slope of the unloading line (3.1.1) decreases when the maximum stress σ_M or the plastic strain increases (JUKOV [1960]). This is even more evident for other kinds of materials, as for instance for rocks. In fig. 3.1.2, for example, there are given two stress-strain curves in compression for schist (cylindrical specimen 10 cm high, 5 cm in diameter) obtained with loading rates of $|\dot{\sigma}| = 300$ kgf cm^{-2}min^{-1} and $|\dot{\sigma}| = 1.33$ kgf cm^{-2}min^{-1} both in loading and unloading. Thus the stress-strain curve is non-linear during loading and during unloading. During successive unloading and reloading pronounced hysterezis loops can also be noticed.

1.2. **Long range tests.** Let us assume now that the experiment previously described which was done to get the stress-strain curve of fig. 3.1.1 with a standard testing machine, is stopped at a certain stress and strain state (point M on fig. 3.1.3*a*, say). With a hard testing machine the strain can further be maintained constant $\varepsilon = \varepsilon_M$ while the stress variation can be measured. The experiments are showing that the stress decreases in time. This decrease which occurs in time intervals much longer than those involved in the previously described experiments is called *relaxation*. Thus during relaxation the stress variation follows the path MR in fig. 3.1.3*a*.Starting from the initial state in M, the simplest mechanical model which can qualitatively describe this phenomenon is the so called Maxwell model

Maxwell Fluid

$$\dot{\sigma} + \frac{\sigma}{\tau} = E\dot{\varepsilon}, \tag{3.1.5}$$

where E and τ are constants. If we put here $\varepsilon = \varepsilon_M =$ const., we get for the variation of the stress starting from state in M

$$\sigma(t) = \sigma_M \exp\left(-\frac{t}{\tau}\right), \tag{3.1.6}$$

where τ is the "relaxation time". This kind of behaviour is shown in fig. 3.1.3*b*. In various experiments two situations may occur, depending on the material and on the initial stress-strain state: The stress may decrease in the long interval which is of interest to us down to nearly zero, or, most likely, (for most materials mainly for relatively high initial stress states) the curve in fig. 3.1.3*b* may possess an horizontal asymptote $\sigma = \sigma_R > 0$, where σ_R depends on the initial stress-strain state. In this last case, after a long period of time, remanent nearly constant stresses will continue to be present in the specimen. As concerns mathematical models, in the

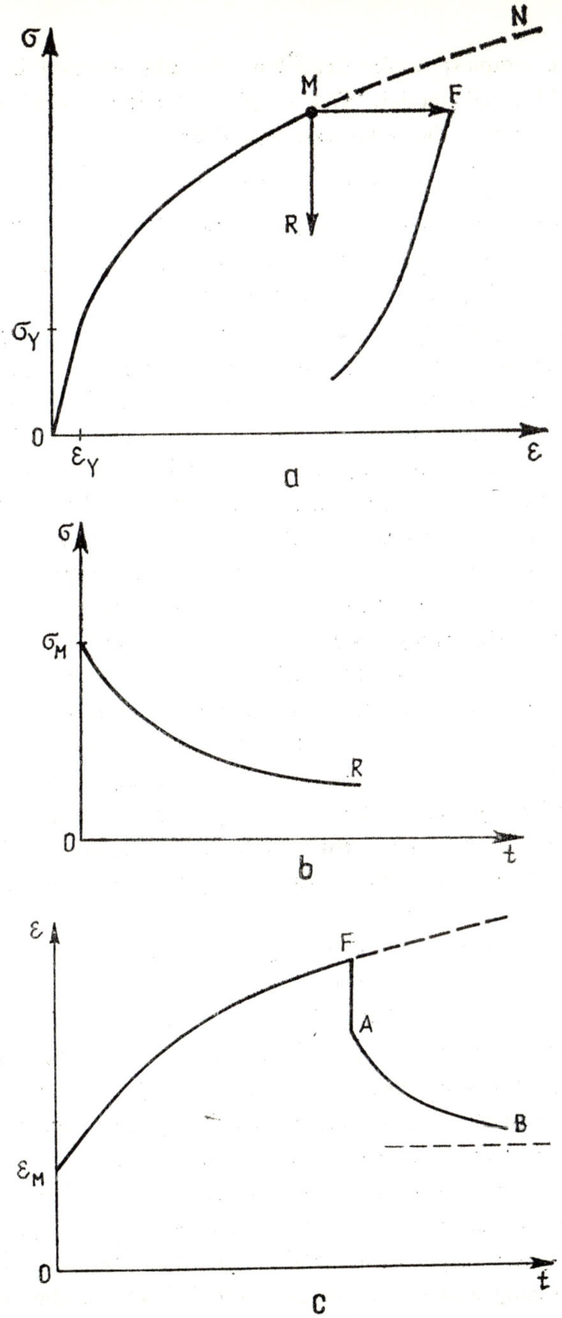

Fig. 3.1.3 a. Standard test followed by a much slower one b. Stress relaxation c. Variation of strain in time during a creep test followed by unloading

first case the simplest model describing the phenomenon is the already mentioned Maxwell model. The simplest possible model which may in principle, describe the latter case would be

Three Param. Solid

$$\dot{\varepsilon} + a\varepsilon = b\dot{\sigma} + c\sigma. \tag{3.1.7}$$

This is known as the "linear standard model". Here the coefficients a, b, c are constant. To describe relaxation, we have to integrate (3.1.7) with $\dot{\varepsilon} = 0$, the initial data being those of the state in M.

Another diagnostic test can be done with a soft testing machine if after a continuously increasing loading to the state M (fig. 3.1.3 *a*) the stress (or force) is further kept constant. In this case, if strain is measured in a long subsequently following interval, a slow increase of the strain is found so that the diagram in fig. 3.1.3 *a* is continuing from M in the direction MF. This is the *creep* phenomenon. Therefore if a "long" time interval is elapsing (by „long" we mean here that the deformation from M to F takes place in a time interval which is usually by several orders of magnitude bigger than the time interval necessary to produce the deformation from 0 to M in a standard testing machine; the rate of strain in such experiments is generally even by several orders of magnitude smaller than 10^{-5} sec^{-1}), the irreversible strain is sometimes considered to be the sum of two components; the one obtained during the deformation from 0 to M is traditionally called *plastic* while the one obtained during the process from M to F is called *creep strain*. The later component is negligible with respect to the first one if the experiment is done in an "ordinary" time interval usually taken to perform quasistatic tests with standard testing machines. On the contrary, during creep tests which run for very long time intervals the plastic component becomes sometimes negligible when compared with the creep component. Note that if in particular $\sigma_M < \sigma_Y$, the plastic component of strain is absent and therefore after a very long time interval only the creep component will be present as irreversible component. Generally the creep phenomenon depends on a variety of parameters such as magnitude of the applied stress, temperature, superimposed vibrations etc. A standard creep strain-time curve is shown on fig. 3.1.3 *c*. If the applied stress is low, then after a long time this curve becomes horizontal (has an horizontal asymptote); we say that the *strain is stabilizing*. However, if the applied stress surpasses a certain limit, then the strain-time curve is strictly increasing and sooner or later the fracture of the specimen will occur.

To give an example, on fig. 3.1.4 there are given three creep strain-time curves for rock-salt (Baroncea et al. [1977]). The three curves were obtained with different initial stress-strain states. The lowest one corresponds to the initial state $\sigma = 520$ N cm^{-2}, $\varepsilon = 0.175\%$ reached in 0.19 min. (loading time); the middle curve corresponds to the initial state $\sigma = 800$ N cm^{-2}, $\varepsilon = 0.6\%$ and was reached in 0.67 min.; finally the upper curve corresponds to the initial state $\sigma = 1200$ N cm^{-2}, $\varepsilon = 0.975\%$ which was reached in 1.085 minutes.

For the purpose of finding the various possible fundamental properties which a certain material may possess it is sometimes useful to study the creep phenomena not only during the loading process (increasing strain) but also during unloading when the applied stress is suddenly removed. For instance, if at the point F (fig. 3.1.3 c) the applied stress σ_M is suddenly removed, then the strain will also decrease. If the strain is suddenly decreasing by a certain part FA, then the model must possess as constituent the elastic instantaneous model, i.e. to the previous mentioned irreversible strains we must also add the elastic reversible one. If the initial stress σ_M is smaller than σ_Y, then generally the segment FA equals ε_Y, but if the applied load surpasses σ_Y, then $FA > \varepsilon_Y$. From the point A onwards the strain con-

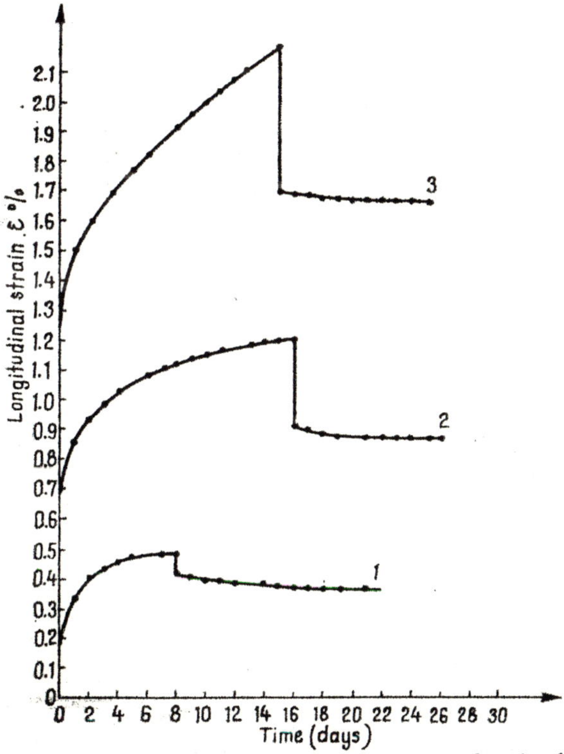

Fig. 3.1.4 Variation of strain in time for creep of rock-salt

tinues to decrease slowly following the path AB. For this phenomenon the terminology *inverse creep* is used (see also fig. 3.1.4). In most of the cases after a very long period of time the strain is not decreasing to zero but towards a certain limit value reached asymptotically. Sometimes however, the strain is decreasing towards zero, and in this case the model to be used for such kind of material is a "viscoelastic" one (the model (3.1.7) is one of the simplest models of this type which may describe an inverse creep down to zero strain).

1.3. **Short range tests.** In the previous paragraph it was assumed that a standard testing is performed up to a certain state corresponding to the point M (fig. 3.1.3 *a*) and subsequently the test is continued for a much longer period of time. It is also of interest to discuss shortly what happens if the standard test is suddenly replaced at M by another one performed in a much shorter period of time (see fig. 3.1.5). Let us assume that the standard test is performed with a certain average rate of strain $\dot{\varepsilon}_0$ (generally in such tests $\dot{\varepsilon}_0$ ranges between 10^{-4} sec^{-1} and 10^{-1} sec^{-1}). If starting from the state in M the rate of strain is suddenly increased to a value $\dot{\varepsilon}_1$ greater than $\dot{\varepsilon}_0$, then the stress-strain curve which is obtained lies above the one

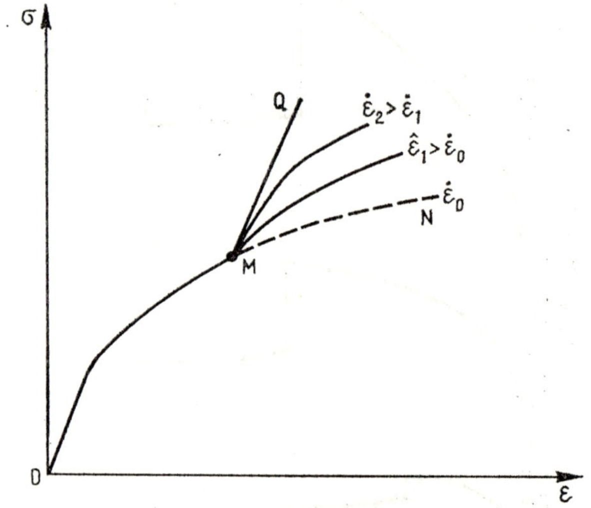

Fig. 3.1.5 Changing the rate of strain during a test showing various possible behaviours

which would be obtained for the rate of strain $\dot{\varepsilon}_0$ (segment MN on fig. 3.1.5). If starting from the state in M the experiment is performed with a rate of strain $\dot{\varepsilon}_2$ still higher, i.e. $\dot{\varepsilon}_2 > \dot{\varepsilon}_1 > \dot{\varepsilon}_0$, then the stress-strain curve which is obtained lies even higher than the previous one, and so on. When $\dot{\varepsilon} \to \infty$ (sudden impacts, explosive loading etc.), then the limit response of the material will be called the *instantaneous response*. For some metals and in certain experiments it can be assumed that the instantaneous response can be approximated by a straight line having the slope of Hooke's line (segment MQ on fig. 3.5.1). However, for other materials (plastics, soils, rocks etc.) the instantaneous response is generally non-linear and sometimes the corresponding stress-strain curve may have a slope higher than the elastic one. Thus, if starting from the state in M the rate of strain is increased we do not obtain a unique response of the material; for each rate of strain we get another stress-strain curve. In particular, the state in M can be the initial stress-free-strain-free state. In this case it was observed that for metals the slope of the elastic portion of the stress-strain curve is little influenced, if at all, by the increase of the rate of strain. For other materials however (plastics, soils, rocks, some metals etc.) Young's modulus is significantly influenced by the rate of strain (see fig. 3.1.2) i.e. the entire stress-strain curve is influenced by the rate of strain. Therefore, for such kind of material for each rate of strain another stress-strain curve is obtained, i.e. there is no unique response of the material for various rates of loadings. But it is certain that within certain ranges of variation of the rates of strains which depend on the material considered, the rate effect on the stress-strain curve can be negligible; this is to be considered separately for each particular case (material, loading rate, geometry of the body etc.).

To obtain very high rates of strain special experimental devices are necessary, while the state at M, which in such experiments plays the role of initial state, is obtained also by a special preloading device (generally not by a standard testing machine). These special devices for "dynamic loadings" may produce rates of strains much higher than obtained with standard machines. Thus rates of strains of the order 10^4sec^{-1} and higher can be produced. To give an example, in figure 3.1.6 are given the stress-strain curves for pure aluminium as obtained for various rates of strains by HAUSER et al. [1960] (reproduced here from THOMSEN et al. [1968]). If the rates of strains are high (over 10^2sec^{-1} say), the inertia forces cannot be neglected and the wave propagation phenomena must be considered in the analysis of the experiment. This will be described in Chapter IV.

The first attempt to find an empirical formula to describe the previously discussed phenomena, seems to be due to LUDWIK [1909]; later the problem was considered also by PRANDTL [1928] and by many other authors. Let us assume that a series of experiments is performed with increasing rates of strains and let us denote by $\dot{\varepsilon}_0^p$ an arbitrary reference plastic rate of strain. Then, for a certain fixed strain, the stresses which

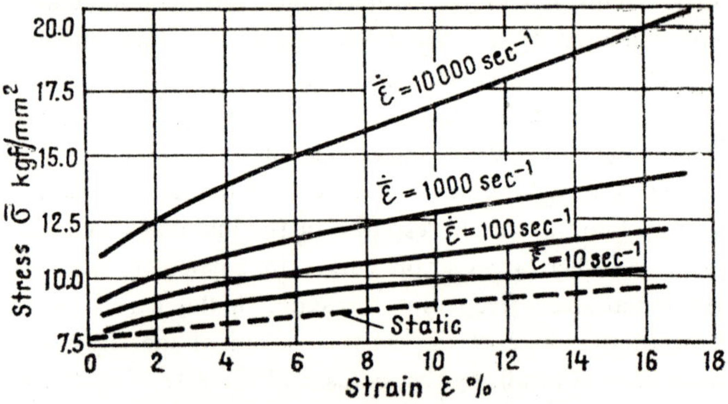

Fig. 3.1.6 Stress-strain curves for pure aluminium obtained for various rates of strains (after HAUSER et al. [1960])

are obtained for various plastic rates of strains $\dot{\varepsilon}^p > \dot{\varepsilon}_0^p$ can be related by the empirical formula

$$\sigma = \sigma_Y + \sigma_0 \ln(\dot{\varepsilon}^p/\dot{\varepsilon}_0^p) \text{ for } \sigma > \sigma_Y > 0, \tag{3.1.8}$$

where σ_Y is a certain "yield stress" corresponding to the reference rate of strain $\dot{\varepsilon}_0^p$ and $\sigma_0 > 0$ is a material constant which is essential in describing the rate effect. Generally the main conclusion resulting from most experimental data is that the stress is strictly monotonically increasing with the rate of strain.

Many simple models describing the rate effect were inspired by formula (3.1.8). If the elastic part of the strain can be neglected with respect to the plastic one, then the simplest of these models can be written as

$$\dot{\varepsilon} = \begin{cases} \dfrac{\sigma - \sigma_Y}{3\eta} & \text{if } \sigma > \sigma_Y, \\ 0 & \text{if } 0 \leqslant \sigma \leqslant \sigma_Y, \end{cases} \tag{3.1.9}$$

where η is the "viscosity coefficient". Since, according to this model, no deformation can occur for stresses not surpassing the yield stress σ_Y, the model is a *rigid viscoplastic* model known also as a *Bingham model* (BINGHAM [1922]). The model can be considered as an extension to viscoplasticity of the rigid/perfectly plastic model of classic plasticity. However, while stresses above the yield stress $\sigma > \sigma_Y$ are not possible within the framework of the perfectly plastic model, according to the model (3.1.9) it is only for such stress states that deformation is possible. The aim of the model (3.1.9) is to describe the experimentally observed fact that to higher rates of strains there correspond during deformation higher stresses. The model (3.1.9) was first used for various pastes, dough, vaseline, paints, mud etc., but also for various solid bodies as metals, rocks etc. In this last case the coefficient η can be found experimentally in the following way. Let us assume that two "standard" tests are performed with

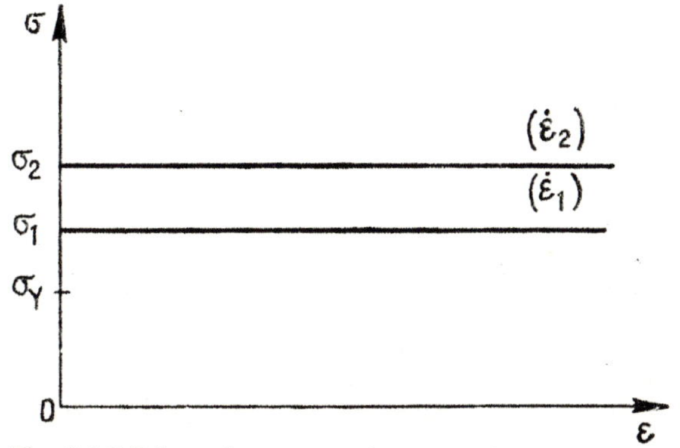

Fig. 3.1.7 Schematic stress-strain curves for two distinct rates of strains

two distinct rates of strains $\dot{\varepsilon}_1$ and $\dot{\varepsilon}_2$ which produce the stresses σ_1 and σ_2 respectively. It is assumed that $\sigma_2 > \sigma_1 > \sigma_Y > 0$ and $\dot{\varepsilon}_2 > \dot{\varepsilon}_1$ (see fig. 3.1.7). Then (3.1.9) yields

$$3\eta = \frac{\sigma_2 - \sigma_1}{\dot{\varepsilon}_2 - \dot{\varepsilon}_1} \tag{3.1.10}$$

which is within the framework of this model a "measure" of the rate of strain effect on the stress magnitude. To keep terminology simple, we will call η the *viscosity coefficient*. This coefficient (defined sometimes as $\eta = \partial\sigma/\partial\dot{\varepsilon}$), and some of its generalizations, play an important role not only in the dynamic deformation theory (when inertia forces are also taken into account) but sometimes also in problems involving medium or even small rates of strains. We observe that for the same material this coefficient might not have a unique value for all possible rates of strains. Thus, if $\dot{\varepsilon}_4 > \dot{\varepsilon}_3$ are two rates of strains by several orders of magnitude greater than $\dot{\varepsilon}_2$ and $\dot{\varepsilon}_1$, then the viscosity coefficient which would be obtained with a formula of the type (3.1.10) is generally distinct from the one obtained for the rates $\dot{\varepsilon}_2$ and $\dot{\varepsilon}_1$. Generally, for most solid bodies, η is decreasing with increasing $\dot{\varepsilon}$, and therefore it is only within certain ranges of variation of $\dot{\varepsilon}$ that η can be considered to be constant. A step function for

$\eta(\dot{\varepsilon})$ can be very useful in practical applications. On the other hand, since for most bodies η is not constant for all ranges of rates of strains (except possibly for very high rates of strains), another non-linear model has to be used whenever possible.

In applying the model (3.1.9) to solid bodies it is also useful to consider the actual yield stress σ_Y as variable with the strain, and therefore to describe the work-hardening process in some other way than in classical plasticity. Then, instead of (3.1.9), we can write

$$\dot{\varepsilon} = \begin{cases} \dfrac{\sigma - f(\varepsilon)}{3\eta} & \text{if } \sigma > f(\varepsilon), \\ 0 & \text{if } 0 < \sigma \leqslant f(\varepsilon), \end{cases} \tag{3.1.11}$$

where $\sigma = f(\varepsilon)$ is the "work-hardening condition" and in some problems the quasistatic stress-strain curve can be considered to be a first guess for this function.

To give an idea about the magnitude of η, we note here that if the model (3.1.11) is applied which describes the creep of rock-salt (see BARONCEA et al. [1977]), then η is of the order of magnitude of 10^{15}Nm^{-2} sec (Poise). For other rocks, also for creep, this value can be smaller; for instance for schist it is 10^{12} Poise, while for still other rocks it may be higher ($10^{17} - 10^{18}$ Poise for argillaceous schist, see VYALOV [1978]). Also Vyalov gives values for ice from 10^{10} to 10^{15} Poise. For quasistatic deformations ($\dot{\varepsilon} \simeq 1 \text{ sec}^{-1}$) for mild steel (CRISTESCU [1977]) η is of the order of magnitude of 10^7Nm^{-2}sec, but for faster deformation ($\dot{\varepsilon} \simeq 100 \text{ sec}^{-1}$) η is of the order 10^6 Nm^{-2} sec; for much higher rates of strains, η is still decreasing with one or more orders of magnitudes in dynamic impact problems. For instance for very high rates of strains ($\dot{\varepsilon} > 10^3 \text{ sec}^{-1}$) CAMPBELL [1973] gives for various metals values of the order 10^3 Poise. Generally η is decreasing when $\dot{\varepsilon}$ is increasing. As was already mentioned, a step function which would describe the variation of η with $\dot{\varepsilon}$ is very useful in applications. The values which were given for η depend strongly on the conditions in which the test is done. The values given above are only illustrative.

If the experiment performed to determine the mechanical properties of various materials involves very high rates of strains (over 10^2 sec^{-1} say), generally the inertia forces cannot be neglected any longer in the analysis of the experiment and in these cases instead of a rigid/viscoplastic model, elastic/viscoplastic models seem to be more suitable. Generalizations of the model (3.1.9) which consider also the elastic properties of the material will be discussed in the next chapter.

Important remark: If with a certain material several experiments are performed in identical conditions but in which the loading rates (or rates of strains) are distinct in various tests, and if in "faster" tests effects of the type shown in fig. 3.1.4 are observed, i.e. if for a given strain the stress is increasing with the loading rate, then the most suitable mathematical model is one of a viscoplastic type.

1.4. **The Portevin-Le Chatelier effect.** This effect which seems to have been discovered already by F. SAVART [1837] and A. P. MASSON [1841] (see BELL [1973] § 4.31), caused many controversial disputes. It was discovered with soft type testing machines on metallic specimens. Most studies have been done on aluminium. Let us assume that a specimen is continuously loaded in tension or compression with a very small constant loading rate. These are the so-called dead weight experiments in which the stress variation is prescribed and the strain is the variable measured. If loaded in tension a bucket of water can be attached directly at the end of the specimen, and the water is left to flow continuously into the bucket. Certainly these experiments are slow and the total duration of the experiment may last several days to several months (or years). The main observed features are the following.

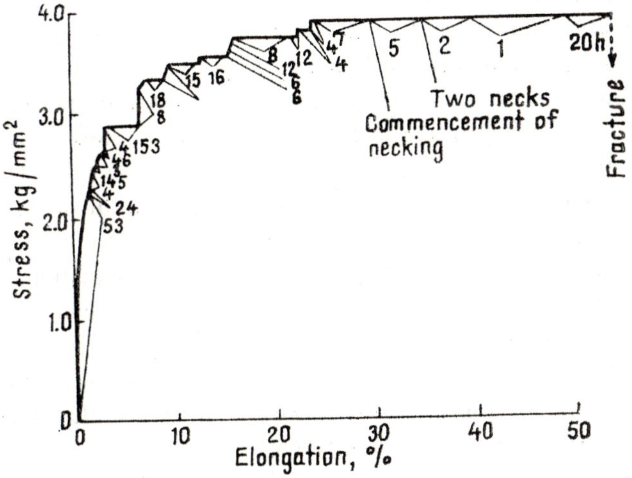

Fig. 3.1.8 Stepwise rising stress-strain curves showing the Masson-Savart effect (after HANSON and WHEELER [1937])

As the stress is increasing, first the response of the material is linear elastic. This is a smooth deformation process in the sense that to a constant loading rate corresponds a deformation with a constant rate. After the plastic range is reached, the deformation process has a "staircase" appearance (see fig. 3.1.8), i.e. when stress is continuously increasing with constant rate, the strain is first increasing very slowly and for a certain value of the stress a fast increase of the strain is triggered off and the strain increases up to an equilibrium strain, and so on. Thus the stress-strain curve has a staircase appearance. During the "slow" increase of the strain the local slope of the stress-strain curve is generally close to (sometimes somewhat smaller than) the slope of Hooke's line. On those portions where the strain is increasing fast, the local slope of the stress-strain curve is nearly horizontal. In fig. 3.1.8 there is shown a stress-strain curve of this type as obtained in 573 days by HANSON and WHEELER [1931] for aluminium. Integers on the figure refer to the number of days involved for each interval. The largest interval is of 153 days.

Thus the phenomenon here described refers to a stepwise increase of the strain when the stress is continuously increasing. The phenomenon was observed by very many experimentalists in the last decades, though all were using soft machines with various kinds of loading methods. PORTEVIN and LE CHATELIER [1923] have systematically studied the phenomenon and have also observed that at each discontinuity point the specimen is producing an audible sound. Some authors showed that the phenomenon can be observed with hard machines too, in which case the continuous variation of the strain is imposed and small oscillations of the stress are recorded.

The phenomenon was attributed to various causes and received various explanations. Some considered it an unstable strain-rate phenomenon, others consider that it is due to the propagation of "slow waves" (speed of propagation ranging between 0.5 m/sec and 800 m/sec) etc. This phenomenon again raised the problem whether the constitutive equations of classical plasticity theories should be improved, by including both "slow" response and "fast" response in a single constitutive equation. Temperature, crystalographic structure, impurities etc. are certainly influencing the phenomenon. This was observed on monocrystals and polycrystals in uniaxial or biaxial tests (KENIG and DILLON [1966], KENIG [1967]). For details and a history of the experiments describing this phenomenon see BELL [1973] (see also SHARPE [1966]).

2. DYNAMIC LONGITUDINAL IMPACT OF BARS

Experiments with longitudinal impact of bars are among the most commonly used experiments in dynamic plasticity. Among the various kinds of experiments which were done we shall discuss only those bearing a relation to the conclusions used in the subsequent chapters for the determination of a constitutive equation. These concern the "symmetric" longitudinal impact of two identical bars. Usually one is at rest and the other one is shot out from an air gun. Further, we will discuss what phenomena can in principle be discovered experimentally and which of them cannot be described with models of classical plasticity.

To begin with, the speed of the projectile bar and then the time interval during which the two bars are in contact can be measured quite accurately by an optical method. The time of contact is a parameter which gives some global information concerning the whole process of propagation of elastic and plastic waves in the bars impacted longitudinally. Another parameter which gives global informations on the whole dynamic loading-unloading process is the final velocity of the specimen which can be measured, for instance, by optical methods; the specimen, initially at rest, is put into motion by the impact and the resulting speed can be measured by successive crossings of some light beams by the moving specimen. Both the time of contact and the final speed give some "final" information at the end of the entire process about the dynamic plastic deformation (the loading-unloading processes in the whole bar are influencing this "final" information).

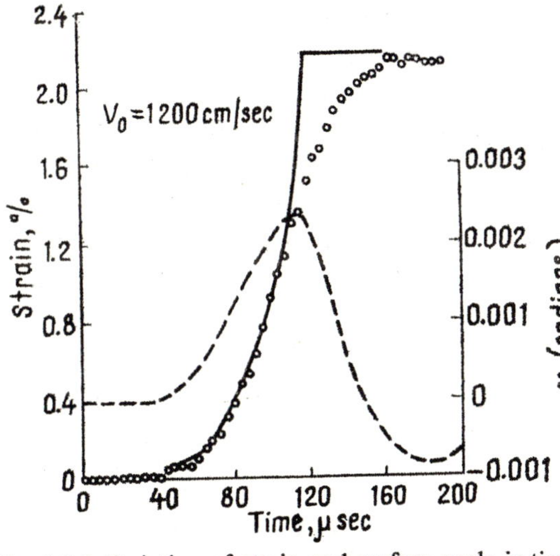

Fig. 3.2.1 Variation of strain and surface angle in time at a certain cross-section for copper (after BELL [1968])

Other important measurements provide data which are fundamental for the formulation of a constitutive equation. Thus with diffraction gratings made on the lateral surface of the specimen at several distances from the impacted end (see BELL [1968] and also the development by SHARPE [1970]) the variation in time of the strain at various cross-sections along the bar can be measured during the process of propagation of elastic and plastic waves. As an illustration, we give on fig. 3.2.1 one of those curves obtained by BELL for copper (see BELL [1968] p. 32); the impact velocity of the specimen was $V_0 = 1200$ cm/sec and the measurement is done at 5.1 cm from the impacted end. Some important features on this experimental curve can be observed. Thus the first rise of the strain-time curve occurs when the first wave arrives at the considered cross-section. According to classical plasticity theory, the magnitude of the strain on this first wave is equal to ε_Y. However, the experiments are showing that in the portion of the bar close to the impacted end the strain on the first wave is sometimes greater than ε_Y and in these cases this magnitude depends on the impact velocity in the sense that for very high impact velocities the strain on the first wave is always higher than ε_Y. Further, in the considered cross-section of the specimen the strain increases and the strain-time curve changes curvature at a certain moment. After reaching its maximum, the strain remains constant for a quite long period of time; this is the so called "plateau" in time of the strain at the considered cross-section. The thick line on fig. 3.2.1 is the prediction of the phenomena according to a classical constitutive equation. One can see that the change of the concavity of the strain-time curve cannot be explained within such a theory. The experiments are also showing that plateau strain at the cross section close to the impacted end is higher than the plateau strain at a certain distance from the impacted end. This experimental result can neither be explained by a classical plasticity theory. In analysing the experimental results near the end of the bar, one has to take into account that the phenomenon is a three-dimensional one, and thus much more difficult to interpret.

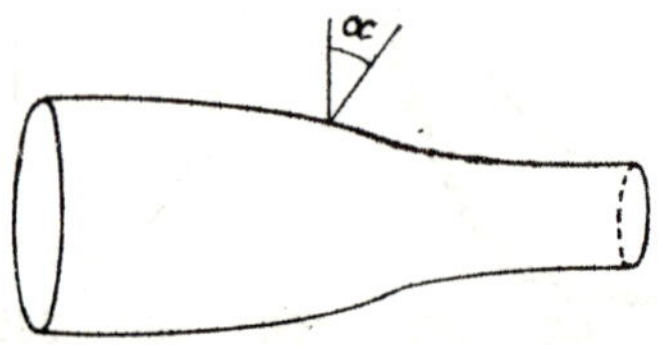

Fig. 3.2.2 Surface angle during longitudinal impact of a bar

During dynamic plastic deformation the lateral surface of the specimen bulges. Let α denote the surface angle, i.e. the angle between the actual position of the normal to the lateral surface and its initial direction at the same material point (fig. 3.2.2); it is easy to see that during the propagation of the plastic waves this angle is initially zero then it increases and afterwards decreases to become zero at the end of the process. This angle too can be measured using the same optical method. An example of variation of this angle at a point is given on fig. 3.2.1 by the dotted line. The maximum of surface angle α corresponds to the point of change of curvature of the strain-time curve.

Curves analogous to those shown on fig. 3.2.1 can be obtained by other techniques. For instance, on fig. 3.2.3 we give after MALVERN [1965] (see also EFRON and MALVERN [1969]) the transient surface particle

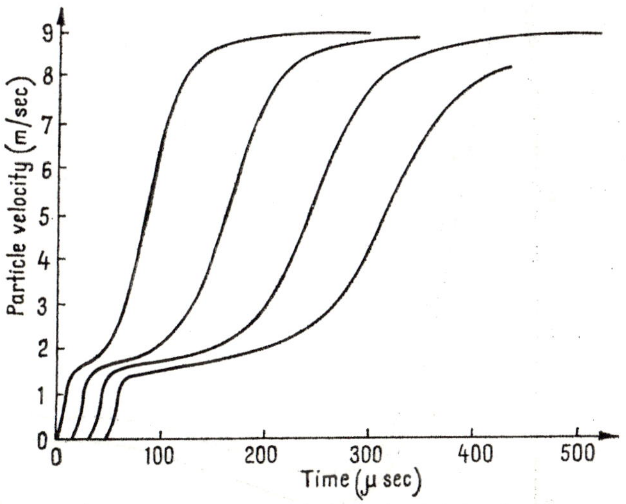

Fig. 3.2.3 Transient surface particle velocity records for aluminium (after MALVERN [1965])

velocity records, as obtained at four stations along the plastically deforming aluminium specimen by means of electromagnetic transducers. The four stations are approximately 7.5 cm apart. Again the initial step is the leading elastic wave, followed by the more slowly rising plastic wave, rising to approximately the same constant plateau velocity at the first three stations. The curve in fig. 3.2.3 has the same shape as the one in fig. 3.2.1.

Both experimental methods, the optical one and the electromagnetic transducer one, are considered to be quite accurate. They can be used along quite a long portion of the bar, enabling us to obtain the overall picture of variation in time of the strain (in the case of the optical method) or of the particle velocity (in the case of the electromagnetic method) from the moment of impact up to the moment when unloading starts, and sometimes even during unloading.

Another parameter which can be measured at the impacted end of the bar is the stress. Generally the stress can be measured with the help of piezoelectric crystal wafers placed at the impacted face. FILBEY [1961]

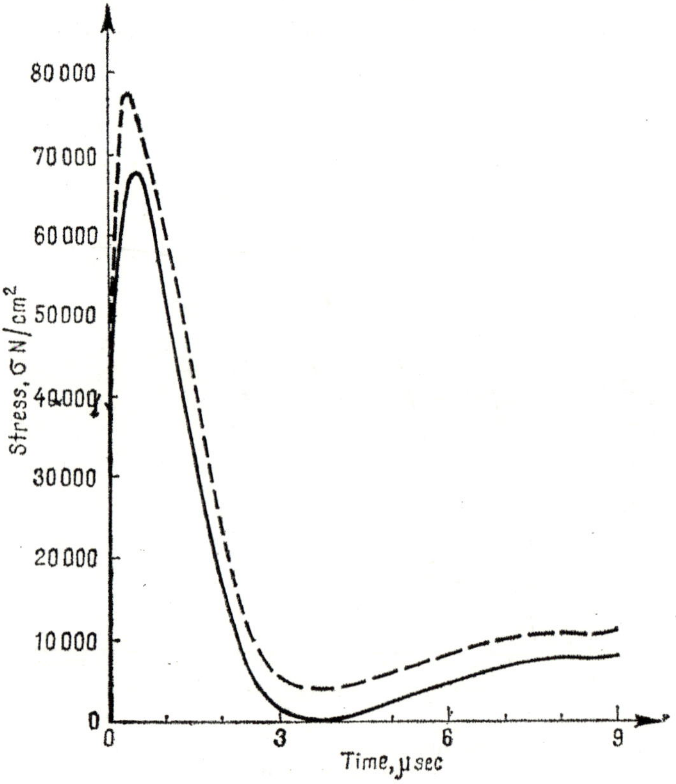

Fig. 3.2.4 Peak stress at impacted end (after FILBEY [1961])

has used an X-cut quartz crystal wafer 2.12 cm in diameter and 0.25 mm thick. A typical result for two experiments with symmetric impact is given in fig. 3.2.4 after FILBEY [1961] who used the impact velocity $V_0 =$ 7100cm/sec. Therefore the stress increases very fast at the impacted end up to a "peak" and then decreases within a few microseconds to become nearly constant for a relatively long period of time (fig. 3.2.5). The medium value of the peak stress is 71043 N cm^{-2}. The stress "plateau" which is reached after several microseconds is at the level 13160 N cm^{-2}. The decrease of the stress which follows afterwards is due to the unloading. The magnitude of the peak stress is much higher than the one which could be calculated by using a classical theory of plasticity or even a linear elastic model. Thus the presence of the peak stress is pointing out to another feature which cannot be described by a classical plasticity theory; the decrease of the stress from the peak down to a plateau can neither be described by a classical plasticity theory. The classical theories of plasticity

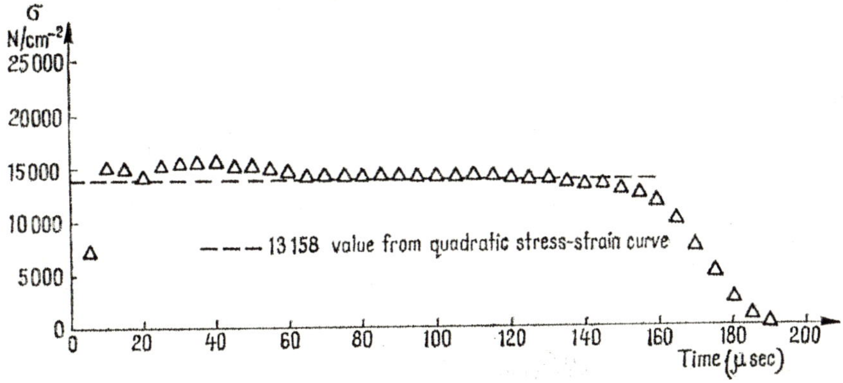

Fig. 3.2.5 Stress plateau (after FILBEY [1961])

can describe only the appearance of a stress plateau at the impacted end but not the stresses above this plateau.

The variation of the stress at the impacted end of the specimen was measured also with a hard "transmitter bar", i.e. a hard bar which remains elastic during the experiment, placed between the projectile and specimen and transmitting the impact to the specimen. In this case the specimen was made from a much softer material than the transmitter bar. During impact the strain is measured on the lateral surface of the transmitter bar with standard strain gages and thus one finds the variation of the stress at the impacted end of the specimen. The conclusions obtained (concerning the peak stress etc.) with this last method generally coincide with those previously mentioned.

3. RELOADING EXPERIMENTS

3.1. **A static loading followed by a dynamic one.** An experiment which played an important role in the development of the theory was the following one carried out by BELL [1951]. A medium carbon steel long bar (about 183 cm in length and consisting of portions 0.64 or 0.35 cm in diameter) was subjected to a static extension.The schema of the experiment is shown on fig. 3.3.1. The specimen *AB* is fixed in *A* and subjected in *B* to a stretching with uniform speed. *BC* is a much harder loaded bar through which the static loading is transmitted and where the applied stress is measured. During the experiment, when the specimen is already in the plastic state, several small dynamic reloadings are produced by the falling of a ring D. Therefore the reloading, producing also an extension, is applied at the end B of the specimen. Velocity determinations were made by measuring the time between impact of the cylinder and initiation of strain on resistance wire strain gages mounted in E at measured distances from the point of impact B. If the stress-strain curve of the material is a priori known and the specimen is already in a plastic static preloading state, then the classic plasticity theory predicts a velocity of propagation defined by

Fig. 3.3.1 Schema of BELL [1951] experiment

$$c = \sqrt{\frac{1}{\rho} \frac{d\sigma}{d\varepsilon}},$$

i.e. established by the slope of the stress-strain curve computed for the strain state corresponding to the static preloading. However, the experimental results obtained by BELL have shown that the first waves which are reaching the point E are propagating much faster with the elastic bar velocity

$$c_0 = \sqrt{\frac{E}{\rho}}.$$

This result cannot be explained by a classical plasticity theory. BELL has performed also some experiments concerning dynamic unloading by dropping a steel sphere on the top A of the specimen already in a plastic static preloading state. This is producing compressive unloading waves propagating in the specimen again with the elastic bar velocity. This last result is in agreement with the classical theory but the first one is not.

Experiments of the same type were afterwards repeated by several experimentalists. The experimental set-ups were sometimes different but the final conclusions were always the same. Thus STERNGLASS and STUART [1953] have used a standard testing machine with a cold-rolled copper strip 302 cm long, 1.27 cm wide, and 0.32 cm thick, as a specimen. Approximately in the middle of the specimen a striking platform was mounted. The dynamic reloading was produced at this platform with the help of a hammer. The impact produced tensional reloading waves on one side of the specimen and a compressive unloading wave on the other side. Again the leading reloading wave, registered by electric resistance gages, was propagating with the elastic bar velocity, though the specimen was already in a plastic static preloading state.

G. BIANCHI [1964] has made also some experiments with long (8 m) copper ribbons fixed at one end and stretched at the other end by a weight. The dynamic reloading was again produced somewhere in the middle of the specimen so that during a single experiment both reloading waves and unloading waves could be observed. Very long specimens (wires of copper or aluminium alloy) were used also by MALYSHEV [1960] [1961]. However he used another method to measure the arrival time of the leading wave at various cross-sections along the rod; the electromagnetic transducers were employed to record the displacement of the various cross-sections of the wire. The conclusions of the last two authors are in agreement with the previously mentioned ones.

For more details concerning the type of experiments which we have briefly described above or for other similar experiments performed by other authors see for instance BELL [1973] Subsection 4.27, CAMPBELL [1973], KLEPACZKO [1973], TING [1975] and the literature mentioned in these and in CRISTESCU [1976] Chapter III, Section 9.

3.2. **A dynamic loading followed by a dynamic reloading.** The previously discussed experiments have suggested the question whether the propagation of the reloading waves with the elastic bar velocity can also be observed if the first loading is a dynamic one. The first experiments in this set-up have been done by ALTER and CURTIS [1956] on lead specimens. The main idea of the experiment, which afterwards was used also by others, is the following (fig. 3.3.2). A lead bar *AB* is in axial collision with

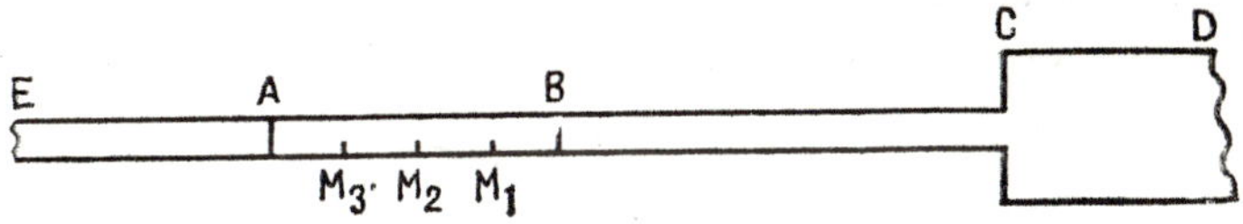

Fig. 3.3.2 Schema of ALTER and CURTIS [1956] experiment

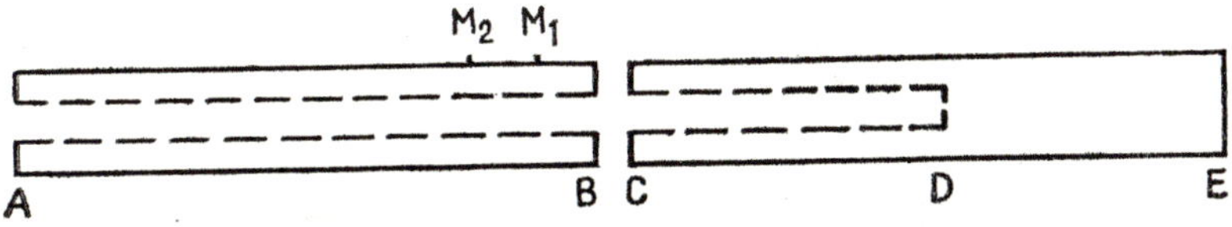

Fig. 3.3.3 Schema of BELL and STEIN [1962] experiment

a stepped steel bar *BCD*. In *C* the cross-section is suddenly changed. The specimen *AB* is continued by an anvil steel bar *AE*. Due to the impact the specimen *AB* undergoes a dynamic plastic deformation, while the strain is measured by electric resistance gages at various points M_1, M_2, . . .

on the lateral surface. The impact produces an elastic wave in the stepped bar, which upon being reflected from the step, returnes as an incremental loading wave, following the dynamic prestress of the specimen from the initial impact. If the length *BC* of the stepped bar is varied, one can place the second loading at any desired time after the first one. The velocity of propagation of the leading reloading wave can be measured at various points M_1, M_2, ..., closer or further away from the impacted end. The impact bar and the anvil bar must be long enough to minimize the unloading effects coming from the free ends of the two hard bars.

Similar experiments were performed by BELL and STEIN [1962]. The specimens were soft aluminium tubes *AB* (fig. 3.3.3) struck longitudinally by a composite hard aluminium hitter *CE*, which was tubular on the portion *CD* and had the form of a plain bar on the portion *DE*. Thus the dynamic reloading was produced by the reflection from the cross-section *D* of the specimen of the leading elastic wave created at the impact, and its propagation backwards up to the cross-section *C*. By varying the length of the tubular portion of the hitter it was possible to introduce the reloading in the specimen at any desired time. On various cross-sections M_1, M_2, ... along the specimen the strain was measured by the diffraction gratings method which gives very accurate results.

Other experimental settings were also used in which the hitter was a composed bar made from two distinct materials (possessing distinct densities). Besides measuring the strain at various cross-sections along the specimen, one can measure also the stress at the impacted end, for instance by using a hard load bar.

The various experiments described above lead to some conclusions

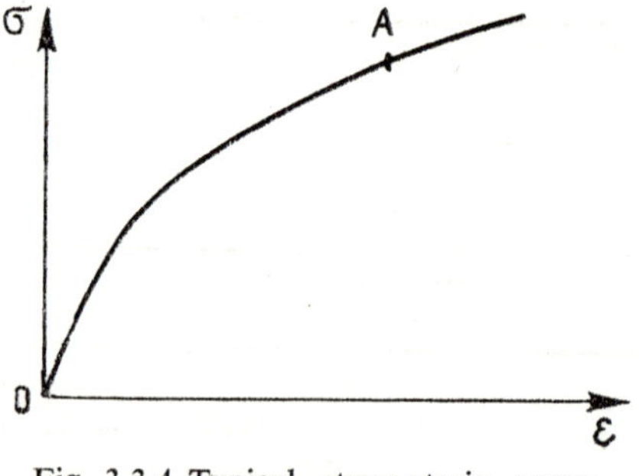

Fig. 3.3.4 Typical stress-strain curve

which can be easily discussed in a characteristic plane using a classic plasticity theory. Thus let us assume that the tested material has a unique stress-strain curve (fig. 3.3.4) obtained by using a standard testing machine, describing the mechanical properties of the material during the whole loading period. Let us assume that at the impacted end of the specimen after a first dynamic loading we reach the stress σ_A and the strain ε_A (point

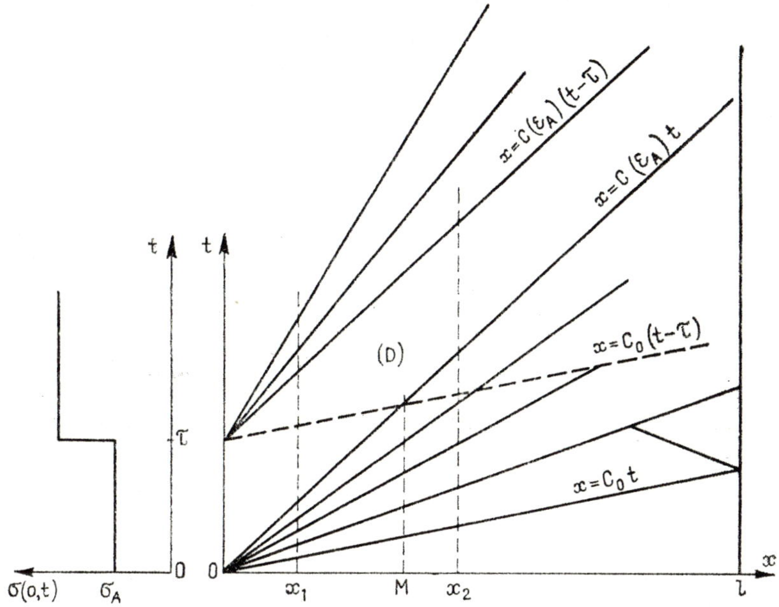

Fig. 3.3.5 Characteristic plane for two successive stepwise increasing loadings

A on the characteristic curve), and that the second dynamic loading produces the stress $\sigma > \sigma_A$. The dynamic loading at the end of the specimen is represented by two steps on the left hand side of the fig. 3.3.5. If $\sigma_A > \sigma_Y$, then the first loading generates a bundle of plastic waves (centered in 0) which will propagate into the specimen according to the classic theory. In the interval of time $0 < t < \tau$ no wave will be generated at the impacted end of the specimen. At time $t = \tau$ the first reloading wave starts from the impacted end of the specimen. According to the classic theory, the slope in the characteristic plane of the last wave in the first bundle of waves is equal to $c^2(\varepsilon_A)$. But the first wave from the second bundle of waves produced by the reloading has also the slope $c^2(\varepsilon_A)$. Therefore these two straight lines are parallel, while domain (D) is a domain of constant state where

$$\sigma = \sigma_A, \qquad \varepsilon = \varepsilon_A, \qquad v_A = \int_0^{\sigma_A} \frac{d\bar{\sigma}}{\rho_0 c(\bar{\sigma})}. \tag{3.3.1}$$

This is the description of a dynamic reloading experiment within a classical plasticity theory.

However the experiments are revealing some facts that cannot be explained by a classical plasticity theory. Let us assume for instance that at the cross-section x_1 quite close to the impacted end either the strain or the particle velocity is measured. One finds that after the passing of the last plastic wave from the first bundle of waves the cross-section x_1 remains in a constant state for a time interval much shorter than τ. The first disturbance due to the reloading reaches the cross-section x_1 at time

$$t' = \frac{x_1}{c_0} + \tau \tag{3.3.2}$$

just as if the leading wave due to reloading would be an elastic one. This conclusion is true for any cross-section $0 \leqslant x \leqslant x_M$, where x_M is obtained as the intersection of the straight lines

$$x = c_0(t - \tau), \quad x = c(\varepsilon_A)t.$$

If the observations are registered at cross-sections $x > x_M$, then the conclusions are different. Let us consider for instance the cross-section $x_2 > x_M$. At time

$$t = \frac{x_2}{x_0} + \tau \tag{3.3.3}$$

the plastic waves from the first bundle still continue to propagate in the considered cross-section and no noticeable influence of the dynamic reloading can be observed experimentally. However, during the time interval

$$\frac{x_2}{c(\varepsilon_A)} + \tau > t > \frac{x_2}{c_0} + \tau \tag{3.3.4}$$

the picture of the plastic wave propagation is somewhat different from that which would apply if the reloading were absent. Thus the leading reloading wave is propagating with the speed c_0 only until this wave catches up the last plastic wave from the first bundle. On the other hand for the cross-section $x > x_M$ there does not exist an interval of time when all variables involved in the problem would be constant (domain (D)), i.e. the reloading is observed much before the time

$$t = \frac{x_2}{c(\varepsilon_A)} + \tau \tag{3.3.5}$$

predicted by the classical plasticity theory.

Therefore dynamic reloading experiments which are following another dynamic loading yield some conclusions which cannot be explained by a classical plasticity theory. This concerns both the speed of propagation of the leading reloading wave and the interaction between the reloading waves and waves produced by the first dynamic loading. Thus, though some aspects of the propagation of plastic waves can be described sometimes even accurately by a classical plasticity theory, the total nature of plastic waves is qualitatively not well described and this suggests the idea that an improvement of the constitutive equation would be desirable.

Similar experiments to the ones described above but performed in torsion were reported by several authors (KLEPACZKO [1967], YEW and RICHARDSON [1969], CAMPBELL and LEWIS [1969], DUFFY et al. [1971], NICHOLAS and CAMPBELL [1971], CAMPBELL and TSAO [1972], DUFFY et al [1972], FRANTZ and DUFFY [1972] etc.). The main conclusions were similar to those obtained for longitudinal impact.

Chapter IV

DETERMINATION OF CONSTITUTIVE EQUATION BY A NUMERICAL APPROACH

1. CONSTITUTIVE EQUATIONS IN THE FORM USED IN NUMERICAL EXAMPLES

Numerical results obtained from examples of rate-type constitutive equations will be compared below with experimental data and with the predicted behaviour of several constitutive equations of the kind used in classical plasticity theory. For this reason besides constitutive equations of the rate-type we shall also use several constitutive equations written in the classical time-independent form. Some classical constitutive equations which have played an important role in the development of the rate-type constitutive equations will be given first. In the present chapter, only a single loading from the virgin state $\sigma = \varepsilon = 0$ will be considered, followed by a single unloading up to $\sigma = 0$. Since all the examples considered in this chapter are for compressive stresses only, it is convenient to consider stresses and strains to be positive in compression.

In the framework of classical plasticity theory one uses the non-linear equation (2.1.3) in the loading domains (further denoted by L) while in the unloading domain (denoted here by U) the equation (2.1.2) is used. Therefore at crossing the loading/unloading boundary L/U the form of the constitutive equation is changed. The L/U boundary is a priori unknown. The shape of this boundary, as well as the distribution of stresses, strains and velocities along this boundary depend on initial and boundary conditions as well as on the constitutive equation. On this boundary, when first passing from loading to unloading we have $\dot{\varepsilon}(X, t) = 0$, i.e. the strain reaches its maximum in each cross-section of the bar. Thus in order to determine this maximum the whole strain history in that particular cross-section X is to be known. Generally the L/U boundary is determined by recording at each point X the maximum stress reached in the whole loading-unloading history at that particular point, as will be described further.

The L/U boundary for a semi-infinite bar was first considered by RAKHMATULIN [1945], KARMAN and DUWEZ [1950] and TAYLOR [1946].

For general boundary and initial conditions, for general constitutive equations and for finite bars (i.e. taking into account the reflections of waves as well) the L/U boundary can be determined only by numerical methods (see CRISTESCU [1970] and the Appendix). Basically the L/U boundary is determined by the condition that the solutions in the L and U domains be compatible with the boundary conditions and with the requirement that σ, ε and v be continuous across this boundary. It is obvious that the determination of this boundary is done simultaneously with the determination of the solution in both neighbouring regions L and U. No further details will be given here; only several revealing aspects will be discussed in order to compare the solution obtained in the framework of classical plasticity theory with those obtained for various rate-type constitutive equations.

Let us recall that in the U-domains the characteristics are straight lines (see further (4.1.3) in which $\Phi = 0$), while in the L-domains the characteristics are in general curved lines of the form (4.1.3). This is one of the major difficulties in the formulation of the numerical method. This difficulty can be overcome (CRISTESCU [1970]) if the constitutive equation (2.1.3) is written in the equivalent differential form

$$\dot{\varepsilon} = \frac{\dot{\sigma}}{E} + \chi\left[(f^{-1}(\sigma))' - \frac{1}{E}\right]\dot{\sigma} = \left[\frac{1}{E} + \chi\Phi(\sigma)\right]\dot{\sigma} \tag{4.1.1}$$

with

$$\chi(\sigma_m, \operatorname{sign}\dot{\sigma}) = \begin{cases} 0 & \text{if} \quad \sigma \leqslant \sigma_Y \text{ or } \sigma_Y \leqslant \sigma < \sigma_m \\ & \quad\text{or} \quad \sigma = \sigma_m \text{ and } \dot{\sigma} < 0, \\ 1 & \text{if} \quad \sigma = \sigma_m \text{ and } \dot{\sigma} \geqslant 0, \end{cases} \tag{4.1.2}$$

where $\sigma_m = \sigma_m(X, t) = \max\limits_{0 \leqslant s \leqslant t} \sigma(X, s)$. The coefficient χ defined by (4.1.2) is useful since the constitutive equation (4.1.1) is now valid in loading and in unloading domains.

According to the form (4.1.1) of the constitutive equation, the equations of the characteristic lines in both domains L and U, can be written as

$$\begin{aligned} \frac{dX}{dt} &= \pm\, c(\sigma) = \pm \sqrt{\frac{E}{\rho_0(1 + E\Phi)}}, \\ dX &= 0. \end{aligned} \tag{4.1.3}$$

The differential relations satisfied along these characteristic lines are

$$\begin{aligned} d\sigma &= \mp\, \rho_0 c \, dv, \\ d\varepsilon^P &= \Phi \, d\sigma, \\ E\, d\varepsilon^E &= d\sigma, \end{aligned} \tag{4.1.4}$$

respectively. We stress that the last characteristic (4.1.3) is a double characteristic in the loading domains and a simple characteristic in unloading domains. Along this characteristic the equations $(4.1.4)_2$ and $(4.1.4)_3$ hold in L but only $(4.1.4)_3$ holds in U.

Thus, in the unloading domain the characteristics are

$$\frac{dX}{dt} = \pm c_0 = \pm \sqrt{\frac{E}{\rho_0}}, \qquad dX = 0, \tag{4.1.5}$$

and the differential relations satisfied along them are

$$d\sigma = \mp \rho\, c_0\, dv, \qquad d\sigma = E\, d\varepsilon^E. \tag{4.1.6}$$

The numerical examples described below have been computed for a certain kind of aluminium (the so called 1100°F aluminium with standard purity and annealed for two hours at 590°C and then furnace cooled — see BELL [1968]). For this material the relations

$$\sigma = E\varepsilon \quad \text{if } \sigma \leqslant \sigma_Y, \qquad \sigma = \beta(\varepsilon + \varepsilon_0)^{1/\alpha} \quad \text{if } \sigma > \sigma_Y \tag{4.1.7}$$

were used, where $E = 7038000$ N/cm² and $\rho_0 = 2.7$ g/cm³, and therefore $c_0 = 5080$ m s^{-1}. Concerning the other constants, the following numerical values were used (see CRISTESCU and BELL [1970] and CRISTESCU [1972b]):

a) for the so called quasistatic stress-strain curve,

$$\sigma_{Y_0} = 703.80 \text{ N/cm}^2, \quad \beta = 22908.0 \text{ N/cm}^2, \quad \alpha = \frac{8}{3}; \tag{4.1.8}$$

b) for the so called "dynamic" stress-strain curve,

$$\sigma_{Y_1} = 212.18 \text{ N/cm}^2, \quad \beta = 38640.0 \text{ N/cm}^2, \quad \alpha = 2; \tag{4.1.9}$$

c) for the translated dynamic stress-strain curve,

$$\sigma_{Y_2} = 759.00 \text{ N/cm}^2, \quad \beta = 38640.0 \text{ N/cm}^2, \quad \alpha = 2, \quad \varepsilon_0 = 0.000278. \tag{4.1.10}$$

The data (4.1.8) were obtained by static tests using standard testing machines. The data (4.1.9) were obtained by dynamic tests in two ways: by measuring the speed with which various magnitudes of the strain are propagating in the bar and by establishing, for various velocities of impact, a one-to-one correspondence between the magnitude of the strain at the strain plateau and the magnitude of the stress at the corresponding stress plateau (see Chapter III, Section 2, Chapter V, Section 5). Since the true yield stress for the aluminium under consideration is in fact σ_{Y2}, the third set of numerical values (4.1.10) has been considered as well; here the stress-strain curve is obtained from the dynamic one by a translation along the strain axis (this defines ε_0). With the exception of the immediate neighbourhood of the yield point, the stress-strain curve obtained by this procedure is not too far from the dynamic one.

In the unloading domain equation (3.1.1) has been used with the value of the constant E given above.

In the numerical examples the rate-type constitutive equations under consideration were used in the form (see Chapter II, Section 3)

$$\dot{\varepsilon} = \left[\frac{1}{E} + \chi\Phi(\sigma, \varepsilon)\right]\dot{\sigma} + \Psi(\sigma, \varepsilon). \tag{4.1.11}$$

The characteristic lines for this constitutive equation were discussed in Chapter II, Section 4. In the form used in the numerical programme they can be written formally as (4.1.3) while the differential relations satisfied along them are

$$\begin{aligned} \mathrm{d}\sigma &= \pm \rho_0 c\, \mathrm{d}v - \rho_0 c^2 \Psi(\sigma, \varepsilon), \\ \mathrm{d}\sigma &= E\, \mathrm{d}\varepsilon^E, \\ \mathrm{d}\varepsilon^P &= \Phi(\sigma, \varepsilon)\, \mathrm{d}\sigma + \Psi(\sigma, \varepsilon)\, \mathrm{d}t. \end{aligned} \tag{4.1.12}$$

The coefficient function $\Psi(\sigma, \varepsilon)$ describing the non-instantaneous response was assumed to depend linearly on the overstress (CRISTESCU [1972 b]):

$$\Psi(\sigma, \varepsilon) = \begin{cases} \dfrac{k(\varepsilon)}{E}[\sigma - f(\varepsilon)] & \text{if} \quad \sigma > f(\varepsilon) \quad \text{and} \quad \varepsilon \geqslant \dfrac{\sigma}{E}, \\ 0 & \text{if} \quad \sigma \leqslant f(\varepsilon). \end{cases} \tag{4.1.13}$$

Formulae (4.1.8) – (4.1.10) suggest several expressions for the relaxation boundary $\sigma = f(\varepsilon)$ which all can be written in the form

$$f(\varepsilon) = \begin{cases} \sigma_Y & \text{if} \quad \varepsilon \leqslant \varepsilon_Y, \\ \beta(\varepsilon + \varepsilon_0)^{1/\alpha} & \text{if} \quad \varepsilon > \varepsilon_Y. \end{cases} \tag{4.1.14}$$

However some of the comparisons with the experimental data have suggested that a slight change of this boundary in the neighbourhood of the yield stress is necessary, thus the following expression was also used

$$f(\varepsilon) = \begin{cases} \sigma_Y & \text{if} \quad \varepsilon \leqslant \varepsilon_Y, \\ \sigma_Y + \dfrac{\beta}{2}\, \varepsilon_z^{-1/2} (\varepsilon - \varepsilon_Y) & \text{if} \quad \varepsilon_Y < \varepsilon < \varepsilon_z, \\ \beta \varepsilon^{1/2} & \text{if} \quad \varepsilon_z \leqslant \varepsilon \end{cases} \tag{4.1.15}$$

with

$$\varepsilon_z = \left(\frac{\beta \varepsilon_Y}{\sigma_Y - \sqrt{\sigma_Y^2 - \varepsilon_Y \beta^2}} \right)^2. \tag{4.1.16}$$

This change of the relaxation boundary produces a raising of this boundary in the neighbourhood of the yield point. Thus in this case the relaxation boundary is approximated by two segments of straight lines continued by a parabola of the form (4.1.9).

In a first approximation we have $k(\varepsilon) \gg 0$ for the coefficient in (4.1.13) which may be considered to be constant, especially for large strains. However, for a more accurate description of the deformation phenomena, especially for strains $\varepsilon > \varepsilon_Y$ close to ε_Y, one has to assume that $k(\varepsilon)$ is an increasing function of ε. Various expressions used for $k(\varepsilon)$, suggested by experimental data, lead to the assumption that k is nearly constant for higher strains but becomes much smaller or even very small for strains close to ε_Y. For instance,

$$k(\varepsilon) = k_0 \left[1 - \exp\left(- \frac{\varepsilon}{\hat{\varepsilon}} \right) \right] \tag{4.1.17}$$

with k_0 and $\hat{\varepsilon}$ material constants, was one of the expressions used. Another expression which can be easily adapted to the experimental data is

$$k(\varepsilon) = \begin{cases} 0 & \text{if} \quad \varepsilon \leqslant \varepsilon_0, \\ k_1 & \text{if} \quad \varepsilon_0 < \varepsilon \leqslant \varepsilon_1, \\ k_2 + \dfrac{k_1 - k_2}{\varepsilon_1 - \varepsilon_2} (\varepsilon - \varepsilon_2) & \text{if} \quad \varepsilon_1 < \varepsilon < \varepsilon_2, \\ k_2 & \text{if} \quad \varepsilon_2 \leqslant \varepsilon \end{cases} \tag{4.1.18}$$

with k_1, k_2, ε_0, ε_1, ε_2 positive constants. For certain particular materials it is possible that $k_1 = 0$ as well as $\varepsilon_0 = \varepsilon_Y$, in general. However, if $\varepsilon_0 > \varepsilon_Y$ and $k_1 \neq 0$, then the corresponding model can describe the property of some materials, the so called "delayed yield". For the aluminium under consideration we have taken $\varepsilon_0 = \varepsilon_Y$, $\varepsilon_1 \simeq 3\varepsilon_Y$, while ε_2 is by one order of magnitude bigger than ε_1 (some of the numerical values used in the numerical examples are given in Table 4.1). We would like to stress here that the experimental data indicate that k depends on the magnitude of the rate of strain; k is increasing when ε is decreasing by several orders of magnitude (see Chapter III, Subsection 1.3). For certain ranges of variation of $\dot{\varepsilon}$ the dependence of k on $\dot{\varepsilon}$ can be disregarded, as it was done in this chapter.

TABLE 4.1

No. of example	Formulas used	Constants used
1	(4.1.11), (4.1.13) (4.1.15), (4.1.18) (4.1.20), (4.1.23) (4.1.2)	$\sigma_Y = 1033$ N/cm², $\varepsilon_Y = 0.0001471$ $\varepsilon_z = 0.002575$, $\alpha = 2$, $\beta = 38622$ N/cm² $k_1 = 10^5$ s^{-1}, $k_2 = 10^6$ s^{-1}, $\varepsilon_1 = 0.0005$, $\varepsilon_3 = 0.004$, $m = 22390$ N/cm², $n =$ $= 62010$ N/cm², $h = 0.00003$, $\lambda =$ $= -0.00019$
2	(4.1.11), (4.1.13) (4.1.15), (4.1.18) (2.3.13), (4.1.23) (4.1.2)	same as in example 1 but h = 0, $\lambda = 0$, $\underset{\vee}{\varepsilon} = 0$
3	same as in example 2	same as in example 2 but $\underset{\vee}{\varepsilon} = 0.0004$

The problem of determining the explicit form for the coefficient Φ from the experimental data is much more difficult. Apparently Φ could be determined from the speed of propagation $(4.1.3)_1$ of the elastic-viscoplastic acceleration waves. However, it is not possible to determine this speed experimentally for all possible dynamic states of stress and strain. Thus Φ has to be determined by an indirect procedure which uses the following experimental data: the variation in time of ε and v for various prescribed cross-sections X (see Chapter III, Section 2) and the variation in time of the stress at the impacted end of the bar. Some simple expressions for the instantaneous curve have been chosen assuming the hypothesis (CRISTESCU [1972 *b*]) that for any $\varepsilon > \varepsilon_Y$ we have

$$E \geqslant \frac{1}{\Phi(\sigma, \varepsilon) + \dfrac{1}{E}} > f'(\varepsilon), \tag{4.1.19}$$

i.e. that the slope of the instantaneous curve lies between the elastic slope and that of the relaxation boundary. Further $\Phi(\sigma, \varepsilon)$ was chosen to match a certain set of experimental data. Other restrictions to be satisfied by other forms of the coefficient function Φ, due to certain experimental evidence, will be discussed in Chapter V. A simple function was chosen for a first approximation to the "instantaneous response curve" whose slope is $1/\left(\Phi + \frac{1}{E}\right)$ and it was found that a good representation of the propagation of the rising part of the strain-time curve could be obtained by adjusting slightly the function Φ. Two forms were used, both initially based on a cubic equation for the instantaneous response curve; $\varepsilon = A\sigma^3$ suggested equation (4.1.22) below while $\sigma - \sigma_Y = B(\varepsilon - \varepsilon_Y)^{1/3}$ suggested equation (4.1.20). In each case the resulting expression for Φ was then further adjusted by making the constant A or B a slowly varying function of ε in order to decrease the influence of Φ at the higher strains.

Besides (2.3.13) in the numerical examples given below other two expressions for Φ have been used. One of them is (CRISTESCU [1972 *b*])

$$\Phi(\varepsilon) = \frac{3\left[\varepsilon - \varepsilon_Y - \varepsilon^* + \left(\frac{a}{3E}\right)^{3/2}\right]^{2/3}}{a} - \frac{1}{E}, \tag{4.1.20}$$

where

$$a = m + n\sqrt{\varepsilon} \tag{4.1.21}$$

with m and n constants. ε^* is a threshold strain and is defined below by (4.1.23).

The other one is

$$\Phi(\sigma, \varepsilon) = \frac{\gamma}{E}\left[3\left(\frac{E}{p + q\sqrt{\varepsilon}}\right)^3\left(\frac{\sigma}{E}\right)^2 - 1\right], \tag{4.1.22}$$

where p, q and γ are constants. In the programs the case of the semilinear constitutive equation was obtained in the above model simply by taking $\gamma = 0$ in (4.1.22).

σ^* is a threshold stress defined in some of the examples so that for $\sigma \leqslant f(\varepsilon) + \sigma^*$ we have $\Phi = 0$. Other examples have been considered with $\sigma^* = 0$; generally, it was assumed that $\sigma^*(\varepsilon)$ is defined by

$$\varepsilon^* = \frac{\sigma^*}{E} = h - \lambda\sqrt{\left|\varepsilon - f^{-1}\left[\frac{\sigma_Y}{E} + h\right]\right|} \tag{4.1.23}$$

with h and λ constants. Formula (4.1.23) also yields the ε^* of (4.1.20). The values of various constants are given in Table 4.1.

2. INITIAL AND BOUNDARY CONDITIONS

All the numerical examples given here concern the so called symmetric impact of two identical bars, since for this case a full set of very many accurate experimental data obtained by J. F. BELL was available (see Chapter III, Section 2). For the purpose of writing a program for a computer the symmetric impact is a simpler case, since due to symmetry it suffices to perform the computation in one of the two bars only. It will be assumed that the bar (called "specimen") for which the computations will be carried out is at the initial moment at rest and undeformed, its initial length being l. Therefore

$$\left.\begin{matrix} t=0 \\ 0<X\leqslant l \end{matrix}\right\}: \sigma=\varepsilon=v=0 \tag{4.2.1}$$

are the adopted initial conditions.

As concerns the boundary conditions, the end $X=l$ of the bar is assumed to be free, therefore the boundary conditions at $X=l$ are

$$\left.\begin{matrix} X=l \\ t\geqslant 0 \end{matrix}\right\}: \sigma=0. \tag{4.2.2}$$

From (4.2.2) together with the constitutive equation (Hooke's law) we get that for any t we also have $\varepsilon=0$ while the particle velocity of this end of the bar must be computed by a special subroutine used at $X=l$.

At the moment $t=0$ the bar is impacted at the end $X=0$ by another identical bar (i.e. of same size and mechanical properties), called hitter, which is moving with a known initial velocity V. Therefore, the initial conditions for the hitter are

$$\left.\begin{matrix} t=0 \\ -l\leqslant X<0 \end{matrix}\right\}: \sigma=\varepsilon=0, \quad v=V. \tag{4.2.3}$$

By comparing conditions (4.2.3) and (4.2.1) one can see that for $t=0$ at the end $X=0$ of the specimen the particle velocity is subjected to a sudden jump. This jump was handled in two ways, by two variants of the program. First, it was assumed that the impact velocity is fast but smoothly transferred to the end of the specimen. A convenient very short time interval $0\leqslant t<t_m$ was chosen, in which this smooth increase of the velocity at the impacted end of the specimen takes place from zero up to v_{max}. For simplicity, it was assumed in the program that the velocity at the end of

the specimen increases according to a linear law. Thus, the first kind of boundary conditions were

$$X = 0 \begin{cases} 0 \leqslant t < t_m: & v = \dfrac{t}{t_m} v_{max}, \\ t_m \leqslant t \leqslant T_c: & v_H = v_S, \\ T_c < t: & \sigma_S = 0, \end{cases} \tag{4.2.4}$$

where t_m is a conveniently chosen time interval (in most examples given here $t_m = 0.5\mu s$) and $v_{max} = V/2$. T_c is a computed time, called *time of contact*. In the time interval $t_m \leqslant t \leqslant T_c$ the ends $X = 0$ of the two bars are moving together, i.e. at $X = 0$ the particle velocity v_H of the hitter and the particle velocity v_S of the specimen are equal. In $(4.2.4)_3$ σ_S denotes the specimen stress. More details concerning the condition (4.2.4) and the way in which this condition was handled in a program subroutine are given in the Appendix. The conditions $(4.2.4)_{1,2}$ are used as long as $\sigma_S(0, t) > 0$. Let us recall that $\sigma_S(0, t)$ is computed by a subroutine of the program. The time when $\sigma_S(0, t)$ first vanishes will be called the time of contact and will be denoted by T_c. For $t \geqslant T_c$ the end $X = 0$ of the specimen becomes stress free and this is the reason why from now on we shall use condition $(4.2.4)_3$ at $X = 0$. Condition $(4.2.4)_3$ in the form of a subroutine is used in order to get the behaviour of the specimen after separation from the hitter.

Since the loading process at the end $X = 0$ is anyway very fast, in order to apply the boundary conditions (4.2.4) at the end $X = 0$, the mesh interval Δt for the characteristic net used in the computations, was sometimes considered variable. Thus for the domain $X > 0$, $X < c_0 t$, $X < -c_0(t - t_1)$ with $0 \leqslant t \leqslant t_1 = 2\ \mu s$, the mesh size used was $\Delta t = \dfrac{1}{200}\ \mu s$ and then, between the characteristics $X = -c_0(t - t_1)$ and $X = -c_0(t - t_2)$ with $2\mu s = t_1 < t \leqslant t_2 = 10\ \mu s$, Δt was steadily increased up to $\Delta t = \dfrac{1}{4}\ \mu s$. This last mesh interval was subsequently used throughout the field for all the computations carried out in the strip $0 < X \leqslant l$, $t > 0$.

3. SHOCK WAVES

Since the conditions (4.2.1) and (4.2.3) prescribe distinct values for v at $X = 0$ and $t = 0$, we get a velocity discontinuity introduced by the boundary conditions, and this discontinuity is propagating into the bar in the form of a shock wave. It was shown in the previous section how this discontinuity can be handled in a computer program by considering boundary conditions of the form (4.2.4), i.e. by replacing the discontinuity in the boundary conditions by a smooth but fast variation of v at $X = 0$. However, this discontinuity can be handled in another way by introducing shock waves (see Chapter I). As was shown in Chapter I, across a shock wave the jump conditions

$$\rho_0 U[v] = [\sigma],$$
$$U[\varepsilon] - [v] = 0 \tag{4.3.1}$$

are satisfied with the sign convention for σ used in this chapter. Since in this chapter only mechanical processes will be considered, the jump conditions $(1.1.11)_1$ and $(1.1.11)_2$ will be disregarded. A short discussion on thermomechanical shock waves will be presented in Chapter VIII.

The compatibility relation $(4.3.1)_1$ is called the *dynamic jump condition*, while $(4.3.1)_2$ is called the *kinematic jump condition*. These conditions are to be satisfied at the crossing of a shock wave (or the front of a shock wave).

Let us consider now our problem by assuming that, due to the sudden impact, first a shock wave will propagate in the bar. Let us first examine the case of the constitutive equation of classical plasticity theory.

In the case of constitutive equation of the form (3.1.4) which is linear for $\sigma \leqslant \sigma_Y$, i.e. $\sigma = E\varepsilon$ for $\sigma \leqslant \sigma_Y$, an elastic compressive shock wave of speed c_0 propagates first in the bar. At time $t = l/c_0$ this wave reflects from the free end $X = l$ and becomes a dilatational shock wave, which is an unloading wave. During propagation this dilatational wave will gradually be absorbed (that is the jump will gradually decrease) by the plastic compressive direct waves generated at the impacted end. Let X_A be the cross-section of the bar where the reflected dilatational shock wave is completely absorbed (see fig. 4.3.1), i.e. this wave is propagating as a shock wave in the portion $X_A < X \leqslant l$ of the bar only. Thus for this portion of the bar the straight line

$$X = -c_0\left(t - \frac{2l}{c_0}\right) \tag{4.3.2}$$

is just the front of the reflected shock wave. Therefore the functions σ, v and ε may suffer jumps across the line (4.3.2). In order to estimate these jumps the solution of the problem just below (4.3.2) is necessary. Using the equation of the characteristics of positive slope

$$X = c(\sigma)t, \tag{4.3.3}$$

the relationship (4.1.4)$_1$

$$v = +\int_0^\sigma \frac{d\sigma}{\rho_0 c(\sigma)} \tag{4.3.4}$$

and (4.3.2) we can express the loading domain along the line (4.3.2) as function of t (see CRISTESCU [1967] Chapter II, say). For instance, if a

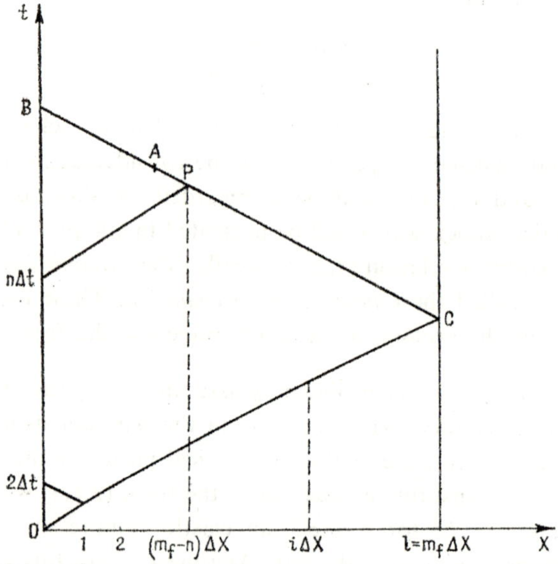

Fig. 4.3.1 Characteristic plane for numerical integration

constitutive equation of the form (4.1.7) is used with $\varepsilon_0 = 0$ and $\alpha = 2$, then (4.3.2) becomes

$$X = \sqrt{\frac{\beta}{2\rho_0}\left(\frac{\sigma}{\beta}\right)^{-1}}\, t$$

and solving this with respect to σ and using (4.3.2) we get (CRISTESCU [1970])

$$\begin{gathered} \sigma_b = \frac{\beta^2}{2\rho_0 c_0^2} \frac{t^2}{\left(\dfrac{2l}{c_0} - t\right)^2}, \\ v_b = \frac{2}{3\beta}\sqrt{\frac{2}{\rho_0}}\,\sigma_b^{3/2} + \left[\frac{1}{\rho_0 c_0} - \frac{2}{3\beta}\sqrt{\frac{2}{\rho_0}}\sqrt{\sigma_Y}\right]\sigma_Y, \end{gathered} \tag{4.3.5}$$

where the subscript b stands for "before", i.e. just below the line (4.3.2).

The solution along (4.3.2) but on the other side in the unloading domain, i.e. above the line (4.3.2), can be obtained by using the jump conditions (4.3.1), where $U = -c_0$. Thus we obtain (LEE [1953])

$$\sigma_a = \sigma_b - \frac{1}{2}[2\sigma_Y - (\rho_0 c_0 v_b - \sigma_b)],$$

$$v_a = v_b - \frac{1}{\rho_0 c_0}(\sigma_a - \sigma_b), \tag{4.3.6}$$

$$\varepsilon_a^E = \frac{\sigma_a}{E}, \quad \varepsilon_a^P = \varepsilon_b^P,$$

where the subscript a stands for "after", i.e. just above the line (4.3.2). The values of all unknown functions along the straight line (4.3.2) in the unloading domain can now be obtained using (4.3.6). All these values are denoted by the subscript a. However, it is still necessary to know the value of one of the functions σ_a, v_a or ε_a^E at a single point of (4.3.2). Since just after reflection at $X = l$ and $t = \frac{l}{c_0} + 0$ we have $\sigma_a = 0$, the formulae (4.3.6) yield σ_a, v_a and ε_a^E along (4.3.2) as long as $\sigma_a \neq \sigma_b$. The computations are carried out at successive points along (4.3.2) starting from $X = l$ and $t = \frac{l}{c_0} + 0$. By this procedure we also obtain the location of the point A on (4.3.2), where the unloading wave is completely absorbed by the direct plastic loading wave and therefore $\sigma_a = \sigma_b$.

Thus along the line (4.3.2), on the upper part of it, a subroutine of the program based on the formulae (4.3.5) and (4.3.6) yields the solution for the portion $X_A \leqslant X < l$ of the bar. For the remaining portion of the bar $0 \leqslant X < X_A$ the solution along (4.3.2) is furnished by a subroutine based on formulae (4.3.5). Thus the solution along (4.3.2) at any cross-section $0 \leqslant X \leqslant l$ can be obtained. These data were used further on for the numerical computations in the strip

$$0 \leqslant X \leqslant l,$$

$$X > -c_0\left(t - \frac{2l}{c_0}\right), \tag{4.3.7}$$

$$t < T_c + t_f,$$

where t_f is a conveniently chosen time, in order to get the elastic vibration of the bar after separation from the hitter, if desired.

If the constitutive equation used is of rate-type, then the problem of propagation of the first shock wave must be considered starting from the point $X = 0$, $t = 0$. For semi-linear constitutive equations the propagation of the first shock wave was studied first by SOKOLOVSKIĬ [1948 a, b] and by MALVERN [1951 a, b]. They observed that for semi-linear constitutive equations (of type (1.2.6)—(1.2.7)) the jump conditions are

$$\begin{aligned} \rho_0 c_0 [v] &= [\sigma], \\ c_0 [\varepsilon] &= [v], \end{aligned} \tag{4.3.8}$$

that is the velocity of propagation of the shock wave coincides with the velocity of propagation $c_0 = \sqrt{E/\rho_0}$ of the acceleration waves. Therefore in the interval $0 \leqslant X \leqslant l$, along the characteristic line

$$X = c_0 t \tag{4.3.9}$$

both the jump conditions (4.3.8) and the differential relations $(4.1.12)_1$ are satisfied (generally $(4.1.12)_1$ are satisfied along the characteristic lines). Since velocities are prescribed by the boundary conditions, one can first compute $[v]$ from (4.2.3) and (4.2.1) and then $[\sigma]$ and $[\varepsilon]$ at the point $X = 0$, $t = 0+$, by using (4.3.8). Now $(4.1.12)_1$ can be integrated along (4.3.9),

$$-\sigma + \sigma_0 = \sigma_0 c_0 (v - v_0) - \int_0^t \psi(\sigma, \varepsilon)\, dt, \tag{4.3.10}$$

where σ_0 and v_0 are the values at $X = 0$, $t = 0+$ while $\psi \leqslant 0$ according to (2.2.9). From (4.3.10) and using (4.3.8) we obtain the equation

$$\varepsilon - \varepsilon_0 = \frac{1}{2E} \int \psi(E\varepsilon, \varepsilon)\, dt, \quad t \in \left[0, \frac{l}{c_0}\right) \tag{4.3.11}$$

with the initial condition $\varepsilon(0+) = \varepsilon_0 = v_0/c_0$. If V is the initial velocity of the hitter according to (4.2.3), then in symmetric impact we get $v_0 = V/2$. From equation (4.3.11) we can get the decreasing behaviour of ε along the characteristic (4.3.9), starting from the value ε_0 at $X = 0$, $t = 0+$. After the determination of ε, the values of σ and v along the same characteristic lines can be obtained from formulae (4.3.8), i.e.

$$\begin{aligned} v(t) &= v_0 + c_0(\varepsilon(t) - \varepsilon_0), \\ \sigma(t) &= E\varepsilon(t). \end{aligned} \tag{4.3.12}$$

At $t = l/c_0$ the shock wave (4.3.9) reflects from the free end $X = l$ as an unloading shock wave with the equation

$$X = -c_0 t + 2l. \tag{3.4.13}$$

Just above this line, at the point C (fig. 4.3.1) the stress satisfies $\sigma_a = 0$ and, since the wave (4.3.13) is an unloading wave, at this point we also have $E\varepsilon_a = \sigma_a = 0$, while v_a can be obtained from (4.3.8),

$$v_a = v_b + c_0 \varepsilon_b = 2c_0 \varepsilon_b. \tag{4.3.14}$$

Here $\varepsilon_b = \varepsilon_-(l/c_0)$ and $v_b = v_-(l/c_0)$ are obtained from (4.3.11) and $(4.3.12)_1$ respectively. The values of $\sigma(t)$, $\varepsilon(t)$ and $v(t)$ determined according to formulae (4.3.11) and (4.3.12) together with the boundary condition $v(0, t) = v_0$ prescribed for $0 < t < 2l/c_0$ are now used to integrate the system (4.1.12) in the triangle OCB

$$\begin{gathered} 0 < X < l, \\ X < c_0 t, \quad X < -c_0 t + 2l. \end{gathered} \tag{4.3.15}$$

Thus, the values of $\sigma_b(t)$, $\varepsilon_b(t)$ and $v_b(t)$ are determined along and just "before" (4.3.13). Using now the jump conditions (4.3.8) and the particular values of v_a, σ_a and ε_a at the point C, we can obtain $\sigma_a(t)$, $\varepsilon_a(t)$ and $v_a(t)$ along (4.3.13) and just "after" it by integration from C to B. The point A is defined as that point on (4.3.13) where for the first time during this integration $\varepsilon_a = \varepsilon_b$ holds and therefore, by (4.3.8) also $\sigma_\alpha = \sigma_b$ and $v_a = v_b$.

Let us examine how one can handle the propagation of a shock wave in a material satisfying a quasilinear constitutive equation of the form (2.2.1)

$$\dot{\sigma} = \varphi(\sigma, \varepsilon)\, \dot{\varepsilon} + \psi(\sigma, \varepsilon). \tag{4.3.16}$$

Let us assume the same initial conditions, i.e. that at the initial moment the bar is at rest, undeformed and stress free. We shall follow a procedure described in detail in Chapter V, Section 2. Let us write (4.3.16) in the functional form

$$\sigma(t) = f(\varepsilon(t), \tau(t)), \tag{4.3.17}$$

where $\tau(t)$ is the history parameter, depending for each fixed X on the history of deformation in the time interval $[0, t)$. The function f in (4.3.17) satisfies the equation

$$\frac{\partial f(\varepsilon, \tau)}{\partial \varepsilon} = \varphi(f(\varepsilon, \tau), \varepsilon).$$

The curve in the σ, ε plane

$$\sigma = f(\varepsilon, 0) \tag{4.3.18}$$

is called the *instantaneous curve* with respect to the natural rest configuration (a configuration for which $\varepsilon = 0$ all the time implies $\sigma = 0$ all the time) and it is determined as the solution of the equation

$$\frac{\partial f(\varepsilon, 0)}{\partial \varepsilon} = \varphi(f(\varepsilon, 0), \varepsilon), \quad f(0, 0) = 0. \tag{4.3.19}$$

Depending on the properties described by the function φ, the solution (4.3.18) of the equation (4.3.19) on some interval on the ε-axis may either coincide with Hooke's straight line $\sigma = E\varepsilon$, or possess a downward or an upward directed concavity (see fig. 4.3.2). Henceforth all these cases will be assumed possible. However, for all these cases one will assume that there is a certain interval $[0, \varepsilon_{YD}]$ in which the function f is linear and therefore (4.3.18) coincides with the Hooke straight line on this interval. Here ε_{YD}

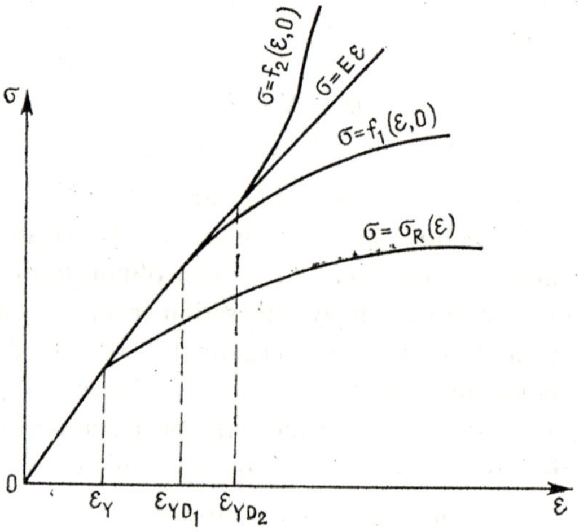

Fig. 4.3.2 Relaxation boundary and three possible instantaneous curves

can have the meaning of a strain corresponding to a certain dynamic limit of nonlinearity. As was previously mentioned, the relaxation boundary $\sigma = \sigma_R(\varepsilon)$ and the three possible cases concerning the instantaneous curve, are represented in fig. 4.3.2.

In order to describe the propagation of the shock wave the jump conditions $(1.1.11)_1$ and (1.1.12) will be used together with (4.3.18). Now, in the cases when f has a concavity directed towards the positive stress axis and therefore a shock wave occurs, $U(\varepsilon)$ becomes a variable velocity of propagation.

All three previously mentioned cases will be considered using the fig. 4.3.1 in which OC is now no longer a straight line. Using the initial conditions (4.2.1) and the boundary conditions

$$\begin{aligned} v(0, t) &= v_0, \quad t > 0, \\ \sigma(l, t) &= 0, \quad t > 0, \end{aligned} \tag{4.3.20}$$

the jump conditions $(1.1.11)_1$ and (1.1.12) and finally (4.3.18) we obtain that along OC

$$\rho_0 U v_a - f(\varepsilon_a, 0) = 0,$$
$$v_a - U\varepsilon_a = 0, \tag{4.3.21}$$
$$\rho_0 U^2 = \frac{f(\varepsilon_a, 0)}{\varepsilon_a}.$$

Considering now the equation of motion $(1.1.17)_1$ and the compatibility equation $(1.1.7)_2$ in which (4.3.17) is also taken into account, we get

$$\left(\frac{\partial v}{\partial t}\right)_a + c^2\left(\frac{\partial \varepsilon}{\partial X}\right)_a = -\frac{1}{\rho_0}\frac{\partial f}{\partial \tau}(\varepsilon, 0)\left(\frac{\partial \tau}{\partial X}\right)_a,$$
$$\left(\frac{\partial \varepsilon}{\partial t}\right)_a + \left(\frac{\partial v}{\partial X}\right)_a = 0. \tag{4.3.22}$$

The derivatives of v and ε along OC are

$$\left(\frac{\partial v}{\partial t}\right)_a + U\left(\frac{\partial v}{\partial X}\right)_a = \frac{\mathrm{d}v_a}{\mathrm{d}t},$$
$$\left(\frac{\partial \varepsilon}{\partial t}\right)_a + U\left(\frac{\partial \varepsilon}{\partial X}\right)_a = \frac{\mathrm{d}\varepsilon_a}{\mathrm{d}t}. \tag{4.3.23}$$

Using (5.3.11) and (5.3.12) as well as the fact that τ is continuous across the shock wave, we can write

$$\left[\frac{\partial \tau}{\partial t}\right] + U\left[\frac{\partial \tau}{\partial X}\right] = \frac{\mathrm{d}[\tau]}{\mathrm{d}t} = 0,$$
$$\frac{\partial f}{\partial \tau}\frac{\partial \tau}{\partial t} = \psi(f(\varepsilon, \tau), \varepsilon). \tag{4.3.24}$$

Since $\left(\frac{\partial \tau}{\partial t}\right)_b = \left(\frac{\partial \tau}{\partial X}\right)_b = 0$ and since $(4.3.24)_2$ yields

$$\frac{\partial f}{\partial \tau}(\varepsilon_a, 0)\left(\frac{\partial \tau}{\partial t}\right)_a = \psi(f(\varepsilon_a, 0), \varepsilon_a), \tag{4.3.25}$$

we get

$$\frac{\partial f}{\partial \tau}(\varepsilon_a, 0)\left(\frac{\partial \tau}{\partial X}\right)_a = -\frac{1}{U}\psi(f(\varepsilon_a, 0), \varepsilon_a). \tag{4.3.26}$$

From $(4.3.23)_2$ and $(4.3.22)_2$ one has

$$\left(\frac{\partial v}{\partial X}\right)_a = -\left(\frac{\partial \varepsilon}{\partial t}\right)_a = -\frac{\mathrm{d}\varepsilon_a}{\mathrm{d}t} + U\left(\frac{\partial \varepsilon}{\partial X}\right)_a$$

and therefore, by $(4.3.23)_1$,

$$\left(\frac{\partial v}{\partial t}\right)_a = \frac{dv_a}{dt} + U\frac{d\varepsilon_a}{dt} - U^2\left(\frac{\partial \varepsilon}{\partial X}\right)_a. \tag{4.3.27}$$

Now $(4.3.22)_1$, (4.3.26) and (4.3.27) imply

$$\frac{dv_a}{dt} + U\frac{d\varepsilon_a}{dt} = \frac{1}{\rho_0 U}\psi(f(\varepsilon_a, 0), \varepsilon_a) + (U^2 - c^2)\left(\frac{\partial \varepsilon}{\partial X}\right)_a, \tag{4.3.28}$$

and if $(4.3.21)_2$ is used to eliminate v_a, (4.3.28) becomes

$$\frac{d}{dt}(U(\varepsilon_a), \varepsilon_a) + U(\varepsilon_a)\frac{d\varepsilon_a}{dt} = \frac{1}{\rho_0 U(\varepsilon_a)}\psi(f(\varepsilon_a, 0), \varepsilon_a) + \\ + (U^2 - c^2)\left(\frac{\partial \varepsilon}{\partial X}\right)_a, \tag{4.3.29}$$

where $U(\varepsilon_a)$ is obtained from $(4.3.21)_3$. If the initial (i.e. at $X = 0$, $t = 0+$) value $v_a(0) = v_0$ is prescribed by (4.3.20), the initial value for ε results from $(4.3.21)_1$ and $(4.3.21)_3$:

$$\rho_0 v_0^2 = f(\varepsilon_0, 0)\,\varepsilon_0. \tag{4.3.30}$$

Therefore we get the initial condition for the equation (4.3.29)

$$\varepsilon_a(0) = \varepsilon_0. \tag{4.3.31}$$

Formula (4.3.11) corresponding to the semi-linear case can be obtained from (4.3.28) for $U = c_0$, However, these two cases are quite distinct, since in the semi-linear case the variation of ε along the shock wave is determined if merely the initial data $\varepsilon_a(0) = \varepsilon_0$ is prescribed ,while in the quasilinear case, in order to determine ε, the derivative of ε with respect to X everywhere along the shock wave OC must also be known. For this purpose the whole solution along OC is to be obtained. For rate-type materials this result was obtained by AHRENS and DUVALL [1966]. Similar conclusions were obtained for other kind of nonlinear materials (see for instance CHEN and GURTIN [1972]).

Equation (4.3.29) does not allow one to determine ε along the shock wave even if the initial value (at the impacted end of the bar) is known (compare with the semi-linear case (4.3.11)). The reason for this is the presence in equation (4.3.29) of terms connected with the instantaneous response of the material (as in $U(\varepsilon)$) and the term connected with the relaxation properties of the material (as in ψ). We would like to recall at this point that there are experimental techniques allowing in various cross-sections of the bar the simultaneous determination of ε and of the derivative $\partial\varepsilon/\partial X$

(by determining the surface angle) during the dynamic process of deformation (see Chapter III, Section 2, and Chapter V, Section 5).

As will be shown in Chapter V, Section 4, the function $f(\varepsilon, 0)$ can be determined in a first approximation by knowing two material constants. Thus equation (4.3.29) can be used in order to determine the function ψ along the curve of instantaneous response of the material, provided that enough experimental data concerning ε_a and $(\partial\varepsilon/\partial X)_a$ are known along the shock wave.

If $\varepsilon_a = \varepsilon_a(t)$ is already known, then $(4.3.21)_2$ yields $v_a = v_a(t)$ and $(4.3.21)_3$ yields $\sigma_a = \sigma_a(t)$. The equation of the shock wave OC is then determined from

$$\frac{dX}{dt} = c(\varepsilon_a(t)),\ X(0) = 0. \qquad (4.3.32).$$

4. NUMERICAL EXAMPLES

We shall give in this section the results of the numerical computations for rate-type models of the form (2.3.1). Three examples, i.e. three kinds of rate-type constitutive equations, will be discussed (CRISTESCU [1974 *a*]); various expressions used for the coefficients as well as the numerical values of the constants involved are given in Table 4.1. Table 4,2. contains the numerical values which were obtained by computations for various parameters as well as the corresponding experimental values. Most of the experimental data used further on in this chapter have been obtained by J. F. BELL for the symmetric longitudinal impact of two aluminium bars.

In order to compare the results obtained by means of various rate-type constitutive equations with the ones obtained within a classical plasticity theory, let us first discuss the description of the symmetric impact within the framework of a classical plasticity theory (CRISTESCU [1970]). The constitutive equation used is (4.1.7) with (4.1.10). One of the bars (the "specimen") is at rest while the other one (the "hitter") is moving with the initial velocity V. A typical description of the interaction between the elastic and plastic waves as well as of the reflection of waves from the free end is given on fig. 4.4.1. In this specific example $V = 34.98$ m/sec, the diameter of the bar is $D = 2.5$ cm, while its length $l = 10\ D$. The first elastic compressive wave $X = c_0 t$ reflects from the free end $X = l$ as an unloading tensile wave. In the neighbourhood of the impacted end $X = 0$ during the time interval $0 \leqslant t < 2l/c_0$ no reflected wave from the free end arrives. Thus, above the characteristic line $X = c(\sigma_{max})t$ there is a region in the characteristic plane where $v = v_{max} = V/2$, $\sigma = \sigma_{max}$ and $\varepsilon = \varepsilon_{max}$. All these values can be easily obtained for the boundary conditions and the differential relations

$$dv = \pm\, c(\varepsilon)\, d\varepsilon \quad \text{or} \quad dv = \pm \frac{d\sigma}{\rho c(\sigma)} \tag{4.4.1}$$

satisfied along the characteristic lines. The presence of such region of constant state is typical for the solution obtained by means of a classical plasticity theory constitutive equation (of the form (2.1.1)). Once the unloading process is reaching this region, the stress begins to decrease here too. Therefore the region of constant state will be bounded by the straight lines $X = 0$, $X = c(\sigma_{max})t$ and a segment of the L/U boundary.

Another region of constant state is comprised between the straight lines

$$X = c_0 t, \quad X = c(\varepsilon_Y)\, t,$$

$$X = -\, c_0 \left(t - \frac{2l}{c_0}\right). \tag{4.4.2}$$

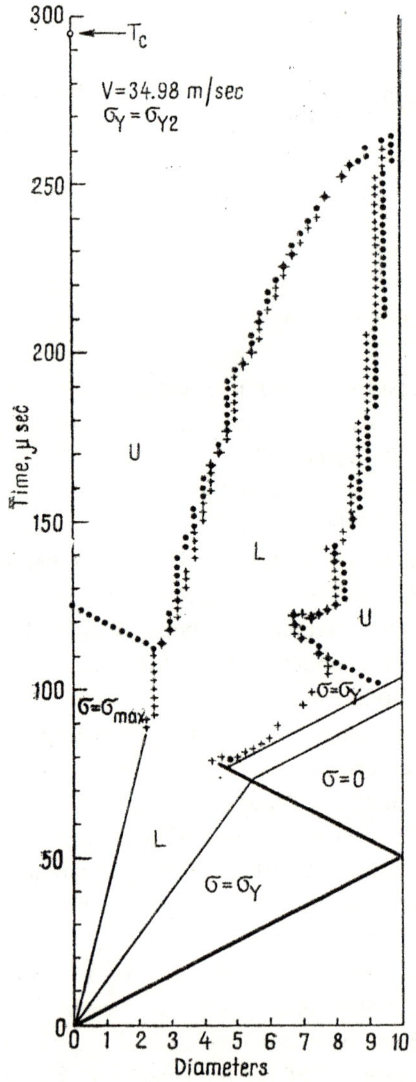

Fig. 4.4.1 Characteristic plane showing the loading/unloading boundary obtained with numerical methods

In domain L only plastic waves are propagating while in domain U there are elastic unloading waves only. The L/U boundary shown by dots was obtained by numerical methods using a computer. T_c is the time of contact between the two bars which was obtained also by computation.

The solution in the loading and unloading domains can be represented in various ways. However, most useful is the representation of those features of the solution which can be compared with the experimental data. For instance the variation in time of the strain at various cross-sections along the specimen is given on fig. 4.4.2. Observe that in the neighbourhood of the impacted end, after reaching its maximum, the strain remains constant for quite a long period of time. This is the so called "plateau" in time of the strain. Besides that, on a portion of the bar near the impacted end ($0 \leqslant X < 2.5\, D$ say)

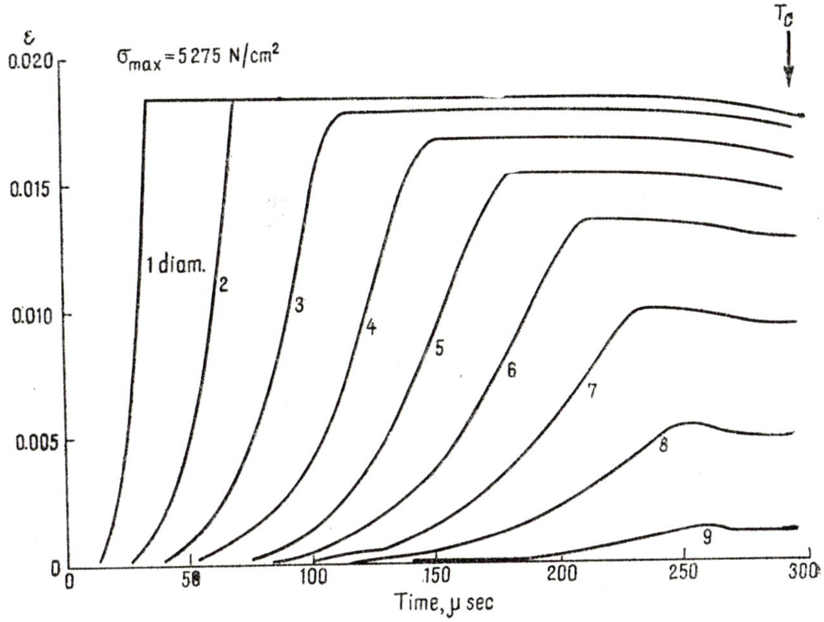

Fig. 4.4.2 Computed strain-time curves at various cross-sections of the bar

the maxima on various cross-sections coincide, that is we have a "plateau in space" of the strain.

For the same impacting velocity as before we show on fig. 4.4.3 the variation in time of the particle velocity. In a similar way we can point out a "plateau" for the particle velocity.

Let us mention that, except on the very last portion (i.e. the portion when the unloading influence is felt), the curves at $X = 1\, D$ and $X = 2\, D$ on both fig.. 4.4.2 and 4.4.3 are typical for the solution in the loading domain obtained with a classical constitutive equation (2.1.1) and which was initially obtained for the "semi-infinite" bar.

Let us discuss now the results which can be obtained with rate-type constitutive equations.

Since the strain variation in time at various cross-sections along the bar was measured by the diffraction grating technique and since this experimental method is a very accurate one for finite strains, great weight was given to these data in the determination of the coefficient functions. These curves, obtained for various cross-sections X along the bar, are given in fig. 4.4.4. In this figure, for instance, 3 D means the cross-section which is

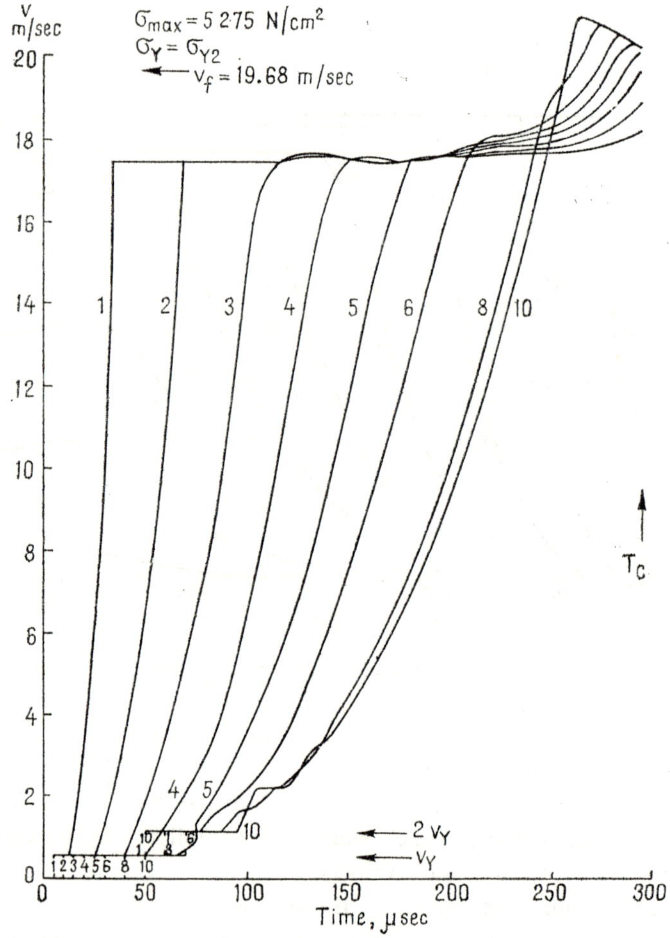

Fig. 4.4.3 Computed velocity-time curves at various cross-sections of the bar

three diameters away from the impacted end of the bar. The small circles are the experimental data while the vertical arrows indicate the times of contact. The beginnings of these curves in fig. 4.4.4 were not represented on the figure. Since it is important for the establishement of the constitutive equation to know the shapes of these curves corresponding to small

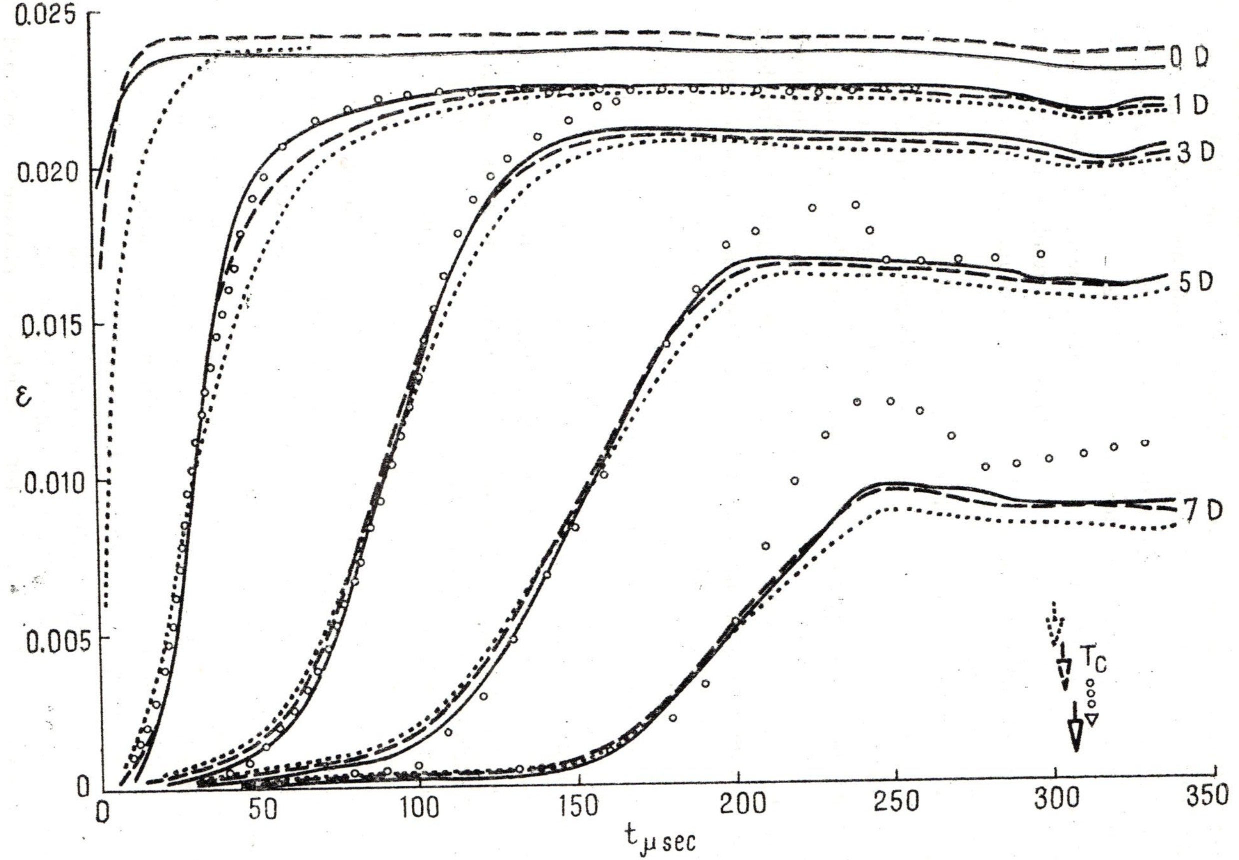

Fig. 4.4.4 Comparison of the computed strain-time curves at various cross-sections of the bar with experiments

strains, mainly for the cross-sections close to the impacted end, these were represented in a separate figure 4.4.5 (the experimental data were not available). One can see that the first fast increase of the strain which is nearly linear (recall that according to the boundary conditions $(4.2.4)_1$, the initial rise was assumed to be linear) results in a strain well above the static yield strain. This first rise of the strain was called the "elastic precursor". From figure 4.4.5 it follows that during propagation the elastic precursor is decaying quite fast towards ε_Y. Qualitatively this result is in agreement with experimental data obtained by several authors and it is described by the model as well.

At the impacted end of the bar this initial rise of the strain corresponds to the nearly linear portion of the stress-strain curve before reaching the peak stress (see fig. 4.4.6 *a*, *b*, *c*). Therefore the decay of the elastic precursor during propagation takes place nearly simultaneously with the decay of the peak stress during propagation. In figure 4.4.6 the full lines are the relaxation boundaries, while the numbers written beside the lines are showing how many microseconds elapsed from the moment of impact $t = 0$ of the two bars. For instance, in fig. 4.4.6 *c*, at the impacted end (i.e. at zero *D*) the peak is reached in 0.5 μs (we recall that in these examples the loading at the end $X = 0$ is handled by means of the relation (4.2.4) in which $t_m = 0.5\ \mu$s); after 40 μs the state of stress and strain is very close to the relaxation boundary but it takes 185 microseconds until this state falls noticeably below the relaxation boundary. In other words the states of stress and strain on the relaxation boundary and in its immediate neighbourhood are quasistable states.

From fig. 4.4.5 (full lines) and fig. 4.4.6, *a* there follows that the expression (4.1.20) for Φ does not describe accurately the peak stress phenomena, though generally from other points of view the solution is in good agreement with the experimental data and in many instances it coincides with the solution obtained when expression (2.3.13) is used for Φ. Thus if we would like to describe more accurately the peak stress phenomena and the initial shape of the strain-time curves we may choose an expression for Φ which is vanishing along $\sigma = E\varepsilon$ and possibly even in the neighbourhood of this straight line. In other words in the neighbourhood of Hooke's line the instantaneous response of the material must be "nearly" elastic. When Φ is a function of ε^P, this requirement is satisfied; $\underset{\vee}{\varepsilon}$ was introduced just to create a strip along the $\sigma = E\varepsilon$ line where $\Phi = 0$.

Therefore, if the expression for Φ satisfies the previously mentioned requirement, the magnitude of the peak stress and that of the elastic precursor will be determined, for boundary conditions of the form (4.2.4),

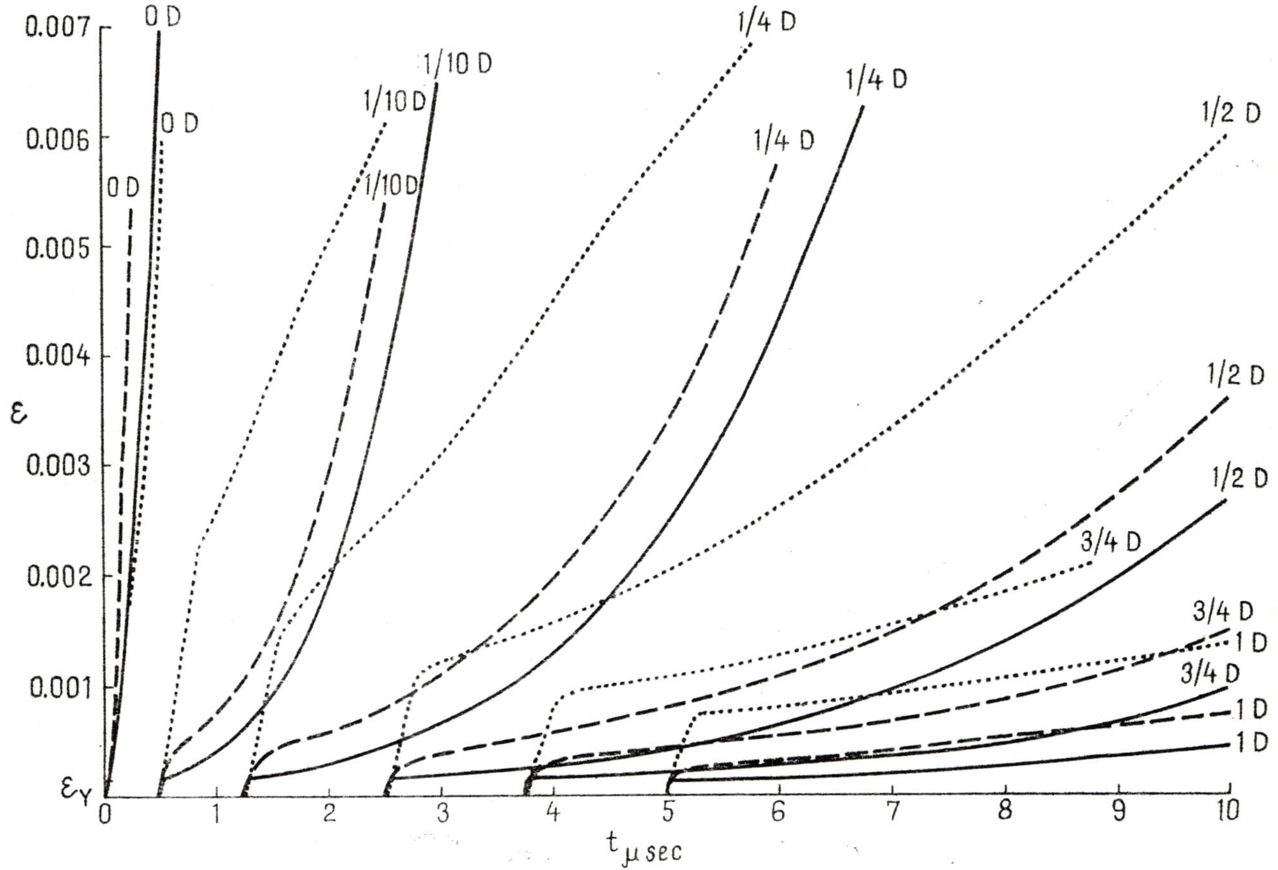

Fig. 4.4.5 Beginning of the strain-time curves at cross-sections close to the impacted end

by the impact velocity only (and certainly also by the speed by which the boundary conditions of the program introduce the loading, i.e. by the magnitude of t_m and possibly also by that of $\check{\varepsilon}$). If the loading law prescribed at the end of the bar is sudden and if Φ satisfies the mentioned requirements, then the magnitude of the peak stress coincides with the one obtained

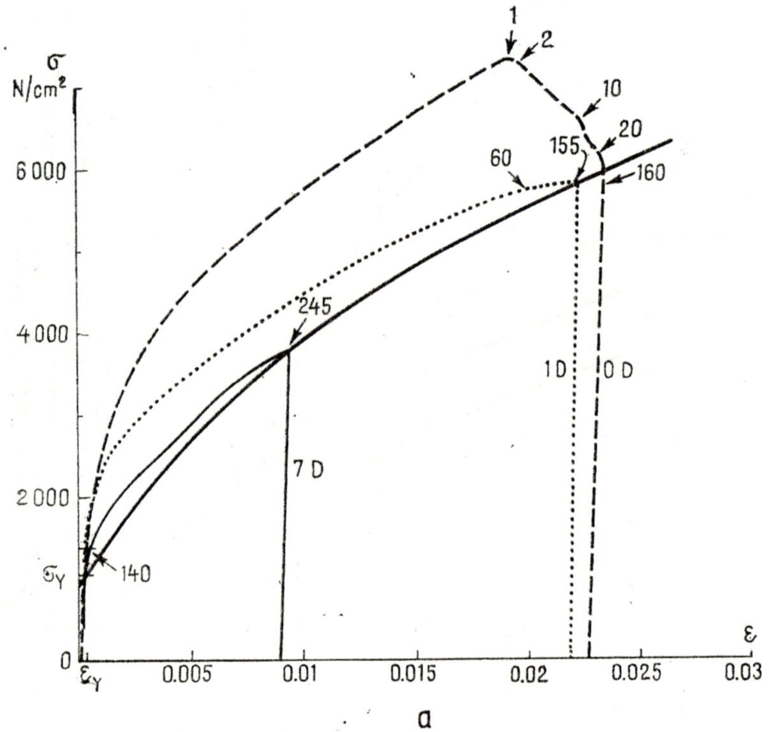

Fig. 4.4.6 a. Stress-strain curves at various cross-sections of the bar for model 1 (Table 4.2)

within the elastic model $\sigma = E\varepsilon$ or within the semi-linear model ((2.2.6)—(2.2.7), say) computed for the same impact velocity. However if the peak stress which results from the computation is still too small when compared with the experimental value, then we have to consider instantaneous curves with an upward directed concavity and possibly with a higher slope than the elastic one; for such a model the velocity of propagation of the waves is expected to be even higher than the elastic bar velocity, mainly in the portion of the bar close to the impacted end.

The explicit expression for the function Φ is determined mainly by the arrival times of various magnitudes of strain at various cross-sections of the bar (see fig. 4.4.4). For instance, at $1D$ the computed arrival times for smaller strains ($\varepsilon < \bar{\varepsilon}$), as they result from example 3 (dotted lines), are too early when compared with experimental data. Here $\bar{\varepsilon}$ is the strain corresponding to the inflection point of the curves of fig. 4.4.4; the same numerical

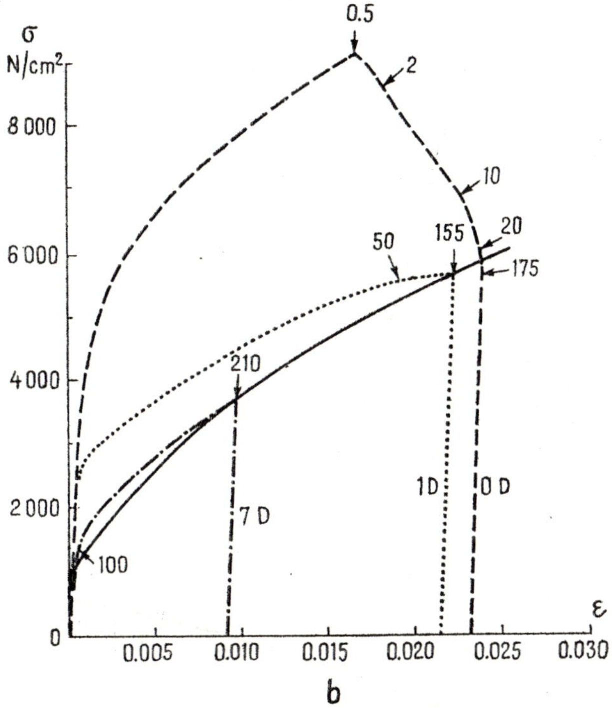

Fig. 4.4.6 b. Stress-strain curves at various cross-sections of the bar for model 2 (Table 4.2)

values for $\bar{\varepsilon}$ are given in Table 4.2. For higher strains, i.e. for $\bar{\varepsilon} > \varepsilon$, the situation is reversed: the computed arrival times are too late. In the case of example 3 both effects are due to the fact that Φ is too small. In order

TABLE 4.2

No. of example	σ_{max} N/cm² peak	plateau	ε_{max} peak	plateau	T_c µs	v_f m/s	$\bar{\varepsilon}$ $x = 1\ D$	Penetration of the first unloading wave
1	7158	5783	0.02359	0.02248	306.5	22.46	0.0108	4 D
2	9284	5763	0.02403	0.02229	303.0	22.42	0.0107	4.25D
3	20999	5738	0.02360	0.02210	301.0	22.41	0.0101	4.25D
exp.	29627	5719	~0.0275	0.0219	310.8	22.63	0.0135	4~5D

to improve the solution, Φ must be increased, for instance by a change of coefficient a or by a decrease of $\check{\varepsilon}$ (example 2, broken line). A converse situation takes place in example 1 (full line) where Φ is slightly too big.

A somehow similar effect on the strain-time curve can be obtained by changing the parameter k entering the expression for Ψ. In other words,

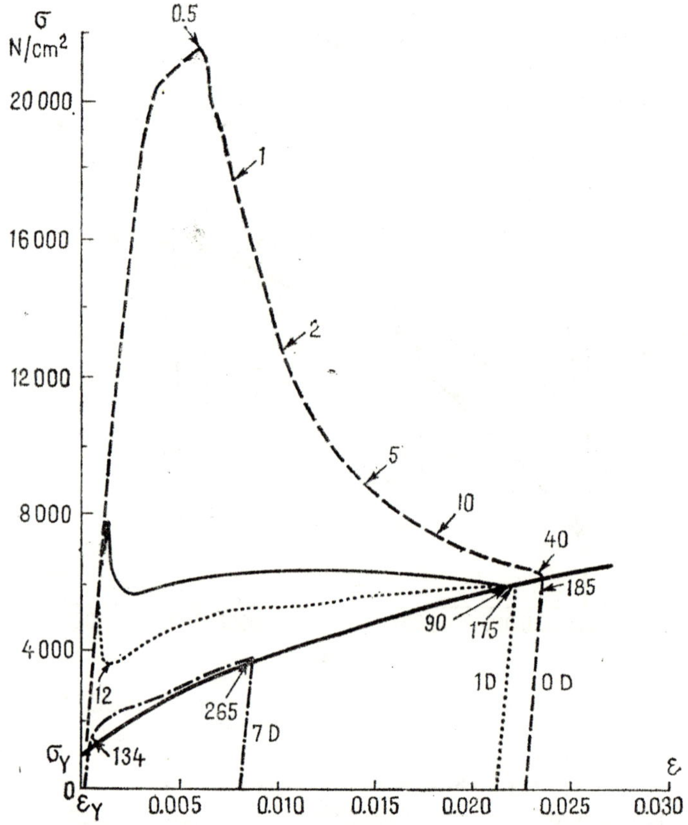

C

Fig. 4.4.6c Stress-strain curves at various cross-sections of the bar for model 3 (Table 4.2)

by increasing k or by increasing Φ one will produce more or less the same effect on the shape of the strain-time curve. However, if k is increased too much, both the peak stress and the initial sharp rise of the strain-time curve will be drastically diminished. This was one of the reasons why for small strains one has to choose a small value for k (see (4.1.18)). A second reason which imposes a small value for k for small strains is the behaviour near the yield point as described by the strain-time curve near the free end of the specimen (where strains are relatively small, i.e. surpassing a few times only the yield strain). Figs. 4.4.7 and 4.4.8, for instance, are showing the variation in time of the strain at the distances of eight and nine diameters from the impacted end. These figures show that the computed

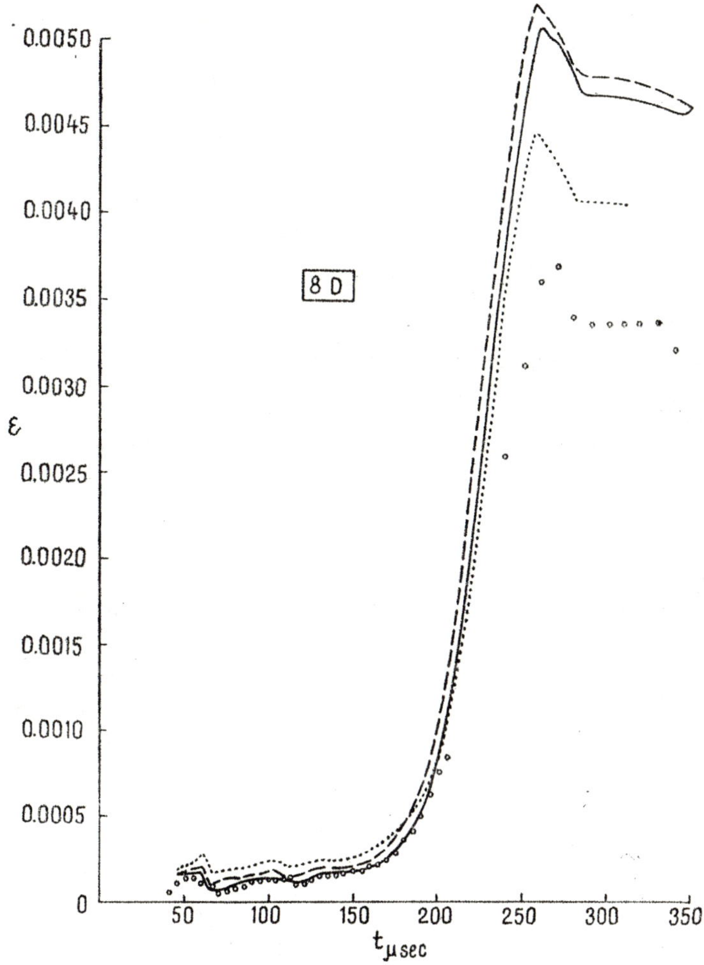

Fig. 4.4.7 Comparison of the computed strain-time curves at $8D$ with experiments

solutions are in excellent agreement with the experimental data. This is greatly due to a correctly chosen value for k. For a higher k the maximum would be higher, portions of the curve would raise too early, etc. (see Cristescu [1972 *b*]). Finally, a third argument suggesting a small value of k for small strains is the variation in time of the displacement at various cross-sections along the bar. Fig. 4.4.9 gives a comparison of such

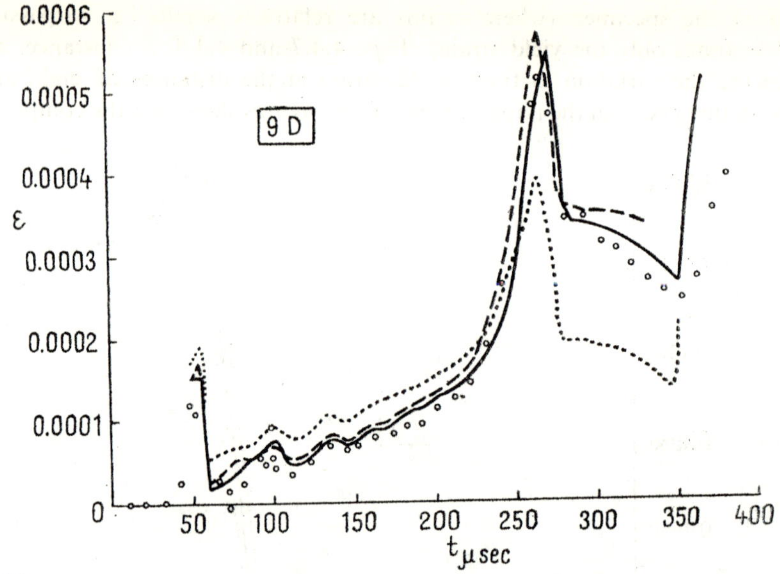

Fig. 4.4.8 Comparison of the computed strain-time curves at $9D$ with experiments

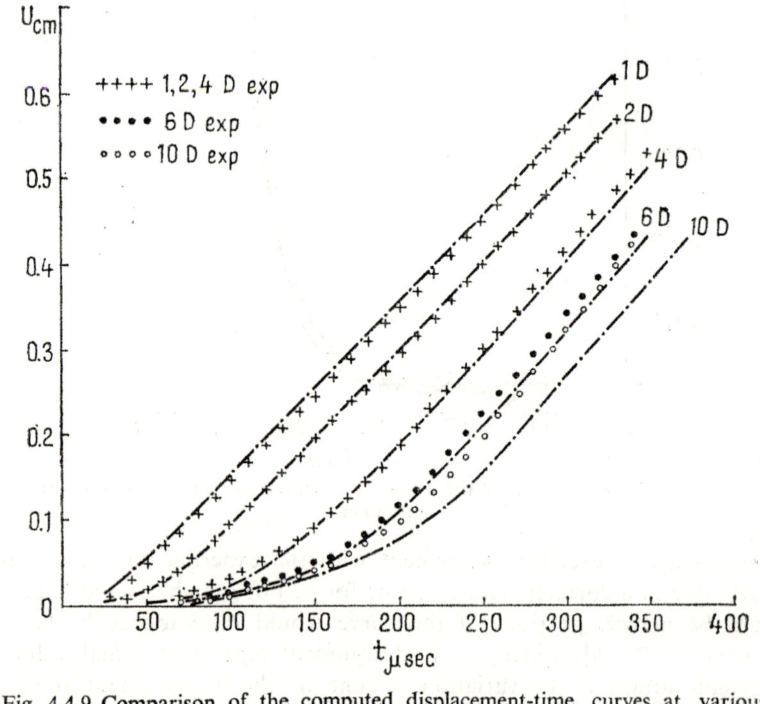

Fig. 4.4.9 Comparison of the computed displacement-time curves at various cross-sections with experimental data

experimentally obtained curves with the numerical solution of example 2. The general agreement seems to be reasonably good. It is important to observe that the agreement for the cross-sections close to the free end is better if for smaller ε the value of k is smaller (see CRISTESCU [1972 *b*]).

Among all parameters and coefficient functions entering the constitutive equation, the most important and the first to be determined from experiments is the equation of the relaxation boundary. The starting point is the experimental observation that in the first few diameters near the impacted end, the strain is constant for a relatively long period of time after reaching its maximum (see fig. 4.4.4) before unloading occurs, and to this strain plateau there corresponds a stress plateau (see fig. 4.4.10 for the

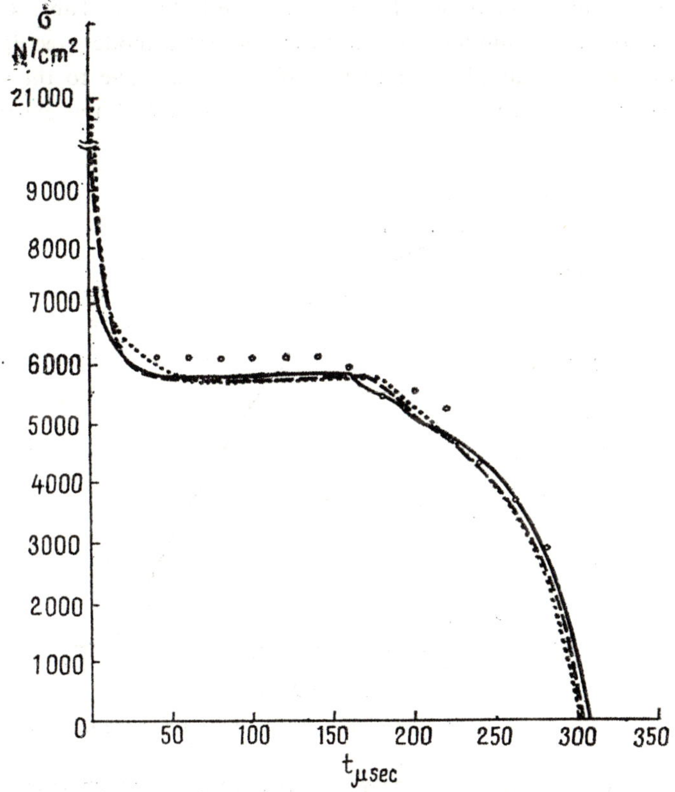

Fig. 4.4.10 Variation of stress at impacted end

variation of stress at the impacted end of the specimen) which again lasts for quite a long time. It is shown in Chapter V, Section 5 that the rate-type constitutive equation of the form (4.1.11) may describe even an absolute plateau, i.e. domains may exist in the characteristic plane where all three functions σ, ε and v are constant. The stress-strain states corresponding to these plateaux (for σ and ε) are, by definition, points on the relaxation boundary. Thus, by making experiments with various velocities of impact and therefore by determining the stress-strain states corresponding to the stress and strain plateaux, one can determine the relaxation boundary for a certain considered material. Note that in fig. 4.4.10 the raising portions of the curves shown have not been represented; this was done in order to simplify the figure.

The experiments (BELL [1961], Lee [1960]) are showing that the maximum strain is constant for a certain portion along the bar (generally until a few diameters away from the impacted end) but very close to the impacted end (one half diameter or so) the strain is slightly higher (Table 4.2). This follows also from computations within the accepted model (see fig. 4.4.11 corresponding to example 2), and the effect is again due to the presence of the peak stress near the impacted end. In fig. 4.4.11 the experimental

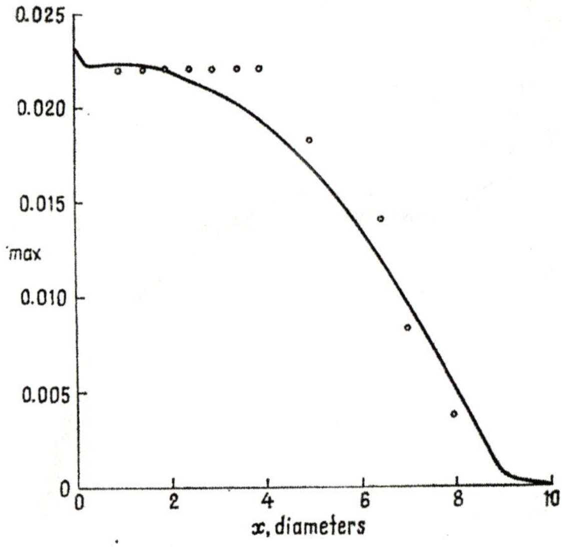

Vig. 4.4.11 Variation of maximum strain along the bar

points showing higher strains near $X = 0$ have not been represented since these were not available.

Another quantity determined experimentally quite accurately is $\bar{\varepsilon}$, i.e. the strain magnitude at the inflection point of the strain-time curves of fig. 4.4.4. Note that due to the kind of boundary conditions used, another much lower point of inflection can be found at the beginning of the strain-time curves (fig. 4.4.5) but this one is of lesser importance. As it is well known, this magnitude $\bar{\varepsilon}$ of the strain corresponds to the maximum of the surface angle α (see fig. 3.2.1 Chapter III, Section 2). In the next section (see also CRISTESCU [1972 *b*]) we shall see that $\bar{\varepsilon}$ is the strain corresponding to the peak stress at the impacted end. In other words, at $X = 0$ the lower part $\varepsilon < \bar{\varepsilon}$ of the strain-time curve up to $\bar{\varepsilon}$ corresponds to the raising portion of the stress-strain curve, while the upper part of the strain-time curve ($\varepsilon > \bar{\varepsilon}$) corresponds to the relaxation of the stress (decreasing portion of the stress-strain curve). Thus all the parameters influencing the peak stress will influence $\bar{\varepsilon}$ as well. Numerical values for $\bar{\varepsilon}$ are given in Table 4.2.

The perfectly elastic unloading mechanism according to (3.1.1) and used in all mentioned examples seems to be in reasonably good agreement with the experimental data. This follows from the comparison of various aspects of the computed solutions with the experimental data. An overall picture of the shape of the loading and unloading regions in the characteristic plane is given in fig. 4.4.12 for the example 2. Between the loading and unloading regions, denoted by L and U respectively, there is a strip or small isolated regions where relaxation takes place, i.e. stress is decreasing though strain increases. These regions were

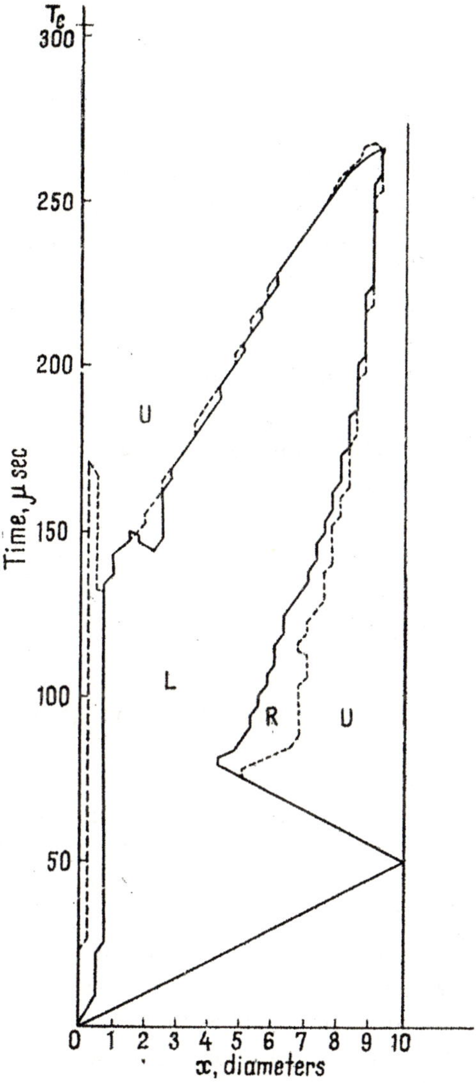

Fig. 4.4.12 Characteristic plane showing loading domains L, unloading domains U and relaxation domains R

denoted by R in fig. 4.4.12. Therefore the exact definition of the three kinds of regions is (see Chapter II, Section 2)

$$\text{in } L\text{: } \Phi > 0 \text{ and } \Psi > 0,$$
$$\text{in } R\text{: } \Phi = 0 \text{ and } \Psi > 0,$$
$$\text{in } U\text{: } \Phi = 0 \text{ and } \Psi = 0.$$

The unloading process starts at time l/c_0 from the free end of the bar. The first penetration of the unloading region into the loading region can again be observed experimentally. The coordinates of the tip of this unloading region, denoted by X_U are given in Table 4.2 (last column). X_U depends on various parameters but mostly on the yield stress. Generally the higher σ_Y, the smaller is X_U, i.e. the unloading region penetrates more into the loading region. The arrival times of the first unloading wave penetrating the loading region as computed, assuming perfect elastic unloading, are again in reasonable agreement with the experimental data. Figures 4.4.7 and 4.4.8 are also showing a very reasonable agreement with the computed solution of the material response in the unloading domain close to the free end, where stresses and strains are relatively small; thus the elastic vibrations in the unloading domain are quite well described by the perfectly elastic unloading assumption. Generally the shape of the lower part of the unloading region near the free end of the bar as well as the overall behaviour of the solution near this end depend mainly on σ_Y and on k.

5. COMPARISON BETWEEN MODELS

In order to gain later a better understanding of the procedure used to determine the rate-type models described in the previous sections of the present chapter, we shall make presently a comparison between the results obtained with various rate-type constitutive equations and some classical time-independent models. The presentation will be in a nearly chronological order; we shall discuss the accepted hypotheses and the computations successively done in order to find the explicit form of the constitutive equation as well as the arguments which have led to the final form of the rate-type constitutive equation.

The first constitutive equation of classical plasticity of the form (4.1.7) was used with the numerical values (4.1.9); it was called the "dynamic" stress-strain curve for the corresponding material. For the aluminium tests under consideration the dynamic stress-strain curve is shown by a full line in fig. 4.5.1. The propagation of (loading and unloading) elastic and plastic waves in a bar initially at rest and symmetrically impacted by a hitter moving with the speed of $V=40.6$ m/s was studied by CRISTESCU and BELL [1970]. The constitutive equations used were (4.1.7), (4.1.9) and some other variants. The results obtained were in reasonably good agreement with the experimental data. For instance, fig. 4.5.2 gives the variation in time of the strain at various cross-sections along the specimen for three variants of the boundary conditions. The computed times of contact shown by vertical arrows are in good agreement with the experimental ones. There are, however, some experimental facts which cannot be described by a classical constitutive equation of the form (4.1.7), for instance the peak stress, the change of curvature on the strain-time curve, the elastic precursor, etc. Various other aspects of this solution were given by CRISTESCU and BELL [1970].

The same computations have been then repeated after replacing the dynamic stress-strain curve by the "quasistatic" one for the same considered material, i.e. by (4.1.7), (4.1.8). This curve is shown on fig. 4.5.1 by dotted lines. But the results thus obtained were in desagreement with the experimental data. For instance in fig. 4.5.3 (CRISTESCU [1971 *b*]) the strain-time curves for various cross-sections (full lines) are given. By comparing these curves with the experimental data (crosses) one can see that the coincidence is poor.

Further, in the first attempts to find a rate-type constitutive equation, it was assumed that the relaxation boundary (4.1.14) coincides for $\varepsilon \geqslant \varepsilon_Y$ with the quasistatic stress-strain curve of the corresponding material (this assumption is made by most of the authors who have considered rate-type models). Therefore for this specific case formulae (4.1.7)—(4.1.8) were used for the relaxation boundary $(4.1.14)_2$. The results obtained were not satisfactory. The strain-time curves for various cross-sections along the bar are shown in fig. 4.5.3 by small circles. Agreement with experimental data is poor again; the maximum strain at the plateau is far from the experimental value, etc. One can observe however, that this solution has some common features with the solution obtained within the time-independent theory (4.1.7)—(4.1.8), i.e. the maxima are close and the overall behaviour of one solution is not too different from that of the other. On the other hand, the "dynamic" constitutive equation yields relative good solutions (dotted lines on fig. 4.5.3). This has suggested the use of the dynamic stress-strain curve, i.e. (4.1.7) combined with (4.1.9) or (4.1.10), as a relaxation boundary in (4.1.13). The results obtained in this way were much better. Therefore it was thought that an improved version of the

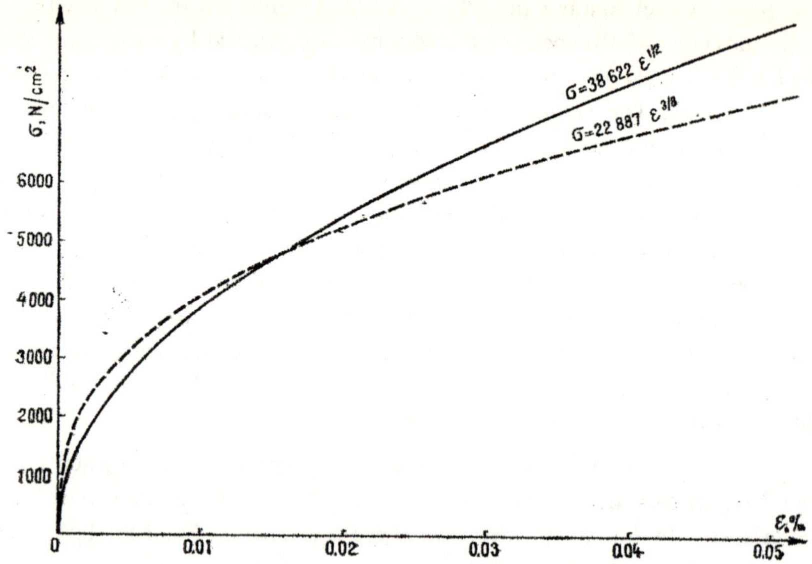

Fig. 4.5.1 Dynamic and static stress-strain curves for aluminium

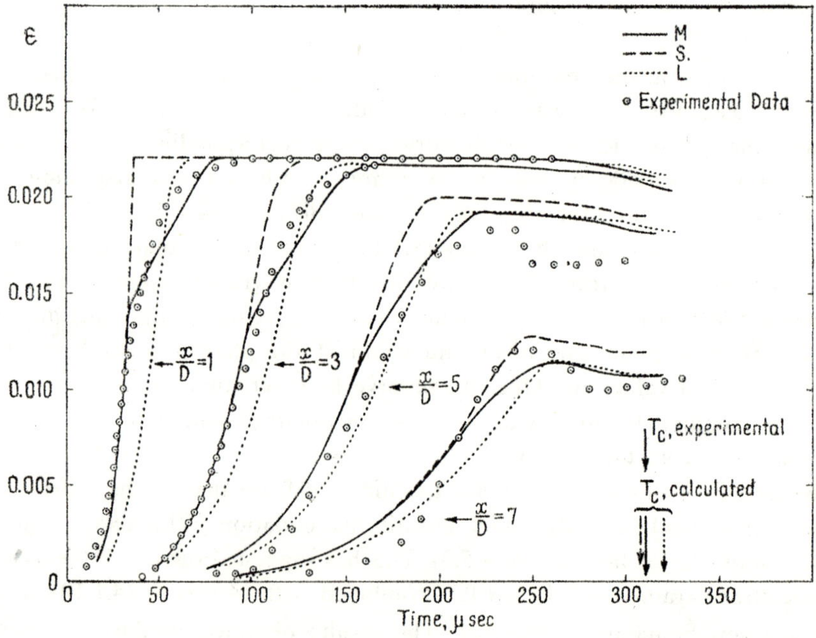

Fig. 4.5.2 Comparison of computed strain-time curves at various cross-sections with experimental data (after CRISTESCU and BELL [1970])

rate-type model could be obtained starting from the assumption that in a first approximation the dynamic stress-strain curve can play the role of a relaxation boundary. This also explains why reasonably good results are obtained by using the dynamic stress-strain curve: This constitutive equation is very close to the relaxation boundary for the corresponding material. As a matter of fact, one of the methods used to obtain experimentally the

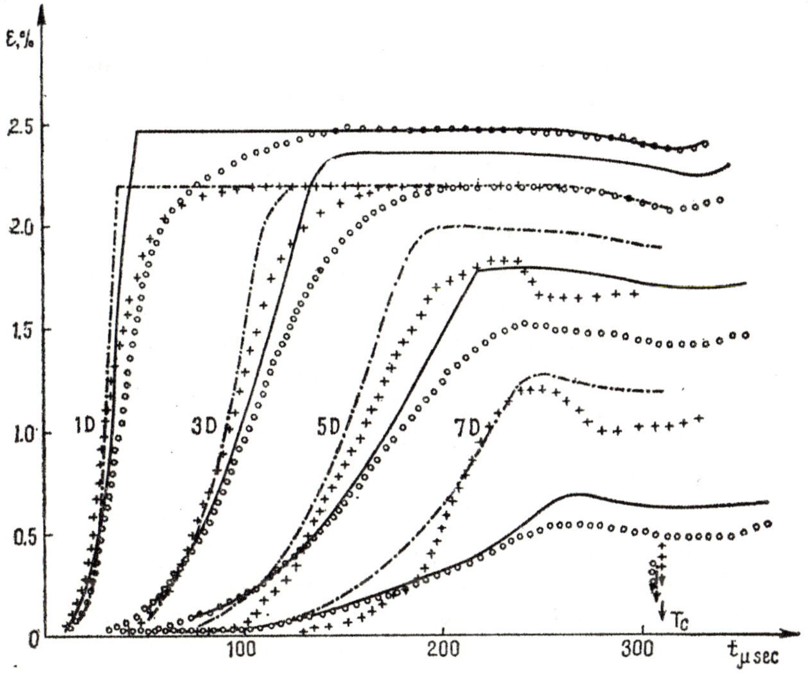

Fig. 4.5.3 Comparison of computed strain-time curves at various cross-sections with experimental data (after CRISTESCU [1971 b])

"dynamic" stress-strain curve consists in obtaining for various velocities of impact the strain and the stress at the plateau (see Chapter III, Section 2). Thus the same experimental method is used in order to get both the "dynamic" stress-strain curve and the relaxation boundary.

6. CONCLUSIONS AND SOME REMARKS

Let us summarize now the procedure followed in order to determine from a set of experimental data a rate-type constitutive equation to be used in dynamic problems.

First one has to find the relaxation boundary. As it was already mentioned, this is a curve in the stress-strain plane which for different impact velocities establishes a one-to-one correspondence between the magnitude of strains at the plateaus and that of the stresses at the corresponding plateaus. It is certainly assumed that for a certain range of variation of the impact velocities such a curve exists and is unique. This curve is called the relaxation boundary since generally a coincidence with the quasistatic stress-strain curve is not expected. Experimentally the relaxation boundary can be obtained by determining the stress plateau at the impacted end with a piezocrystal or with a hard transmitter bar (see Chapter III, Section 3) and the strain (or velocity) at the corresponding plateau along several diameters near the impacted end. Therefore relatively stable quantities must be measured: a stress which stays constant for quite a long period of time and a strain (or velocity) which is constant on a part of the bar. Thus, in dynamic problems it is the relaxation boundary which is of significance and not the quasistatic stress-strain curve. Generally one cannot know a priori if for a certain material under consideration the two curves coincide or are well apart. For the aluminium considered in the present examples they do not coincide. No attempt was made to establish a rate-type model which would describe the whole range of loadings between the very fast loadings and the slow quasistatic ones as well.

The level of the stress plateau is established mainly by the function $f(\varepsilon)$ entering in the expression for $\Psi(\sigma, \varepsilon)$. Certainly other parameters are also influencing the plateau level, but to a lesser degree. For instance the decrease of Φ or of k lifts the plateau.

The behaviour of the solution may essentially depend on the order of magnitude of k. If k is "big" (for the aluminium considered in examples in Chapter IV, $k > 10^7\,s^{-1}$ would be already "big"), the dynamic stress-strain curve is for all cross-sections very close to the relaxation boundary and above it. In order to avoid the accumulation of computational errors it is useful in such cases to replace the rate-type constitutive equation by a classical stress-strain curve, the equation of the relaxation boundary being used as the classic stress-strain curve. Certainly one has to change accordingly the computation program. However if k is very "small" (for the examples considered $k < 10^4\,s^{-1}$ would be already "small"), then the dynamic stress-strain curves mainly in the cross-sections close to the impacted end are not too distant from the instantaneous response curve (2.3.4) and they approach it from below. Again one can be tempted to switch to a classical constitutive equation (for instance an integrated form of (2.3.4) could be such a constitutive equation). In this latter case, however, one can well use a "rate" type program in which the instantaneous response curve would replace the classic stress-strain curve for the corresponding material. This procedure is attractive since the program for a rate-type constitutive equation is much easier to write than a program for a constitutive equation written in finite form. For uniaxial stress or strain problems it is immaterial if in the program the classical constitutive equation is replaced by a relaxation boundary by making $k \to \infty$ or by the curve of instantaneous response by making $k \to 0$; the choice of the method used is a matter of convenience only. For problems of combined stress states, however, the situation changes in a fundamental manner; an introduction of the yield condition in the Φ-terms couples the several possible existing plastic waves while an introduction of the yield condition in Ψ-terms does not couple them (see for details CRISTESCU [1971 a] *). However one has to observe that even for one-dimensional problems the two limit cases describe the unloading phenomena in a significantly different manner so that experimentally one can distinguish easily which of the two procedures is to be followed if a classical theory is desired as a model. The same is true for any generalization of the model to two- or three-dimensional cases. From a rational point of view it is the instantaneous response curve obtained for $k \to 0$ which would correspond to the classical "time-independent" approach of plasticity theory and not the relaxation boundary.

The previous discussion was done for say, an average range of rate of strain involved in certain kinds of dynamical problems. Higher strain rates lead to stress-strain curves closer to the instantaneous response curve while smaller strain rates to curves closer to the relaxation boundary. For instance in the previously mentioned examples, in the sections near the impacted end (with higher strain rates), the stress-strain curves are higher and are followed by a significant decreasing portion where stress is decreasing but the strain is increasing, while in the sections near the free end a

---

*) The coupling of plastic waves, sometimes called plastic waves of combined stresses, has been studied since 1955 by CRISTESCU [1955, 1956] for plane shearing-dilatational waves within classical plasticity theories, for waves produced by oblique impact of two plates by RAKHMATULIN [1958] and CRISTESCU [1959], for cylindrical waves producing shearing in two directions by CRISTESCU [1960].

Combined longitudinal and torsional waves in thin-walled tubes have been studied within classical plasticity theory by CLIFTON [1966], TING [1968], GOEL and MALVERN [1970, 1972]. For the same problem CRISTESCU [1968, 1971a] has shown that the yield condition is responsible for the coupling of plastic waves; various kind of coupling may result from the kind of yield condition used; rate type constitutive equations have been considered as well.

Many other authors also considered this problem. Further informations concerning this subject may be found in CRISTESCU [1958] Chapter VII, RAKHMATULIN and DEMIANOV [1961], CRISTESCU [1967], GOEL and MALVERN [1972], NOWACKI [1978], JUBAEV [1979].

converse situation occurs. Thus one has to be careful when describing the behaviour of the whole specimen during the entire experiment by a single "higher" stress-strain curve. It is important to observe that when the rate of strain is decreasing, a limit procedure does not lead to the static stress-strain curve representing the behaviour of the material but to the relaxation boundary (which, for such cases, may be or not close to the static stress-strain curve).

Let us comment on the coefficient function Φ. This is the coefficient controlling the velocity of propagation of the first part of the raising portion of the strain-time curves in the loading domain. Therefore Φ may be determined by measuring the arrival time of various levels of strain in the loading domain (before any reflection occurs). This procedure will not furnish directly an expression for Φ since in the framework of rate-type theories the arrival time of various levels of strain at successive cross-sections does not correspond to the arrival time of some specific waves (strain is not constant during the propagation of a certain wave). However it is possible to determine Φ, i.e. the instantaneous response curve, by stress measurements. The following indirect procedure can also be suggested. The computed results showed in every case that at $X = 0$ the upper part of the strain-time curve (where $\partial^2\varepsilon/\partial t^2 < 0$) corresponds to the decreasing portion of the stress-strain curve at $X = 0$, when a dynamic creep and relaxation take place. For instance in fig. 4.6.1 several strain-time curves at $X = 0$ are given. The point P where $\partial^2\varepsilon/\partial t^2 = 0$, i.e. where the strain-time curve changes curvature, corresponds to the peak stress on the stress-time curve. Fig. 4.6.1, *a* was obtained using a rate-type model of the form (4.1.11) (with (4.1.20)) (see CRISTESCU [1972 *b*]), fig. 4.6.1, *b* corresponds to a semi-linear model (i.e. $\Phi = 0$) while fig. 4.6.1, *c* corresponds to a rate-type model of the form (4.1.11) with (4.1.22) in which k has a small value ($k = 10^4 s^{-1}$). In this last case there is practically no portion on the strain-time curve where $\partial^2\varepsilon/\partial t^2 < 0$ and the strain-time curve has the same features as the curves obtained by using classical constitutive equations (no change of curvature in the loading portion). For the case $\Phi = 0$ the point where the curvature changes is very low (fig. 4.6.1, *b*) in the sections of the bar very close to the impacted end. Thus the shapes of the upper portion of the strain-time curves (above the point P) are to a great extent governed by the parameters $k(\varepsilon)$ and $\Phi(\sigma, \varepsilon)$; high values for k or small values for Φ increase the length of this upper portion of the curve and vice versa. Thus from experimental strain-time curves obtained in some sections of the bar close to the impacted end one can determine both Φ and k. Finally we would like to remark that the change of the curvature of the strain-time curves and the reason for this change was a long debated subject in the literature.

The conclusion is that some experimentally observed effects which cannot be described by using classical time-independent constitutive equations, can be described with the rate-type constitutive equations. The main such effects are: the presence of a peak stress which decays within a few

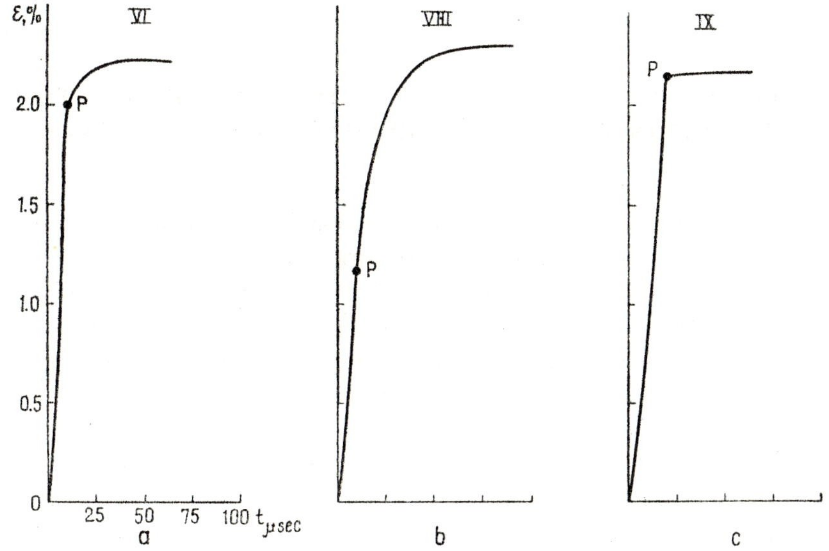

Fig. 4.6.1 Variation of the strain in time at the impacted end according to three models

microseconds to the plateau in the first half diameter, the change of curvature on the strain-time curves, the presence of an "elastic precursor" which decays during propagation, higher maximum strains very close to the impacted end while the plateau follows afterwards etc. Certainly some of these effects may be also influenced by some other phenomena, such as lateral inertia etc.

Chapter V

RATE-TYPE CONSTITUTIVE EQUATIONS FOR THE ONE-DIMENSIONAL CASE

1. CONDITIONS IMPOSED ON A RATE-TYPE CONSTITUTIVE EQUATION BY DYNAMIC RELOADING EXPERIMENTS

We shall show in this section that the conclusions obtained from reloading experiments (see Chapter 3, Section 3) may be used as mathematical assumptions in order to obtain restrictions for the functions present in the rate-type constitutive equations. The results presented here are due to SULICIU [1974 *a*] and SULICIU, MALVERN and CRISTESCU [1974].

Let us consider a one-dimensional motion $x = \chi(X, t)$ of a body $\mathscr{B}$, with $X \in I_1 \subset R$ and $t \in I_2 \subset R$. $I_1 = (a_1, b_1)$ is called the *reference configuration* of the body $\mathscr{B}$ (X is called the coordinate of the material point in the reference configuration or Lagrangeian coordinate, while x is the actual coordinate or Eulerian coordinate, of the material point). $I_2 = (\alpha_2, \beta_2)$ and $\beta_2 - \alpha_2$ is called the *duration of the motion* and it may be finite or not (see chapter I, section 1).

Let us suppose that χ is a continuous motion on $D = I_1 \times I_2$ and the functions

$$V(X, t) = \chi_t(X, t) = \dot{x},$$
$$\varepsilon(X, t) = \chi_X(X, t) - 1 > -1 \tag{5.1.1}$$

exist and are piecewise continuous and differentiable on D. $V(X, t)$ is called the *velocity of the material particle* and $\varepsilon(X, t)$ is called the *strain of the particle*. The stress $\sigma = \sigma(X, t)$ is piecewise continuous and differentiable on D.

We have $\varepsilon \in (-1, \infty)$ and $\sigma \in (-\infty, \infty)$.

A *mechanical process* of a particle with coordinate X that belongs to the body $\mathscr{B}$, is a curve $(\varepsilon(t), \sigma(t)) \in (-1, \infty) \times (-\infty, \infty)$, where t runs over a subinterval of I_2 and both ε and σ are piecewise continuous and differentiable.

Constitutive Assumption I (C.A.I). Let us consider a domain $\mathscr{D} \subseteq (-1, \infty) \times (-\infty, \infty)$ which contains the origin $(0, 0)$; the body $\mathscr{B}$

will be said to have a *rate-type behaviour* in the domain $\mathscr{D}$ if there exist two functions $\varphi, \psi: \mathscr{D} \to R$, $\varphi > 0$, such that

$$\dot{\sigma} = \varphi(\varepsilon, \sigma)\dot{\varepsilon} + \psi(\varepsilon, \sigma), \tag{5.1.2}$$

for any mechanical process $(\varepsilon(t), \sigma(t)) \in \mathscr{D}$. One assumes that $\varphi \in C^s(\mathscr{D})$ and $\psi \in C^r(\mathscr{D})$, with $s, r \geqslant 0$ (see Chapter II, Subsection 1.1).

A piecewise continuous and differentiable curve $(\varepsilon(t), \sigma(t)) \in \mathscr{D}$, $t \in I = (\alpha, \beta) \subseteq I_2$, will be called a *rate-type mechanical process* of duration $\beta - \alpha$, if (5.1.2) holds all along this curve.

A rate-type material is said to possess a *natural equilibrium configuration* if $\varepsilon(t) = 0$ for any $t \in [\alpha, \beta]$, and $\sigma(\alpha) = 0$ implies $\sigma(t) = 0$ for all $t \in [\alpha, \beta]$, for any given $\alpha < \beta$, $\alpha, \beta \in R$.

A necessary condition for a rate-type material to possess a natural equilibrium configuration, is that

$$\psi(0, 0) = 0. \tag{5.1.3}$$

Let $\mathscr{B}$ be a body possessing a natural equilibrium configuration; for such a body consider all strain functions

$$\varepsilon(t) = \begin{cases} 0 & \text{for } t < 0, \\ \varepsilon_* = \text{const.} & \text{for } t \geqslant 0, \end{cases} \tag{5.1.4}$$

for which the rate-type process $(\varepsilon(t), \sigma(t))$ belongs to $\mathscr{D}$ for all $t \geqslant 0$, where

$$\sigma(t) = \begin{cases} 0 & \text{for } t < 0, \\ \tilde{\sigma}(t) & \text{for } t \geqslant 0, \end{cases} \tag{5.1.5}$$

and $\tilde{\sigma}$, for $t \geqslant 0$, is a smooth function determined from the relations:

$$\begin{aligned} &\dot{\tilde{\sigma}} = \psi(\varepsilon_*, \tilde{\sigma}), \qquad \tilde{\sigma}(0) = \sigma_I(\varepsilon_*), \\ &\frac{\mathrm{d}\sigma_I(\varepsilon)}{\mathrm{d}\varepsilon} = \varphi(\varepsilon, \sigma_I(\varepsilon)), \quad \sigma_I(0) = 0. \end{aligned} \tag{5.1.6}$$

The curve $(\varepsilon, \sigma_I(\varepsilon)) \in \mathscr{D}$, where σ_I is the solution of the initial value problem $(5.1.6)_2$, will be called the *instantaneous response curve with respect to the natural equilibrium configuration*.

Let $\mathscr{P}_0$ denote the set of all strain functions (5.1.4) for which there exists a stress function $\sigma(t)$ of the form (5.1.5), determined by (5.1.6) and such that $(\varepsilon(t), \sigma(t)) \in \mathscr{D}$ for all $t \geqslant 0$.

A curve $(\varepsilon, \sigma_R(\varepsilon)) \in \mathscr{D}$ is called a *relaxation* (or *equilibrium*) *curve* if, all along it, one has

$$\psi(\varepsilon, \sigma_R(\varepsilon)) = 0. \tag{5.1.7}$$

Constitutive Assumption II (C.A.II). Any $\varepsilon \in \mathscr{D} \cap R$ belongs to $\mathscr{P}_0$ and, for each such strain function ε, there exist a finite time $t_\varepsilon > 0$ and a function $\varphi_0 : (-1, \infty) \cap \mathscr{D} \to (0, \infty)$ of class C^k, $k = 0, 1, \ldots,$ such that

$$\varphi(\varepsilon_*, \sigma(t)) = \varphi_0(\varepsilon_*) \quad \text{for} \quad t \geqslant t_\varepsilon > 0. \tag{5.1.8}$$

Let us remark that SULICIU, MALVERN and CRISTESCU [1974] have chosen φ_0 as a constant function, i.e. $\varphi_0(\varepsilon) = \varphi(0, 0) = E$, where E is the Young modulus.

The following theorem (see SULICIU [1974 a]) shows the restrictions imposed upon the constitutive equation (5.1.2) by the above constitutive assumption.

Theorem 5.1.1. Let us assume that: i) C.A.I and C.A.II hold for $s = 0$, $r = 0$, $k = 0$; ii) for any ε, $y = \varphi(\varepsilon, \sigma)$ is invertible with respect to σ, on $\mathscr{D}$, iii) $\mathscr{D}$ has the property that if $(\varepsilon, \sigma_1), (\varepsilon, \sigma_2) \in \mathscr{D}$, then $(\varepsilon, \lambda\sigma_1 + (1 - \lambda)\sigma_2) \in \mathscr{D}$ for any $\lambda \in [0, 1]$.

Then: a) there exists a unique relaxation curve (at least locally); b) all along this curve, the propagation velocity is given by $c_0 = (\varphi_0(\varepsilon)/\rho_0)^{1/2}$ (where ρ_0 is the mass density in the reference configuration); c) for $\varepsilon < 0$, $\sigma_I(\varepsilon) < \sigma_R(\varepsilon)$ if and only if $\psi(\varepsilon, \sigma_I(\varepsilon)) \geqslant 0$ and therefore $\psi(\varepsilon, \sigma) > 0$ for $\sigma \in I_1^+ = (\sigma_I(\varepsilon), \sigma_R(\varepsilon))$ (or $\sigma_R(\varepsilon) < \sigma_I(\varepsilon)$ if and only if $\psi(\varepsilon, \sigma_I(\varepsilon)) < 0$ and therefore $\psi(\varepsilon, \sigma) < 0$ for $\sigma \in I_1^- = (\sigma_R(\varepsilon), \sigma_I(\varepsilon)))$; for $\varepsilon > 0$, $\sigma_I(\varepsilon) > \sigma_R(\varepsilon)$ if and only if $\psi(\varepsilon, \sigma_I(\varepsilon)) < 0$ and therefore $\psi(\varepsilon, \sigma) < 0$ for $\sigma \in I_2^+ = (\sigma_R(\varepsilon), \sigma_I(\varepsilon))$ (or $\sigma_I(\varepsilon) < \sigma_R(\varepsilon)$ if and only if $\psi(\varepsilon, \sigma_I(\varepsilon)) > 0$ and therefore $\psi(\varepsilon, \sigma) > 0$ for $\sigma \in I_2^- = (\sigma_I(\varepsilon), \sigma_R(\varepsilon)))$; d) $1/\psi(\varepsilon, \sigma)$ is Lebesgue integrable with respect to σ on the corresponding intervals $I_1^+, I_1^-, I_2^+, I_2^-$ (and, as a consequence of *d*) and *e*), $\psi(\varepsilon, \sigma)$ cannot be locally Lipschitz continuous with respect to σ on the whole domain $\mathscr{D}$).

Proof. a) Let us consider an $\varepsilon \in \mathscr{P}_0$; then, from C.A.II one has

$$\varphi(\varepsilon_*, \sigma(t)) = \varphi_0(\varepsilon_*) \quad \text{for} \quad t \geqslant t_\varepsilon.$$

ii) implies that $\sigma(t) = \sigma_* = \text{const.}$ for $t \geqslant t_\varepsilon$ and

$$\sigma_* = \sigma_R(\varepsilon_*). \tag{5.1.9}$$

σ_R is a continuous function by i) and ii). Now, from (5.1.2) and (5.1.9), one gets

$$\psi(\varepsilon, \sigma_R(\varepsilon)) = \dot{\sigma} = 0 \quad \text{for} \quad t \geqslant t_\varepsilon, \tag{5.1.10}$$

that is $(\varepsilon, \sigma_R(\varepsilon))$ is a unique continuous relaxation (equilibrium) curve (at least locally). (The uniqueness follows from ii).

b) The propagation velocity c for the one-dimensional motion of a rate-type material is given by

$$c = \left(\frac{\varphi(\varepsilon, \sigma)}{\rho_0}\right)^{1/2}$$

(see Chapter II, Subsection 4.1, formula (2.4.12) with $\mathscr{E} = 0$ and λ denoted by c). According to the C.A.II and the above formula, one has

$$c_0 = c(\varepsilon, \sigma_R(\varepsilon)) = \left(\frac{\varphi_0(\varepsilon)}{\rho_0}\right)^{1/2}.$$

c) and d). Let us assume that

$$\psi(\varepsilon_*, \sigma_I(\varepsilon_*)) = 0 \qquad \text{for an} \quad \varepsilon^* \neq 0.$$

Then

$$\tilde{\sigma}(t) = \sigma_I(\varepsilon_*) \qquad \text{for} \quad t \geqslant 0$$

is a solution of (5.1.6). Hence $\varepsilon \in \mathscr{P}_0$, which contradicts the C.A.II since t_ε has to be greater than zero. Therefore $\psi(\varepsilon_*, \sigma_I(\varepsilon_*)) \neq 0$ for any $\varepsilon_* \neq 0$.

Consider now $\varepsilon_* < 0$ and assume that $\sigma_I(\varepsilon_*) < \sigma_R(\varepsilon_*)$; then there exists a $\bar{\sigma} \in I_1^+$ such that $\psi(\varepsilon_*, \sigma) > 0$ for $\sigma \in [\sigma_I(\varepsilon_*), \bar{\sigma}]$ and $\psi(\varepsilon_*, \bar{\sigma}) = 0$, since $\psi(\varepsilon_*, \sigma) < 0$ implies $\dot{\sigma}(t) < 0$ and this leads to $\sigma(t) < \sigma(0) = \sigma_I(\varepsilon_*)$ for $t > 0$. Now, for any $\sigma_1 \in (\sigma_I(\varepsilon_*), \bar{\sigma})$, one has

$$\int_{\sigma_I(\varepsilon_*)}^{\sigma_1} \frac{d\sigma}{\psi(\varepsilon_*, \sigma)} = t_1 < t_\varepsilon < \infty.$$

Therefore, $t_1 \to \bar{t} \leqslant t_\varepsilon$ when $\sigma_1 \to \bar{\sigma}$,. Let us apply ii) and iii) to the mechanical process $(\varepsilon(t), \sigma(t)) \in \mathscr{D}$, with

$$\sigma(t) = \begin{cases} 0 & \text{for} \quad t < 0, \\ \tilde{\sigma}(t) & \text{for} \quad 0 \leqslant t \leqslant \bar{t}, \\ \tilde{\sigma}(\bar{t}) & \text{for} \quad t \geqslant \bar{t}, \end{cases} \qquad \varepsilon(t) = \begin{cases} 0 & \text{for} \quad t < 0, \\ \varepsilon_* & \text{for} \quad t \geqslant 0 \end{cases}$$

where $\dot{\tilde{\sigma}}(t) = \psi(\varepsilon_*, \tilde{\sigma}(t))$, $\tilde{\sigma}(0) = \sigma_I(\varepsilon_*)$. One concludes that $\bar{\sigma} = \sigma_R(\varepsilon_*)$ and $\bar{t} = t_\varepsilon$.

Conversely, let us assume that $0 < \psi(\varepsilon, \sigma_I(\varepsilon_*))$; then $\psi(\varepsilon_*, \sigma) > 0$ for all $\sigma \in [\sigma_I(\varepsilon_*), \bar{\sigma}]$, $\bar{\sigma} > \sigma_I(\varepsilon_*)$ and, just as above, one concludes that $\bar{\sigma} = \sigma_R(\varepsilon_*)$.

The other cases are dealt with in the same way.

e) Suppose ψ is locally Lipschitz continuous with respect to σ on $\mathscr{D}$. Then, for any point $(\varepsilon_0, \sigma_0) \in \mathscr{D}$, $\sigma_0 = \sigma_R(\varepsilon_0)$, there exists a constant $M = \bar{M}(\varepsilon_0, \sigma_0) > 0$, such that

$$|\psi(\varepsilon, \tau)| \leqslant M |\tau - \sigma_0|$$

for all (ε_0, τ) in a neighborhood of $(\varepsilon_0, \sigma_0)$. For $\tau_1 < \sigma_0$ it follows that

$$\int_{\tau_1}^{\sigma_0} \frac{d\tau}{|\psi(\varepsilon_0, \tau)|} \geqslant -\frac{1}{M} \log|\tau - \sigma_0| \Big|_{\tau_1}^{\sigma_0} = \infty,$$

i.e. t_ε cannot be finite.

Remark 5.1.1. If $\sigma_I(\varepsilon) < \sigma_R(\varepsilon)$ and $\psi(\varepsilon, \sigma) \leqslant 0$ for $\sigma \geqslant \sigma_R(\varepsilon)$ (for $\varepsilon < 0$ and $\sigma < 0$), then the solution $\sigma(t) = \sigma_R(\varepsilon)$ for $t \geqslant t_\varepsilon$ of the equation $\dot{\sigma} = \psi(\varepsilon, \sigma)$ (ε = const. for $t \geqslant t_\varepsilon$), $\sigma(t_\varepsilon) = \sigma_R(\varepsilon)$, is unique (see HARTMAN [1964], p. 33).

Remark 5.1.2. If in the previous theorem one makes the additional assumption that C.A.II holds for $r = 1$, $\psi \in C^1(\mathscr{D})$ and $\frac{\partial \varphi}{\partial \sigma}(\varepsilon, \sigma) \neq 0$ for any $(\varepsilon, \sigma) \in \mathscr{D}$, then σ_R is a smooth function (see SULICIU, MALVERN and CRISTESCU [1974]) and one has

$$\frac{d\sigma_R}{d\varepsilon} = \frac{\dfrac{d\varphi_0(\varepsilon)}{d\varepsilon} - \dfrac{\partial \varphi}{\partial \varepsilon}(\varepsilon, \sigma_R(\varepsilon))}{\dfrac{\partial \varphi}{\partial \sigma}(\varepsilon, \sigma_R(\varepsilon))}. \tag{5.1.11}$$

Two examples of functions ψ defined for $\varepsilon \leqslant 0$ and $\sigma \leqslant 0$ and giving finite relaxation time are furnished by

$$\psi(\varepsilon, \sigma) = \begin{cases} 0 & \text{if } \sigma \geqslant \sigma_R(\varepsilon), \\ k(\sigma_R(\varepsilon) - \sigma)^\alpha & \text{if } \sigma < \sigma_R(\varepsilon), \quad \alpha \in (0, 1) \end{cases} \tag{5.1.12}$$

and

$$\psi(\varepsilon, \sigma) = \begin{cases} 0 & \text{if } \sigma \geqslant \sigma_R(\varepsilon), \\ -\dfrac{k_1}{\log(\sigma_R(\varepsilon) - \sigma) - \log(k_2 - \sigma)} & \text{if } \sigma < \sigma_R(\varepsilon), \end{cases} \tag{5.1.13}$$

where $\sigma = \sigma_R(\varepsilon)$ is an increasing smooth function and k, k_1 and k_2 are positive constants.

Functions ψ of the first kind presented above have been considered by KUKUDJANOV [1967] and [1977] in certain numerical investigations of the strain-rate influence. This example has also been carried over by PIAU [1978] to the framework of a three-dimensional viscoplasticity theory with internal state variables.

The above analysis leads to the following conclusion: rate-type constitutive equations that give a finite relaxation time cannot be linearized by power expansions in the neighborhood of a relaxed state $(\varepsilon_0, \sigma_R(\varepsilon_0))$ even if the strain ε is very close to ε_0.

2. STEADY MOTIONS OF RATE-TYPE MATERIALS

This section contains the results obtained by GREENBERG [1968] for rate-type materials with visco-elastic behaviour and modified by SULICIU [1974 *b*] in order to make them applicable also to elastic-viscoplastic materials and to make them compatible with the C.A.II (see SULICIU [1974 *b*]).

We assume that the body $\mathscr{B}$ has the reference configuration $I_1 = R$ (see Section 1) and $I_2 = R$, while the motion χ and the stress σ satisfy the same hypotheses on $D = R \times R$ as in Section 1.

A motion χ is said to be a *steady motion* if there are two piecewise continuous functions $\hat{\varepsilon}$ and $\hat{V}$, such that

$$\frac{\partial \chi}{\partial X}(X, t) = \hat{\varepsilon}(x) + 1 > 0, \qquad \frac{\partial \chi}{\partial t}(X, t) = \hat{V}(x). \tag{5.2.1}$$

GREENBERG [1968] has shown that χ is steady if and only if there exist a positive number V_0 and a continuous function f with a piecewise continuous derivative and $f' > 0$, such that

$$\chi(X, t) = f(\xi), \qquad \xi = \frac{X}{V_0} + t, \tag{5.2.2}$$

for any $(X, t) \in R \times R$. For steady motions with $\lim_{\xi \to -\infty} f'(\xi) = V_0$, the parameter V_0 can be interpreted as the velocity at great distances, i.e. at the spatial point $x = -\infty$.

The *stress field* is said to be *steady* if there exists a function $\hat{\sigma}$ such that

$$\sigma(X, t) = \hat{\sigma}(x). \tag{5.2.3}$$

If the mass density ρ_0 in the reference configuration is assumed to be constant and body forces to be absent, then the stress σ and the motion χ are related by the equation of balance of momentum

$$\frac{\mathrm{d}}{\mathrm{d}t} \int_{X_1}^{X_2} \rho_0 \frac{\partial \chi}{\partial t}(X, t)\,\mathrm{d}X = \sigma(X_2, t) - \sigma(X_1, t), \tag{5.2.4}$$

for all X_1, X_2 and t. If one takes into account that $\frac{\partial}{\partial X} = \frac{1}{V_0}\frac{\mathrm{d}}{\mathrm{d}\xi}$ and $\frac{\partial}{\partial t} = \frac{\mathrm{d}}{\mathrm{d}t}$, then (5.2.4) reduces to

$$\sigma(\xi_2) - \sigma(\xi_1) = v(\varepsilon(\xi_2) - \varepsilon(\xi_1)), \tag{5.2.5}$$

for all $\xi_1, \xi_2 \in R$, where

$$v = \rho_0 V_0^2. \tag{5.2.26}$$

Let us remak that $\lim_{\xi \to -\infty} f'(\xi) = V_0$ implies $\lim_{\xi \to -\infty} \varepsilon(\xi) = 0$.

For steady motions and stresses the constitutive equation (5.1.2) becomes

$$\frac{d\sigma}{d\xi} = \varphi(\varepsilon, \sigma)\frac{d\varepsilon}{d\xi} + \psi(\varepsilon, \sigma). \tag{5.2.7}$$

Before starting a discussion on the solutions of the system (5.2.5) and (5.2.7), let us adopt the following assumptions (see SULICIU [1974 b]).

Constitutive Assumption III (C.A.III).

a) The domain $\mathscr{D} = \{(\varepsilon, \sigma);\ -1 < \varepsilon \leqslant 0,\ -\infty < \sigma \leqslant 0\}$ contains a curve $\mathbf{R} = (\varepsilon, \sigma_R(\varepsilon))$, $\varepsilon \in (-1, 0)$, of class C^2, with the following properties (see fig. 5.2.1):

$$0 < \sigma_R'(\varepsilon) < \varphi_0(\varepsilon), \quad \sigma_R(0) = 0, \tag{5.2.8 a}$$

$$\sigma_R''(\varepsilon) < 0, \tag{5.2.8 b}$$

$$\varphi(\varepsilon, \sigma_R(\varepsilon)) = \varphi_0(\varepsilon), \tag{5.2.8 c}$$

$$\lim_{\varepsilon \to -1} \sigma_R(\varepsilon) = -\infty. \tag{5.2.8 d}$$

b) C.A.I and C.A.II hold for $s = 1$ and $k = 1$ respectively and, moreover, one has on $\mathscr{D}$

$$\varphi(\varepsilon, \sigma) > 0, \quad (\varepsilon, \sigma) \in \mathscr{D}, \tag{5.2.9 a}$$

$$\frac{\partial \varphi}{\partial \varepsilon} + \varphi \frac{\partial \varphi}{\partial \sigma} < 0, \quad \sigma < \sigma_R(\varepsilon), \tag{5.2.9 b}$$

$$\frac{\partial \varphi}{\partial \sigma} < 0, \quad \sigma < \sigma_R(\varepsilon), \tag{5.2.9 c}$$

$$\varphi(\varepsilon, \sigma) = \varphi_0(\varepsilon), \quad \sigma \geqslant \sigma_R(\varepsilon), \tag{5.2.9 d}$$

$$\varphi_0'(\varepsilon) \leqslant 0, \quad -1 < \varepsilon < 0. \tag{5.2.9 e}$$

c) The function $\psi \in C^\circ(\mathscr{D})$ has the following properties:

$$\psi \text{ is of class } C^1 \text{ on } \mathscr{D} - \mathbf{R}, \tag{5.2.10 a}$$

$$\frac{\partial \psi}{\partial \varepsilon} \geqslant 0, \quad \frac{\partial \psi}{\partial \sigma} \leqslant 0 \quad \text{on } \mathscr{D} - \mathbf{R}. \tag{5.2.10 b}$$

The function $\dfrac{1}{\psi(\varepsilon, \sigma)}$ is integrable on any segment PP^* of the straight line, where $P = (\varepsilon, \sigma)$, $\varepsilon < 0$ and $\sigma < \sigma_R(\varepsilon)$ and $P^* = (\varepsilon_*, \sigma_R(\varepsilon_*))$, i.e.

$$\left| \int_{PP^*} \frac{1}{\psi(\varepsilon, \sigma)} \right| < \infty. \tag{5.2.10 c}$$

It is possible to establish now several consequences of this constitutive assumption. Most of them have a similar form to those discussed by Greenberg [1968] although his hypothesis differs from the C.A.III.

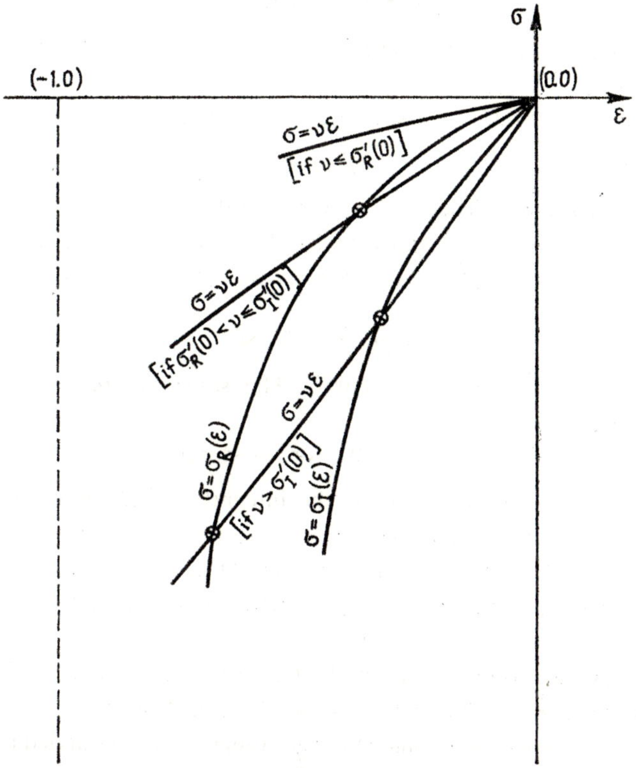

Fig. 5.2.1 The position of the instantaneous and relaxation curves ensuring existence of steady waves

Under the C.A.IIIb), the theory of differential equations asserts that the initial value problem

$$\frac{d\sigma}{d\varepsilon} = \varphi(\varepsilon, \sigma), \qquad \sigma(0) = 0, \tag{5.2.11}$$

has a unique global solution

$$\sigma = \sigma_I(\varepsilon), \tag{5.2.12}$$

of class C^2 for $\varepsilon \in (-1, 0]$.

From (5.2.8 a) and (5.2.8 c) it follows that

$$\sigma_I(\varepsilon) < \sigma_R(\varepsilon), \qquad \varepsilon \in (-1, 0]. \tag{5.2.13}$$

Indeed, since $\sigma_R'(0) < \varphi_0(0) = \varphi(0, 0) = \sigma_I'(0)$, there is an $\varepsilon_0 < 0$ such that $\sigma_R'(\varepsilon) < \sigma_I'(\varepsilon)$ for $\varepsilon \in (\varepsilon_0, 0)$, hence $\sigma_R(\varepsilon) > \sigma_I(\varepsilon)$ for $\varepsilon \in (\varepsilon_0, 0)$ (and $\sigma_R(\varepsilon_0) > \sigma_I(\varepsilon_0)$); (5.2.9 c) implies that

$$\varphi(\varepsilon, \sigma_R(\varepsilon)) < \varphi(\varepsilon, \sigma_I(\varepsilon)), \tag{5.2.14}$$

that is $\sigma_R'(\varepsilon_0) < \sigma_I'(\varepsilon_0)$ and therefore the above procedure can be extended to the left side of ε_0.

From (5.2.9 *b*), (5.2.11) and (5.2.12) one gets

$$\sigma_I''(\varepsilon) < 0. \tag{5.2.15}$$

By applying now Theorem 5.1.1, where the assumption that φ is invertible with respect to σ is replaced by the assumption that there exists a global curve $\sigma = \sigma_R(\varepsilon)$ with the property that (5.2.8 *c*) is satisfied all along it, one obtains

$$\psi(\varepsilon, \sigma_R(\varepsilon)) = 0, \tag{5.2.16}$$

i.e. $\sigma = \sigma_R(\varepsilon)$ is a relaxation curve. The same theorem, together with (5.2.13), leads to

$$\psi(\varepsilon, \sigma) > 0 \quad \text{for} \quad \sigma < \sigma_R(\varepsilon). \tag{5.2.17}$$

According to Theorem 5.1.1, the integral in (5.2.10 *c*) is finite when PP^* is a vertical segment. However it seems that (5.2.10 *a*), (5.2.10 *b*) and Theorem 5.1.1 do not generally imply (5.2.10 *c*); still, the examples (5.1.12) and (5.1.13) satisfy the requirements of the C.A.III *c*), as one can easily verify.

Subsequently we shall adopt the C.A.III and we shall work with the steady fields (ε, σ) satisfying (5.2.5) and (5.2.7). A pair of functions (ε, σ) is said to be a *compressive loading* if both functions are bounded and monotone decreasing on $\boldsymbol{R}$ and if they satisfy the condition

$$\lim_{\xi \to -\infty} \varepsilon(\xi) = 0, \qquad \lim_{\xi \to -\infty} \sigma(\xi) = \sigma_R(0) = 0. \tag{5.2.18}$$

One can view (5.2.18) as the assertion that the material is in equilibrium at $-\infty$.

A *compressive loading solution* is a pair of functions (ε, σ) which is a compressive loading and verifies (5.2.5) everywhere and (5.2.27) almost everywhere on $\boldsymbol{R}$. Let us remark that $(\varepsilon, \sigma) = (0, 0)$ is a trivial compressive loading solution.

A compressive loading (ε, σ) defines a compressive loading solution if and only if it satisfies

$$\sigma = \nu\varepsilon \quad \text{for all } \xi \in \boldsymbol{R}, \tag{5.2.19}$$

as well as (5.2.7) almost everywhere on $\boldsymbol{R}$.

The necessity of this statement follows from (5.2.5) and (5.2.18); sufficiency is immediate.

For compressive loadings one introduces the following notations:

$$C_R = \left(\frac{1}{\rho_0}\frac{d\sigma_R}{d\varepsilon}\bigg|_{\varepsilon=0}\right)^{1/2}, \qquad C_I = \left(\frac{1}{\rho_0}\varphi(0,0)\right)^{1/2}. \quad (5.2.20)$$

The significance of C_I is that of a sound velocity with respect to the natural equilibrium configuration while C_R has no such significance and therefore it is sometimes improperly called the equilibrium sound velocity.

Since $v = \rho_0 V_0^2$, one has

$$\begin{aligned} &V_0 \leqslant C_R \Leftrightarrow v \leqslant \sigma_R'(0), \\ &C_R < V_0 < C_I \Leftrightarrow \sigma_R'(0) < v < \varphi(0,0), \\ &C_I \leqslant V_0 \Leftrightarrow \varphi(0,0) \leqslant v. \end{aligned} \quad (5.2.21)$$

The following results are obvious:

i) If $V_0 > C_R$, then there exists a unique number $\varepsilon_R \in (-1, 0)$, such that

$$\sigma_R(\varepsilon_R) = v\varepsilon_R \quad (5.2.22)$$

and

$$\sigma_R'(\varepsilon_R) - v > 0. \quad (5.2.23)$$

ii) If $V_0 > C_I$, then there exists a unique number $\varepsilon_I \in (-1, 0)$, such that

$$\sigma_I(\varepsilon_I) = v\varepsilon_I \quad (5.2.24)$$

and

$$\sigma_I'(\varepsilon_I) - v > 0. \quad (5.2.25)$$

iii) Under the assumption in ii), one has

$$-1 < \varepsilon_R < \varepsilon_I < 0. \quad (5.2.26)$$

These statements are visualized on fig. 5.2.1.

Another useful remark is the following: from (5.2.10 *c*), (5.2.17) and from the fact that $V_0 > C_R$ implies $v\varepsilon < \sigma_R(\varepsilon)$ for $\varepsilon \in (\varepsilon_R, 0)$, where ε_R satisfies the relation $v\varepsilon_R = \sigma_R(\varepsilon_R)$, one obtains that

$$0 < \int_{\varepsilon_R}^{0} \frac{d\varepsilon}{\psi(\varepsilon, v\varepsilon)} < \infty. \quad (5.2.27)$$

One can now give the analogue to GREENBERG's Theorem 3.1. [1968] (see also SULICIU [1974 *b*]).

Theorem 5.2.1. i) If $V_0 \leqslant C_R$, then the only compressive loading solution is the trivial one. ii) If $C_R < V_0 < C_I$, then there exists a non-trivial compressive loading solution of class C^1. This solution has the property that for any $\xi_* \in R$ and any $\varepsilon_* \in (\varepsilon_R, 0)$, there exists a finite interval (a, b), $a < b$, containing ξ_*, such that

$$\frac{d\varepsilon}{d\xi}(\xi) < 0, \qquad \xi \in (a, b),$$

$$\frac{d\varepsilon}{d\xi}(\xi) = 0, \qquad \xi \in (-\infty, a] \subset [b, +\infty),$$

$$\varepsilon(b) = \varepsilon_R, \qquad \varepsilon(a) = 0.$$

The difference

$$b - a = -\int_{\varepsilon_R}^{0} \frac{[\nu - \varphi(\varepsilon, \nu\varepsilon)]\, d\varepsilon}{\psi(\varepsilon, \nu\varepsilon)} \tag{5.2.28}$$

does not depend on $\xi_* \in R$ or on $\varepsilon_* \in (\varepsilon_R, 0)$ and may be called the *thickness* of the steady wave.

iii) If $C_I < V_0$, then the only compressive loading solution of class C^1 is the trivial one, but there exists a non-trivial compressive loading solution with the following properties:

a) both functions σ and ε are continuous and their first derivatives are continuous on S_0, where

$$S_0 = R - \{\xi_0\} \tag{5.2.29}$$

and $\xi_0 \in R$ is a fixed point;

b) $\varepsilon(\xi) = 0$ and $\sigma(\xi) = 0$ for $\xi \in (-\infty, \xi_0)$;

c) $\varepsilon(\xi_0) = \varepsilon_I$ and $\sigma(\xi_0) = \nu\varepsilon_I$,

where ε_I is given by (5.2.24)] and

d) there exists a number $b > \xi_0$, such that $\dfrac{d\varepsilon}{d\xi}(\xi) < 0$, for $\xi \in (\xi_0, b)$, $\varepsilon(\xi) = \varepsilon_R$ for $\xi \geqslant b$, $\sigma(\xi) = \nu\varepsilon(\xi)$ for $\xi \geqslant \xi_0$, while ε_R satisfies (5.2.22) and (5.2.26). The difference $b - \xi_0$ is determined by

$$b - \xi_0 = -\int_{\varepsilon_R}^{\varepsilon_I} \frac{[\nu - \varphi(\varepsilon, \nu\varepsilon)]\, d\varepsilon}{\psi(\varepsilon, \nu\varepsilon)}, \tag{5.2.30}$$

and does not depend on the choice of ξ_0.

iv) If $V_0 = C_I$, there exists a non-trivial compressive loading solution with the following properties:

a) ε and σ are continuous functions on R while their first derivatives are continuous on S_0, where S_0 is defined by (5.2.29);

b) $\varepsilon(\xi) = 0$ and $\sigma(\xi) = 0$ for $\xi \in (-\infty, \xi_0)$;

c) $\lim_{\xi \to \xi_0^+} \frac{d\varepsilon}{d\xi}(\xi) = -\infty$;

d) there exists a number $b > \xi_0$ such that $\frac{d\varepsilon}{d\xi}(\xi) < 0$ for any $\xi \in (\xi_0, b)$; $\varepsilon(\xi) = \varepsilon_R$ for $\xi \geqslant b$, and $\sigma(\xi) = v\varepsilon(\xi)$ for $\xi \geqslant \xi_0$ while ε_R satisfies (5.2.22). The difference $b - \xi_0$ is determined by

$$b - \xi_0 = -\int_{\varepsilon_R}^{0} \frac{[v - \varphi(\varepsilon, v\varepsilon)]}{\psi(\varepsilon, v\varepsilon)} \, d\varepsilon. \tag{5.2.31}$$

Proof. i) Let us assume that the statement i) is false; thus we assume that there exists a pair of functions $(\varepsilon, \sigma) \neq (0, 0)$ which are bounded, decreasing and satisfy the conditions (5.2.7), (5.2.18) and (5.2.19). Then, there is a number $\bar{\xi}_0 \in R$ such that

$$\varepsilon(\xi_0) < 0, \qquad \sigma(\xi_0) = v\varepsilon(\xi_0) < 0$$

and

$$[v - \varphi(\varepsilon(\xi_0), v\varepsilon(\xi_0))] \frac{d\varepsilon(\xi_0)}{d\xi} = \psi(\varepsilon(\xi_0), v\varepsilon(\xi_0)). \tag{5.2.32}$$

Since $V_0 < C_R(<C_I)$, one has $v\varepsilon > \sigma_R(\varepsilon)$ for any $\varepsilon \in (-1, 0)$ and (5.2.9 *d*) will imply

$$\varphi(\varepsilon(\xi_0), v\varepsilon(\xi_0)) = \varphi_0(\varepsilon(\xi_0)) > \varphi_0(0) > v. \tag{5.2.33}$$

Let us introduce the function

$$g(\lambda) = \psi(\varepsilon(\xi_0), (1 - \lambda)(\sigma_R(\varepsilon(\xi_0)) + \alpha) + \lambda v\varepsilon(\xi_0)), \ \lambda \in (0, 1), \tag{5.2.34}$$

where $\alpha > 0$ is arbitrarily small. Then, according to (5.2.10 *a*) and (5.2.10 *b*), one may write

$$g(1) = g(0) + \frac{\partial\psi}{\partial\sigma}(\varepsilon(\xi_0), \sigma^*)[v\varepsilon(\xi_0) - \sigma_R(\varepsilon(\xi_0)) - \alpha], \qquad \lambda^* \in (0,1)$$

$$\sigma^* = (1 - \lambda^*)[\sigma_R(\varepsilon(\xi_0)) + \alpha] + \lambda^* v\varepsilon(\xi_0),$$

which implies

$$g(1) \leqslant g(0)$$

or

$$\psi(\varepsilon(\xi_0), v\varepsilon(\xi_0)) \leqslant \psi(\varepsilon(\xi_0), \sigma(\varepsilon(\xi_0)) + \alpha).$$

For $\alpha \to 0$, one obtains

$$\psi(\varepsilon(\xi_0), v\varepsilon(\xi_0)) \leqslant 0. \tag{5.2.35}$$

This last relation together with (5.2.32) and (5.2.33) leads to

$$\frac{d\varepsilon}{d\xi}(\xi_0) \geqslant 0$$

which contradicts the assumption that ε is a decreasing function.

ii) Since $C_R < V_0 < C_I$, we have from (5.2.8 *a*), (5.2.14) and (5.2.21) that

$$\varphi(\varepsilon, v\varepsilon) > \varphi(\varepsilon, \sigma_R(\varepsilon)) > \varphi_0(0) > v \tag{5.2.36}$$

for $\varepsilon \in (\varepsilon_R, 0)$; on the other hand, as $v\varepsilon < \sigma_R(\varepsilon)$ for $\varepsilon \in (\varepsilon_R, 0)$, we get from (5.2.17)

$$\psi(\varepsilon, v\varepsilon) > 0, \qquad \varepsilon \in (\varepsilon_R, 0). \tag{5.2.37}$$

The initial value problem

$$[v - \varphi(\varepsilon, v\varepsilon)] \frac{d\varepsilon}{d\xi} = \psi(\varepsilon, v\varepsilon), \qquad \varepsilon(\xi_*) = \varepsilon_* \in (\varepsilon_R, 0),\ \xi_* \in R, \tag{5.2.38}$$

has a unique solution through (ξ_*, ε_*), provided ξ remains in a sufficiently small interval (a_*, b_*) containing ξ_*. Relations (5.2.36) and (5.2.37) imply that $\frac{d\varepsilon}{d\xi}(\xi) < 0$ for $\xi \in (a_*, b_*)$. According to the theory of differential equations, this solution can be extended to the right up to a number b such that $\lim_{\substack{\xi \to b \\ \xi < b}} \varepsilon(\xi) = \varepsilon_R$ and to the left down to a number a such that $\lim_{\substack{\xi \to a \\ \xi > a}} \varepsilon(\xi) = 0$. Since we have

$$a_* - \xi_* = \int_{\varepsilon_*}^{\varepsilon(a_*)} \frac{[v - \varphi(\varepsilon, v\varepsilon)]\, d\varepsilon}{\psi(\varepsilon, v\varepsilon)} \tag{5.2.39}$$

and

$$\xi_* - b_* = \int_{\varepsilon(b_*)}^{\varepsilon_*} \frac{[v - \varphi(\varepsilon, v\varepsilon)]\, d\varepsilon}{\psi(\varepsilon, v\varepsilon)} \tag{5.2.40}$$

and both integrals are finite for $\varepsilon(a_*) \to 0$ and $\varepsilon(b_*) \to \varepsilon_R$ by (5.2.10 *c*), it follows that a and b are finite. Since $\psi(\varepsilon_R, \sigma_R(\varepsilon_R)) = 0$ and $\psi(0, 0) = 0$, we get $\frac{d\varepsilon}{d\xi}(\xi) = 0$ for $\xi \in (-\infty, a) \cup (b, \infty)$, by (5.2.38). Adding now (5.2.39) and (5.2.40) we obtain (5.2.28) for $a_* \to a$ and $b_* \to b$.

iii) The proof that there is no compressive loading solution of class C^1 but the trivial one follows the same lines as the proof of i).

Now let $\xi_0 \in R$ be a fixed number. Then, if we take $\varepsilon(\xi) = 0$ and $\sigma(\xi) = 0$ for $\xi < \xi_0$ and we attach the initial condition $\varepsilon(\xi_0) = \xi_I$ to the equation (5.2.32), then there exists a number $b \in R$, $b > \xi_0$ such that the solution of this initial value problem satisfies all the conditions required in iii). The proof follows the same lines as the proof of ii).

iv) We have to prove that the solution $\varepsilon(\xi) = 0$ and $\sigma(\xi) = 0$ for $\xi \in (-\infty, \xi_0)$ can be extended for $\xi \geqslant \xi_0$ so that it verifies a), c) and d). Therefore we consider the following initial condition for equation (5.2.32): $\varepsilon(\xi_*) = \varepsilon_*$, where $\xi_* \in (\xi_0, b)$, b is determined by (5.2.31) and $\varepsilon_* \in (\varepsilon_R, 0)$. ξ_* and ε_* are chosen to verify

$$\xi_* - \xi_0 = -\int_{\varepsilon_*}^{0} \frac{[v - \varphi(\varepsilon, v\varepsilon)]\, \mathrm{d}\varepsilon}{\psi(\varepsilon, v\varepsilon)}.$$

The solution of equation (5.2.32) through (ξ_*, ε_*) may be extended for all $\xi \in (\xi_0, b)$ and, moreover, it has the property that $\dfrac{\mathrm{d}\varepsilon}{\mathrm{d}\xi} < 0$ on this interval. For $\xi = b$ one has $\varepsilon(b) = \varepsilon_R$ and $\dfrac{\mathrm{d}\varepsilon}{\mathrm{d}\xi}(b) = 0$. If one takes $\varepsilon(\xi) = \varepsilon_R$ for $\xi \geqslant b$ then d) holds. On S_0, $\varepsilon(\xi)$ and $\sigma(\xi) = v\varepsilon(\xi)$ are of class C^1. Since $\lim_{\xi \to \xi_0^+} \varepsilon(\xi) = 0$, a) holds too.

Relation (5.2.32) implies

$$\lim_{\xi \to \xi_0^+} \frac{\mathrm{d}\varepsilon}{\mathrm{d}\xi} = \lim_{\xi \to 0^-} \frac{\psi(\varepsilon, v\varepsilon)}{v - \varphi(\varepsilon, v\varepsilon)}.$$

Now, since $\psi(\varepsilon, v\varepsilon) > 0$ for $\varepsilon < 0$ and $\psi(0, 0) = 0$, and ψ is continuous at $(0, 0)$ but not Lipschitz continuous (see Theorem 5.1.1) while $v - \varphi(\varepsilon, v\varepsilon)$ is of class C^1 and $v = \varphi(0, 0) = \varphi_0(0)$, it follows that c) holds too.

3. ON THE SOLUTION OF THE DIFFERENTIAL EQUATION (5.1.2) FOR MATERIALS WITH RATE-TYPE BEHAVIOUR

We shall need in the following some special classes of functions.

A function f: $[a, b] \to R^n$, $a < b$, $a, b \in R$, is called a *regulated* function on $[a, b]$ if it has both one-side limits at any $t \in (a, b)$ and

$$\lim_{\substack{t \to a \\ t > a}} f(t) = f^+(a), \qquad \lim_{\substack{t \to b \\ t < b}} f(t) = f^-(b)$$

exist. The set of all regulated functions on a given interval will be denoted by $\mathbf{R}^0$. $\mathbf{R}^1$ will denote the set of all regulated functions on a given interval with the additional property that the following limits also exist:

$$\lim_{\substack{t\to t_0\\ t<t_0}} \frac{f(t)-f^-(t_0)}{t-t_0} = f_s'(t_0), \quad \lim_{\substack{t\to t_0\\ t>t_0}} \frac{f(t)-f^+(t_0)}{t-t_0} = f_d'(t_0)$$

for any $t_0 \in (a, b)$, and

$$\lim_{\substack{t\to a\\ t>a}} \frac{f(t)-f^+(a)}{t-a} = f_d'(a), \quad \lim_{\substack{t\to b\\ t<b}} \frac{f(t)-f^-(b)}{t-b} = f_s'(b),$$

and f_s', f_d' are regulated functions on $[a, b]$.

The set of all continuous functions that have one-side derivatives as regulated functions will be denoted by C^{01}, i.e. $C^{01} = C^0 \cap \mathbf{R}^1$. (For details on these classes of functions see, for instance, Nicolescu [1958], Dieudonné [1968], Suliciu [1973, 1974 *c*]).

In this section we shall discuss, following Suliciu [1973, 1974 *c*] the solutions σ of equation (5.1.2) when ε is given and belongs to some class of regulated functions. These solutions will obviously depend on the form of the domain $\mathscr{D}$ as well as on the class of functions where φ and ψ are chosen from.

Assume that $\mathscr{D}$ is a simple and bounded domain in the ε, σ plane. $\mathscr{D}$ is said to be simple with respect to the axes if, for any $(\varepsilon_1, \sigma), (\varepsilon_2, \sigma) \in \mathscr{D}$ one has $(\lambda\varepsilon_1 + (1-\lambda)\varepsilon_2, \sigma) \in \mathscr{D}$ while $(\varepsilon, \sigma_1), (\varepsilon, \sigma_2) \in \mathscr{D}$ implies $(\varepsilon, \lambda\sigma_1 + (1-\lambda)\sigma_2) \in \mathscr{D}$, for any $\lambda \in [0, 1]$. This section generally deals with functions $\varepsilon, \sigma: [t_0, t_1] \to R$, $\varepsilon, \sigma \in \mathbf{R}^1[t_0, t_1]$ or functions belonging to some subclasses of $\mathbf{R}^1[t_0, t_1]$.

One will usually take $t_0 = 0$ and, moreover, ε (respectively σ) will be assumed to be defined on $[0, t_1]$ such that for any t, $\varepsilon(t) = \varepsilon(t-0) = \varepsilon^-(t)$ or $\varepsilon(t) = \varepsilon(t+0) = \varepsilon^+(t)$, i.e. ε (respectively σ) will be assumed right or left continuous. The subset of $\mathbf{R}^1[0, t_1]$ containing all functions with this property, is denoted by $\mathbf{R}^1_*[0, t_1]$. Let us remark that these assumptions, far from being essential for the discussions of the present section, are only intended to simplify them.

Let us consider that the functions φ and ψ in (5.1.2) are defined on the closure of $\mathscr{D}$, i.e. on $\overline{\mathscr{D}}$, and let us consider all functions $\varepsilon = \varepsilon(t)$, $\sigma = \sigma(t)$ of C^1 class, which are solutions of the initial value problem

$$\begin{gathered} \dot{\sigma} = \varphi(\varepsilon, \sigma)\, \dot{\varepsilon} + \psi(\varepsilon, \sigma), \\ \varepsilon(0) = \varepsilon_0, \quad \sigma(0) = \sigma_0, \quad (\varepsilon_0, \sigma_0) \in \mathscr{D}, \end{gathered} \tag{5.3.1}$$

with ε_0 and σ_0 fixed. It is quite obvious that, for a given $\varepsilon \in C^1[0, t_1]$, the natural class of functions where one may look for the solution $\sigma(t)$ of the problem (5.3.1) will be the class of smooth functions on a certain interval $[0, \omega_\varepsilon)$, with $\omega_\varepsilon \in (0, t_1]$. If $\varepsilon \in \mathbf{R}^1[0, t_1]$ then, according to Denjoy's theorem, the set of all points of discontinuity of ε is an at most countable set in $[0, t_1]$. Due to this theorem, the initial value problem (5.3.1) can be rephrased as follows

$$\begin{aligned} &\dot{\sigma}_d = \varphi(\varepsilon^+, \sigma^+)\, \dot{\varepsilon}_d + \psi(\varepsilon^+, \sigma^+), \\ &\sigma^+(0) = \sigma_0, \qquad \varepsilon^+(0) = \varepsilon_0, \\ &\dot{\sigma}_s = \varphi(\varepsilon^-, \sigma^-)\, \dot{\varepsilon}_s + \psi(\varepsilon^-, \sigma^-), \qquad (\varepsilon_0, \sigma_0) \in \mathscr{D}. \end{aligned} \tag{5.3.2}$$

Thus, at all points where ε and σ are smooth, (5.3.2) reduces to (5.3.1).

We shall define a mapping that associates to each $\varepsilon \in \mathbf{R}^1_*[0, t_1]$ (or $\varepsilon \in C^1[0, t_1]$ or $\varepsilon \in C^{01}[0, t_1]$) a function defined on a certain subinterval $[0, \omega_\varepsilon] \subset [0, t_1]$ (where ω_ε is the largest possible t in $[0, t_1]$, with this property) and belonging to the same class of functions as ε, on this subinterval. We shall also give some existence, uniqueness, continuity and differentiability conditions for this mapping with respect to the topology of uniform convergence, when ε runs over $\mathbf{R}^1_*[0, t_1]$ (or $C^1[0, t_1]$ or $C^{01}[0, t_1]$).

A pair of functions $(\varepsilon(t), \sigma(t)) \in \mathscr{D}$ for $t \in [0, \omega)$ will be called an $\mathbf{R}^1_*$-class solution for the initial value problem (5.3.2) if, for any $\varepsilon \in \mathbf{R}^1_*$ $[0, t_1]$, $\varepsilon^+(0) = \varepsilon_0$, there exist an $\omega_\varepsilon \in (0, t_1]$ and a function $\sigma : [0, \omega_\varepsilon) \to R$ of class $\mathbf{R}^1_*$, such that the pair $(\varepsilon(t), \sigma(t))$ verifies (5.3.2) for all $t \in [0, \omega_\varepsilon)$ where for a given $\varepsilon \in \mathbf{R}^1_*[0, t_1]$, ω_ε is the largest t with this property.

In order to find a solution of the problem (5.3.1), we shall apply the Lagrange method of the variation of parameters. With this aim let us consider first the problem

$$\dot{\sigma} = \varphi(\varepsilon, \sigma)\dot{\varepsilon}, \quad \varepsilon(0) = \varepsilon_0, \quad \sigma(0) = \sigma_0, \quad (\varepsilon_0, \sigma_0) \in \mathscr{D}. \tag{5.3.3}$$

If φ is "sufficiently good" and $\varepsilon \in C^1[0, t_1]$, $\dot{\varepsilon}(t) \neq 0$ on $[0, t_1]$, one can find a "sufficiently good" global solution

$$\sigma = f(\varepsilon, \varepsilon_0, \sigma_0), \quad \varepsilon \in (\sigma_-, \omega_+), \quad \omega_\pm = \omega_\pm(\varepsilon_0, \sigma_0); \ (\varepsilon_0, \sigma_0) \in \mathscr{D} \tag{5.3.4}$$

of the problem

$$\frac{d\sigma}{d\varepsilon} = \varphi(\varepsilon, \sigma), \quad \sigma(\varepsilon_0) = \sigma_0, \quad (\varepsilon_0, \sigma_0) \in \mathscr{D} \tag{5.3.5}$$

(see Hartman [1964], Chapters II, III, V, Coddington and Levinson [1955], Chapters I, II).

The solution (5.3.4) has the following properties

$$\frac{\partial f}{\partial \varepsilon} = \varphi(\varepsilon, f), \quad \frac{\partial f}{\partial \sigma_0} > 0, \quad \frac{\partial f}{\partial \sigma_0}(\varepsilon_0, \varepsilon_0, \sigma_0) = 1,$$

$$(\varepsilon, f(\varepsilon, \varepsilon_0, \sigma_0)) \in \mathscr{D} \quad \text{for} \quad \varepsilon \in (\omega_-, \omega_+), \tag{5.3.6}$$

$$(\omega_\pm, \lim_{\varepsilon \to \omega_\pm} f(\varepsilon, \varepsilon_0, \sigma_0)) \in \mathscr{D}.$$

It is obvious now that one can omit the restriction $\dot{\varepsilon}(t) \neq 0$; moreover, for ε belonging to $\mathbf{R}^1_*[0, t_1]$, the function f in (5.3.4) is a solution of the problem

$$\begin{aligned} \dot{\sigma}_d &= \varphi(\varepsilon^+, \sigma^+)\, \dot{\varepsilon}_d, \quad \sigma^+(0) = \sigma_0, \quad \varepsilon^+(0) = \varepsilon_0, \\ \dot{\sigma}_s &= \varphi(\varepsilon^-, \sigma^-)\, \dot{\varepsilon}_s, \quad (\varepsilon_0, \sigma_0) \in \mathscr{D}, \end{aligned} \tag{5.3.7}$$

if $\omega \in (0, t_1]$ is chosen so that $(\varepsilon(t), f(\varepsilon(t), \varepsilon_0, \sigma_0)) \in \mathscr{D}$ for $t \in [0, \omega)$. In fact, ω has to be selected so that $\varepsilon(t) \in (\omega_-, \omega_+)$ for $t \in [0, \omega)$.

We shall prove the following theorem.

Theorem 5.3.1. Let $\varphi: \overline{\mathscr{D}} \to R$, φ and $\dfrac{\partial \varphi}{\partial \sigma}$ be continuous on $\overline{\mathscr{D}}$. Assume that $\mathscr{D}$ is a simple and bounded domain in the $\varepsilon 0 \sigma$-plane. Let $\varepsilon \in R^1_*[0, t_1]$, $\sigma_0 \in R$ and $\varepsilon^+(0) = \varepsilon_0$ be such that $(\varepsilon_0, \sigma_0) \in \mathscr{D}$. Further assume that there exists a continuous function k, depending on ε, $k:[0, t_1] \to R$, $k(t) > 0$, such that for any solution $\sigma \in \mathbf{R}^1_*$ of the problem (5.3.7) and for fixed $\varepsilon \in \mathbf{R}^1_*[0, t_1]$ one has

$$[\sigma] = k[\varepsilon], \tag{5.3.8}$$

where $[\sigma](t) = \sigma(t + 0) - \sigma(t - 0)$.
Then the problem (5.3,7) has a unique solution.

Proof. Only uniqueness need to be proved. Suppose there exist $\omega_1 > 0$, $\omega_2 > 0$ and $\sigma_1, \sigma_2 \in \mathbf{R}^1_*$ such that $(\varepsilon(t), \sigma_1(t)) \in \mathscr{D}$ for $t \in [0, \omega_1)$ and $(\varepsilon(t), \sigma_2(t)) \in \mathscr{D}$ for $t \in [0, \omega_2)$ are solutions of the problem (5.3.7). Then, for $t \in [0, \omega)$, $\omega = min(\omega_1, \omega_2)$ we have

$$\dot{\sigma}_{1d} - \dot{\sigma}_{2d} = (\varphi(\varepsilon^+, \sigma_1^+) - \varphi(\varepsilon^+, \sigma_2^+))\, \dot{\varepsilon} = \frac{\partial \varphi}{\partial \sigma}(\varepsilon^+, \tilde{\sigma})\, (\sigma_1^+ - \sigma_2^+)\, \dot{\varepsilon}_d.$$

Let $\sigma = \sigma_1 - \sigma_2$. According to (5.3.8), σ is continuous. Since $\dot{\varepsilon}_d$ is bounded on $[0, t]$ and $\partial \varphi / \partial \sigma$ is bounded on $\mathscr{D}$, we have

$$|\dot{\sigma}_d| \leqslant M |\sigma|, \quad \sigma(0) = 0. \tag{5.3.9}$$

This inequality implies $\sigma(t) \equiv 0$ for $t \in [0, \omega)$ and therefore $\sigma_1(t) = \sigma_2(t)$ for $t \in [0, \omega)$. If $\omega = \omega_1$, then $\sigma_2(t)$ can be considered as an extension of $\sigma_1(t)$ for $t \in [\omega_1, \omega_2]$; this contradicts the assumption that ω_1 is maximal, hence $\omega_1 = \omega_2$.

If one looks for solutions in C^{01} or C^1 then (5.3.8) is trivially satisfied as ε and σ are continuous.

In order to obtain a solution of the problem (5.3.2) we shall use now Lagrange's method. We choose an arbitrary but fixed $\varepsilon \in \mathbf{R}^1_*[0, t_1]$ with $\varepsilon^+(0) = \varepsilon_0$. σ_0 is replaced by an unknown function of t, say $\tau(t)$, and we have to determine this function so that (5.3.2) is verified. We have

$$\dot{\sigma}_d = \frac{\partial f}{\partial \varepsilon}(\varepsilon^+, \varepsilon_0, \tau)\, \dot{\varepsilon}_d + \frac{\partial f}{\partial \tau}(\varepsilon^+, \varepsilon_0, \tau)\, \dot{\tau}_d =$$

$$= \varphi(\varepsilon^+, f(\varepsilon^+, \varepsilon_0, \tau))\, \dot{\varepsilon}_d + \psi(\varepsilon^+, f(\varepsilon^+, \varepsilon_0, \tau)).$$

According to (5.3.6), we can write

$$\dot{\tau}_d = \mu(\varepsilon^+, \tau), \qquad \tau(0) = \sigma_0 \tag{5.3.10}$$

or

$$\sigma(t) = \sigma_0 + \int_0^t \mu(\varepsilon^+(s), \tau(s))\, ds, \tag{5.3.11}$$

where

$$\mu(\varepsilon, \tau) = \frac{\psi(\varepsilon, f(\varepsilon, \varepsilon_0, \tau))}{\dfrac{\partial f}{\partial \tau}(\varepsilon, \varepsilon_0, \tau)}. \tag{5.3.12}$$

Due to the equivalence between (5.3.10) and (5.3.11), the problem (5.3.2) may be solved by using functions $\tau \in C^{01}$ for $\varepsilon \in \mathbf{R}^1_*$, or $\tau \in C^1$ for $\varepsilon \in C^{01}$, and this means that "roughly speaking" τ is one class "better" than ε (and σ). τ will be called the *history parameter* and it plays the role of an internal state variable.

The existence and uniqueness of the solution of the problem (5.3.10) (or (5.3.11)) for a fixed $\varepsilon \in \mathbf{R}^0_*[0, t_1]$ follow from rather weak assumptions on the function μ. We shall deal here with the continuity and differentiability of this solution when ε runs over $\mathbf{R}^0_*[0, t_1]$.

In fact, we shall prove the following statement.

Theorem 5.3.2. Assume that the following conditions are satisfied:
i) $\mathscr{D}_0$ is a simple and bounded plane domain;

ii) $\mu : \overline{\mathscr{D}}_0 \to R$ is a continuous and differentiable function and $\frac{\partial \mu}{\partial \sigma}$, $\frac{\partial \mu}{\partial \varepsilon}$ are bounded on $\mathscr{D}_0$;

iii) $\varepsilon_0, \sigma_0, \varepsilon_n^0, \sigma_n^0$ are real numbers, with the properties

$$(\varepsilon_0, \sigma_0), (\varepsilon_n^0, \sigma_n^0) \in \mathscr{D}_0, \quad \lim_{n\to\infty} \varepsilon_n^0 = \varepsilon_0, \quad \lim_{n\to\infty} \sigma_n^0 = \sigma_0;$$

iv) $\varepsilon_n, \varepsilon \in \mathbf{R}_*^0[0, t_1]$, $\varepsilon_n^+(0) = \varepsilon_n^0$, $\varepsilon^+(0) = \varepsilon_0$ and $\varepsilon_n \to \varepsilon$ as $n \to \infty$, with respect to the norm $\|\varepsilon\| = \max_{t\in[0, t_1]} |\varepsilon(t)|$.

Then: I) for each problem

$$\tau(t) = \sigma_0 + \int_0^t \mu(\varepsilon(s), \tau(s))\, \mathrm{d}s$$

$$\tau_n(t) = \sigma_n^0 + \int_0^t \mu(\varepsilon_n(s), \tau_n(s))\, \mathrm{d}s,$$

there exist maximal intervals $[0, \omega) \subseteq [0, t_1]$, $[0, \omega_n) \subseteq [0, t_1]$ $(\omega, \omega_n \in \in (0, t_1))$ and uniquely determined functions $\tau(t), \tau_n(t)$ of class C^{01}, such that $(\varepsilon(t), \tau(t)) \in \mathscr{D}_0$ for $t \in [0, \omega)$, $(\varepsilon_n(t), \tau_n(t)) \in \mathscr{D}_0$ for $t \in [0, \omega_n)$ and $(\varepsilon(\omega - 0), \tau(\omega - 0)) \in \overline{\mathscr{D}}_0$, $(\varepsilon_n(\omega_n - 0), \tau_n(\omega_n - 0)) \in \overline{\mathscr{D}}_0$ and, if $\omega < t_1$ and $\omega_n < t_1$, then $(\varepsilon(\omega + 0), \tau(\omega - 0)) \notin \overline{\mathscr{D}}_0$, $(\varepsilon_n(\omega_n + 0), \tau_n(\omega_n - 0)) \notin \overline{\mathscr{D}}_0$. As $n \to \infty$, we have

$$\omega \leqslant \liminf_{n\to\infty} \omega_n, \quad \tau(t) = \lim_{n\to\infty} \tau_n(t), \quad t \in [0, \omega).$$

II) The mapping $\varepsilon(\cdot) \to \tau(\cdot)$ is continuous and Fréchet differentiable with respect to the topology of uniform convergence.

Proof. The first part of this theorem has a similar proof to those used for classical theorems establishing the continuous dependence of a solution on the initial data (see, for instance, HARTMAN [1964], Chapters II, III, V, CODDINGTON and LEVINSON [1955], Chapters I, II). We shall prove here only the last part of the theorem.

Let $\varepsilon, l \in \mathbf{R}_*^1 [0, t_1]$ (i.e. let ε, l be right or left continuous functions in $\mathbf{R}^0[0, t_1]$) be such that $(\varepsilon_0, \sigma_0), (\varepsilon_0 + \lambda l_0, \sigma_0) \in \mathscr{D}_0$, $(\varepsilon_0 = \varepsilon^+(0), l_0 = l^+(0))$ for a sufficiently small λ. Then there exist $\omega, \omega_\lambda \in (0, t_1]$, $\omega \leqslant \liminf_{\lambda\to 0} \omega_\lambda$ and functions $\tau(t), \tau_\lambda(t)$, such that

$$\tau_\lambda(t) - \tau(t) = \int_0^t [\mu(\varepsilon + l\lambda, \tau_\lambda) - \mu(\varepsilon, \tau_\lambda) + \mu(\varepsilon, \tau_\lambda) - \mu(\varepsilon, \tau)]\, \mathrm{d}s =$$

$$= \int_0^t \left[\lambda \frac{\partial \mu}{\partial \varepsilon} (\varepsilon + l\theta_1\lambda, \tau_\lambda)\, l + \frac{\partial \mu}{\partial \tau} (\varepsilon, \theta_2 (\tau_\lambda - \tau)) (\tau_\lambda - \tau)\right] \mathrm{d}s, \quad (5.3.13)$$

for $t \in [0, \tilde{\omega})$, with $\tilde{\omega} = \min(\omega, \omega_\lambda)$, $\theta_1 = \theta_1(s)$, $\theta_2 = \theta_2(s)$, $0 \leqslant \theta_1, \theta_2 \leqslant 1$.

Let us introduce the notation

$$V_\lambda = \frac{\tau_\lambda(t) - \tau(t)}{\lambda}, \qquad C = \max_{(\varepsilon, \tau) \in \mathscr{D}_0} \left| \frac{\partial \mu}{\partial \varepsilon}(\varepsilon, \tau) \right|,$$

$$P = \max_{(\varepsilon, \tau) \in \mathscr{D}_0} \left| \frac{\partial \mu}{\partial \tau}(\varepsilon, \tau) \right|,$$

and let us show that $|V_\lambda(t)| \leqslant M < \infty$ for any $|\lambda| \leqslant a$ and $t \in [0, \tilde{\omega})$. Indeed, we get from (5.3.13) that

$$|V_\lambda(t)| \leqslant C\|l\|\, \tilde{\omega} + P \int_0^t |V_\lambda(s)|\, \mathrm{d}s$$

and, by applying Gronwall's lemma, it follows that

$$|V_\lambda(t)| \leqslant C\|l\|\, \tilde{\omega} \exp(\tilde{\omega} P), \qquad t \in [0, \tilde{\omega}). \tag{5.3.14}$$

Relation (5.3.13) may also be written in the form

$$\begin{aligned} V_\lambda(t) - \int_0^t \left[\frac{\partial \mu}{\partial \varepsilon}(\varepsilon, \tau)\, l(s) + \frac{\partial \mu}{\partial \tau}(\varepsilon, \tau)\, V_\lambda(s) \right] \mathrm{d}s = \\ = \int_0^t \left[\frac{\partial \mu}{\partial \varepsilon}(\varepsilon + \lambda\theta_1 l, \tau_\lambda) - \frac{\partial \mu}{\partial \varepsilon}(\varepsilon, \tau) \right] l(s)\, \mathrm{d}s + \\ + \int_0^t \left[\frac{\partial \mu}{\partial \tau}(\varepsilon, \tau + \lambda\theta_2 V_\lambda) - \frac{\partial \mu}{\partial \tau}(\varepsilon, \tau) \right] V_\lambda(s)\, \mathrm{d}s. \end{aligned} \tag{5.3.15}$$

Let us consider now the initial value problem

$$V(t) - \int_0^t \left[\frac{\partial \mu}{\partial \varepsilon}(\varepsilon, \tau)\, l + \frac{\partial \mu}{\partial \tau}(\varepsilon, \tau)\, V \right] \mathrm{d}s = 0.$$

This problem has a unique continuous solution $V \in C^{01}[0, \omega]$ for $\varepsilon, l \in \mathbf{R}^0_*[0, t_1]$ and $\tau \in C^{01}[0, \omega)$, and this solution is given by

$$V(t) = \exp\left(\int_0^t \frac{\partial \mu}{\partial \tau}(\varepsilon, \tau)\, \mathrm{d}s \right) \left(\int_0^t \exp\left(-\int_0^s \frac{\partial \mu}{\partial \tau}(\varepsilon, \tau)\, \mathrm{d}s_1 \right) \frac{\partial \mu}{\partial \varepsilon}(\varepsilon, \tau)\, l\, \mathrm{d}s \right) \tag{5.3.16}$$

Since τ_λ is continuous with respect to $\varepsilon + \lambda l$, and $\dfrac{\partial \mu}{\partial \varepsilon}, \dfrac{\partial \mu}{\partial \tau}$ are continuous, it follows according to (5.3.14) that the relation (5.3.15) implies

$$\lim_{\lambda \to 0} V_\lambda(t) = V(t), \qquad t \in [0, \omega).$$

By the first part of this theorem we may introduce the functional $\mathscr{F}$ which associates to each function $\varepsilon \in \mathbf{R}^0_*[0, t_1]$ and to each number $\sigma_0((\varepsilon_0, \sigma_0) \in \in \mathscr{D}_0)$, the real valued function defined on an interval $[0, \omega_\varepsilon)$:

$$\tau(t) = \mathscr{F}[\varepsilon(\cdot), t, \sigma_0], \qquad t \in [0, \omega_\varepsilon) \tag{5.3.17}$$

Taking into account the assumption ii) of the theorem as well as formula (5.3.16) we conclude that $\mathscr{F}$ is a continuous and differentiable mapping and

$$\delta\mathscr{F}[\varepsilon(\cdot), t, \sigma_0 \,|\, l(\cdot)] = V(t) = \left(\exp \int_0^t \frac{\partial \mu}{\partial \tau}(\varepsilon, \tau)\, ds\right) \times \\ \times \left[\int_0^t \exp\left(-\int_0^s \frac{\partial \mu}{\partial \tau}(\varepsilon, \tau)\, ds_1\right) \frac{\partial \mu}{\partial \varepsilon}(\varepsilon, \tau)\, l\, ds\right]. \tag{5.3.18}$$

Let us return now to the problem (5.3.2) and prove the following theorem.

Theorem 5.3.3. Assume that the following conditions are satisfied for the problem (5.3.2):

i) $\mathscr{D}$ is a simple and bounded plane domain;

ii) $\varphi, \psi : \bar{\mathscr{D}} \to R$ are continuous on $\bar{\mathscr{D}}$ and $\dfrac{\partial \varphi}{\partial \varepsilon}, \dfrac{\partial \varphi}{\partial \sigma}, \dfrac{\partial^2 \varphi}{\partial \sigma^2}, \dfrac{\partial \psi}{\partial \varepsilon}, \dfrac{\partial \psi}{\partial \sigma}$ exist and are continuous and bounded on $\mathscr{D}$;

iii) condition (5.3.8) is satisfied.

Then, for any $\varepsilon \in \mathbf{R}^1_*[0, t_1]$, $\varepsilon_+(0) = \varepsilon_0$:

I) there exist an $\omega \in (0, t_1]$ and a function $\sigma : [0, \omega) \to R$ such that $(\varepsilon(t), (\sigma(t))$, with $\sigma(t)$ given by

$$\begin{aligned} \sigma(t) &= f(\varepsilon(t), \varepsilon_0, \tau(t)), \qquad t \in [0, \omega), \\ \tau(t) &= \mathscr{F}[\varepsilon(\cdot), t, \sigma_0], \\ (\dot{\tau}(t) &= \mu(\varepsilon(t), \tau), \qquad \tau(0) = \sigma_0) \end{aligned} \tag{5.3.19}$$

is the unique solution (in the sense of the definition given at the beginning of this section) of the problem (5.3.2), where f and $\mathscr{F}$ are determined by (5.3.4) and (5.3.17) respectively;

II) the mapping $\varepsilon \to \sigma$ defined by (5.3.19) is continuous and differentiable (for any $t \in [0, \omega)$) with respect to the topology of uniform convergence.

Proof. Let $\varepsilon \in \mathbf{R}^1_*[0, t_1]$ and $\sigma_0 \in R$ be such that $\varepsilon^+(0) = \varepsilon_0$ and $(\varepsilon_0, \sigma_0) \in \in \mathscr{D}$. One can follow now the same procedure as the one used to obtain the formulas (5.3.3)—(5.3.7) and (5.3.10)—(5.3.11), (5.3.13).

Let us apply Theorem 5.3.2 in order to obtain a solution of the problem (5.3.2). We need to specify the definition domain of $\mu(\varepsilon, \tau)$, given by (5.3.12), for a fixed ε_0. The function $f(\varepsilon, \varepsilon_0, \tau)$ is defined for any τ with $(\varepsilon_0, \tau) \in \mathscr{D}$ and $\omega_-(\varepsilon_0, \tau) < \varepsilon < \omega_+(\varepsilon_0, \tau)$ and it establishes a one-to-one correspondence between τ and σ, for fixed ε_0 and ε.

Let τ_+ and τ_- denote those values of τ for which the segment (ε_0, τ) intersects $\partial\mathscr{D}$, i.e. $(\varepsilon_0, \tau) \in \mathscr{D}$ for $\tau \in (\tau_-, \tau_+)$. We consider domains $\mathscr{D}_n \subset \mathscr{D}$, with $\bar{\mathscr{D}}_n \subset \mathscr{D}_{n+1}$ and $\bigcup_1^\infty \mathscr{D}_n = \mathscr{D}$ (for instance $\mathscr{D}_n = \{(\varepsilon, \sigma);\ (\varepsilon, \sigma) \in \mathscr{D}, |\varepsilon| < n, |\sigma| < n, \operatorname{dist}((\varepsilon, \sigma), \partial\mathscr{D}) > 1/n\}$; see for instance HARTMAN [1964], Chapter II) and we denote by τ_{n-}, τ_{n+} the values of τ for which (ε_0, τ) intersects $\mathscr{D}_n$. Then, the distance from $(\varepsilon_0, \tau_{n\pm})$ to $\partial\mathscr{D}$ is $\geqslant 1/n$ and $(\varepsilon_0, \tau) \in \mathscr{D}$ for $\tau \in [\tau_{n-}, \tau_{n+}]$. According to the properties of $\omega_\pm(\varepsilon_0, \tau)$, for any $\eta > 0$ and any $\bar{\tau} \in [\tau_{n-}, \tau_{n+}]$, there exist two neighborhoods $V'(\bar{\tau})$, $V''(\bar{\tau})$ such that

$$\omega_+(\varepsilon_0, \bar{\tau}) + \eta < \omega_+(\varepsilon_0, \tau), \qquad \tau \in V'(\bar{\tau}),$$

$$\omega_-(\varepsilon_0, \tau) < \omega_-(\varepsilon_0, \bar{\tau}) + \eta, \qquad \tau \in V''(\bar{\tau}).$$

Let us choose $\eta = 1/n$ and $V_n(\bar{\tau}) = V'(\bar{\tau}) \cap V''(\bar{\tau})$; then $\omega_-(\varepsilon_0, \tau) - \frac{1}{n} < \omega_-(\varepsilon_0, \bar{\tau}) < \varepsilon_0 < \omega_+(\varepsilon_0, \bar{\tau}) < \omega_+(\varepsilon_0, \tau) - \frac{1}{n}$ for all $\tau \in V_n(\bar{\tau})$, and there exists a finite covering of $[\tau_{n-}, \tau_{n+}]$, say $V_n(\tau_1), \ldots, V_n(\tau_m)$. If we set $\omega'_-(\varepsilon_0) = \max_{i=1,\ldots,m} \omega_-(\varepsilon_0, \tau_i)$ and $\omega'_+(\varepsilon_0) = \min_{i=1,\ldots,m} \omega_+(\varepsilon_0, \tau_i)$, then $\omega'_-(\varepsilon_0) < \varepsilon_0 < \omega'_+(\varepsilon_0)$. Let us write

$$\Delta(\varepsilon_0, \sigma_0) = \{(\varepsilon, \tau);\ \varepsilon \in (\omega'_-(\varepsilon_0),\ \omega'_+(\varepsilon_0)),\ \ \tau \in (\tau_{n-}, \tau_{n+})\}$$

so that $(\varepsilon, \sigma) \in \mathscr{D}$, where $\sigma = f(\varepsilon, \varepsilon_0, \tau)$, for any $(\varepsilon, \tau) \in \Delta(\varepsilon_0, \sigma_0)$.

According to Theorem 5.3.2, for $\mu: \Delta(\varepsilon_0, \sigma_0) \to R$ and $\varepsilon \in \mathbf{R}^1_*[0, t_1] \subset \mathbf{R}^0_*[0, t_1]$, with $\varepsilon^+(0) = \varepsilon_0$, there exists an $\omega_1 \in [0, t_1]$ such that

$$\tau(t) = \mathscr{F}[\varepsilon(\cdot), t, \sigma_0], \qquad t \in [0, \omega_1], \tag{5.3.20}$$

is a solution of the problem (5.3.11), and

$$\sigma(t) = f(\varepsilon(t), \varepsilon_0, \tau(t)), \qquad t \in [0, \omega_1] \tag{5.3.21}$$

is a solution of the problem (5.3.2).

If $\omega_1 = t_1$, existence is proved. Assume $\omega_1 < t_1$. We have $(\varepsilon(\omega_1 - 0), \tau(\omega_1 - 0)) \in \bar{\Delta}(\varepsilon_0, \sigma_0)$ and $(\varepsilon(\omega_1 + 0), \tau(\omega_1 - 0)) \notin \Delta(\varepsilon_0, \sigma_0)$. There are two possibilities: Either $(\varepsilon(\omega_1 - 0), \sigma(\omega_1 - 0)) \in \bar{\mathscr{D}}$ and $\varepsilon(\omega_1 + 0)$ does not belong to the segment $(\omega_-(\varepsilon_0, \tau(\omega_1 - 0)), \omega_+(\varepsilon_0, \tau(\omega_1 - 0)))$, hence $\omega_1 = \omega$

(and this concludes the proof), or $(\varepsilon(\omega_1 + 0), \sigma(\omega_1 + 0)) \in \mathscr{D}$ (we use here the fact that τ is continuous). If $(\varepsilon(\omega_1 + 0), \sigma(\omega_1 + 0)) \in \mathscr{D}_n$ then this point is used instead $(\varepsilon_0, \sigma_0)$ to construct another $\Delta(\varepsilon(\omega_1 + 0), \sigma(\omega_1+0))$ and applying the same procedure as above, we obtain an $\omega_1' > \omega_1$. If we are still in $\mathscr{D}_n$ the procedure is to be applied again and again until either $\omega_1' = t_1$ or $(\varepsilon(\omega_1' + 0), \sigma(\omega_1' + 0))$ is not in $\mathscr{D}_n$ while still being in $\mathscr{D}$. Then there is an $n_1 > n$ such that $(\varepsilon(\omega_1' + 0), \sigma(\omega_1' + 0)) \in \mathscr{D}_{n_1}$, etc.

Finally we find an $\omega \leqslant t_1$ such that $(\varepsilon(t), \sigma(t)) \in \mathscr{D}$ for $t \in [0, \omega)$ and, if $\omega < t_1$, then $(\varepsilon(\omega - 0), \sigma(\omega - 0)) \in \bar{\mathscr{D}}$ but $\varepsilon(\omega + 0)$ does not belong to the segment $(\omega_-(\varepsilon_0, \tau(\omega - 0)), \omega_+(\varepsilon_0, \tau(\omega - 0)))$.

In order to prove that the solution is unique, let us assume that there are two solutions of the problem (5.3.2), say $(\varepsilon(t), \sigma_1(t)) \in \mathscr{D}$ for $t \in [0, \omega^1)$ and $(\varepsilon(t), \sigma_2(t)) \in \mathscr{D}$ for $t \in [0, \omega^2)$, with $\varepsilon^+(0) = \varepsilon_0$, $\sigma_1^+(0) = \sigma_2^+(0) = \sigma_0$.

The same argument as was employed to prove Theorem 5.3.2 will lead here to $\sigma_1(t) = \sigma_2(t)$ and $\omega^1 = \omega^2$.

The continuity and differentiability of the mapping $\varepsilon \to f(\varepsilon, \varepsilon_0, \tau)$ are obtained as consequences of the properties of φ and ψ; these imply that $\dfrac{\partial f}{\partial \varepsilon}, \dfrac{\partial f}{\partial \tau}$ are continuous functions. On the other hand, $\mathscr{F}[\varepsilon(\cdot), t, \sigma_0]$ is continuous and differentiable by Theorem 5.3.2. The differential of f can be written in the form

$$\delta f(\varepsilon(t), \varepsilon_0, \tau(t) \mid l(t)) = \frac{\partial f}{\partial \varepsilon} (\varepsilon(t), \varepsilon_0, \tau(t))\, l(t) +$$

$$+ \frac{\partial f}{\partial \tau} (\varepsilon(t), \varepsilon_0, \tau(t)) \left(\exp \left(\int_0^t \frac{\partial \mu}{\partial \tau} (\varepsilon(s), \tau(s))\, \mathrm{d}s \right) \right) \times \tag{5.3.22}$$

$$\times \left[\int_0^t \exp \left(- \int_0^s \frac{\partial \mu}{\partial \tau} (\varepsilon(s_1), \tau(s_1))\, \mathrm{d}s_1 \right) \frac{\partial \mu}{\partial \varepsilon} (\varepsilon(s), \tau(s))\, l(s)\, \mathrm{d}s \right],$$

where $\mu(\varepsilon, \tau)$ is given by (5.3.12). The proof is complete.

A consequence of Theorems 5.3.3 and 5.3.2 is that, for $\varepsilon \in \mathbf{R}^1_* [0, t_1]$, $\tau(t)$ is of class C^{01} on $t \in [0, \omega)$ and therefore all the discontinuities of σ at the moment t are given only by the discontinuities of ε at the same time t. The discontinuities of the derivative of σ will also depend on τ.

If $\varepsilon \in C^{01}[0, t_1]$, τ is of class C^1, hence all the discontinuities of the derivative of σ at a moment $t \in [0, \omega)$ depend only on the discontinuities of the derivative of ε at the same t, while the discontinuities of the derivatives of ε on the interval $[0, t)$ have no influence.

4. CONSEQUENCES OF THE CONSTITUTIVE ASSUMPTIONS I AND II

We have discussed in Section 1 of this chapter the restrictions that reloading experiments impose upon the rate-type constitutive equation. On the basis of the results obtained in Sections 1 and 3 we shall present now some further information that can be obtained concerning the constitutive equation. The results presented here are contained in SULICIU, MALVERN and CRISTESCU [1974]; the main task of that paper was to find a number of properties of the functions φ and ψ present in the constitutive equation (5.1.2), when the relaxation curve is identified with an experimental curve. Only case b) of the above mentioned paper will be discussed here.

We assume that the experiments refer to a material which is initially in its natural equilibrium configuration, i.e. $\varepsilon_0 = 0$, $\sigma_0 = 0$. Therefore the solution (5.3.18) can be written in the form

$$\begin{aligned}\sigma(t) &= f(\varepsilon(t), \tau(t)),\\ \tau(t) &= \mathscr{F}[\varepsilon(\cdot), t], \qquad t \in [0, \omega).\end{aligned} \tag{5.4.1}$$

We also assume that φ is a sufficiently smooth function so that relation (5.4.1) can be written by using power series expansions as follows:

$$\begin{aligned}\sigma = f(\varepsilon, \tau) = {} & A\varepsilon^{1/2} + \tau + E\varepsilon + E_1\varepsilon^{1/2}\tau + E_2\tau^2 + C_1\varepsilon^{3/2} + C_2\varepsilon\tau + \\ & + C_3\varepsilon^{1/2}\tau^2 + C_4\tau^3 + D_1\varepsilon^2 + D_2\varepsilon^{3/2}\tau + D_3\varepsilon\tau^2 + \\ & + D_4\varepsilon^{1/2}\tau^3 + D_5\tau^4 + \dots\end{aligned} \tag{5.4.2}$$

The function f verifies $(5.3.6)_1$, that is

$$\frac{\partial f}{\partial \varepsilon}(\varepsilon, \tau) = \varphi(\varepsilon, f(\varepsilon, \tau)).$$

Assuming that $\varphi_0(\varepsilon) = E$, where E is the Young modulus, we apply the C.A.II to any $\varepsilon \in \mathscr{P}_0$ (see Section 1) to conclude that f verifies the relation

$$\frac{\partial f}{\partial \varepsilon}(\varepsilon_*, \tau(t)) = \varphi(\varepsilon_*, f(\varepsilon_*, \tau(t))) = E \quad \text{for} \quad t \geqslant t_\varepsilon > 0. \tag{5.4.3}$$

According to Theorem 5.1.1, the C.A.II implies in general the existence of at least one relaxation curve, i.e. a curve $\sigma = \sigma_R(\varepsilon)$ with the property that ψ vanishes along it.

If we apply $(5.4.1)_1$ to any $\varepsilon \in \mathscr{P}_0$, then due to Theorem 5.1.1, we get that

$$\sigma(t) = f(\varepsilon_*, \tau(t)) = \sigma_R(\varepsilon_*) \quad \text{for} \quad t \geqslant t_\varepsilon > 0. \tag{5.4.4}$$

Now $(5.3.6)_1$ leads to

$$\tau = \tau_R(\varepsilon_*) \quad \text{for} \quad t \geqslant t_\varepsilon. \tag{5.4.5}$$

i.e. the history parameter τ also admits a relaxation curve.

Let us consider the approximations of order 1,2,3,... in the expansion (5.4.2) and impose on them (5.4.3).

The first approximation

$$\sigma = f_1(\varepsilon, \tau) = A\varepsilon^{1/2} + \tau$$

cannot satisfy (5.4.3).

For the second approximation

$$\sigma = f_2(\varepsilon, \tau) = A\varepsilon^{1/2} + \tau + E\varepsilon + E_1\varepsilon^{1/2}\tau + E_2\tau^2 \tag{5.4.6}$$

we have

$$\frac{\partial f_2}{\partial \varepsilon}(\varepsilon, \tau) = \frac{1}{2}\frac{A}{\varepsilon^{1/2}} + E + \frac{1}{2}\frac{E_1\tau}{\varepsilon^{1/2}} = E, \qquad t \geqslant t_\varepsilon.$$

This approximation leads to the general semilinear model proposed by MALVERN [1951 *a*, *b*], if one takes

$$A = 0, \qquad E_1 = 0, \tag{5.4.7}$$

and

$$\tau_1 = \tau + E_2\tau^2, \tag{5.4.8}$$

where τ_1, as a function of t, has to verify the equation

$$\dot{\tau}_1(t) = \psi(\tau_1(t) + E\varepsilon(t), \varepsilon), \qquad \tau_1(0) = 0, \tag{5.4.9}$$

where $\psi \leqslant 0$ for $\varepsilon \geqslant 0$, $\sigma \geqslant 0$, which means that τ_1 has to be negative in this domain. As one can easily see, replacement of ε by $\varepsilon^{1/2}$ induces a change of τ according to formula (5.4.8).

Before starting to discuss the next approximations, let us add another condition to those imposed on f, namely

$$\left.\frac{\partial f(\varepsilon, \tau(\tau_1))}{\partial \tau_1}\right|_{\tau=\tau_R(\varepsilon)} = 1. \tag{5.4.10}$$

This condition expresses the continuous transition of the quasilinear model into the semilinear model, i.e. the semilinear model can approximate the quasilinear one in the neighborhood of the relaxation boundary (since in this neighborhood one has $f(\varepsilon, \tau(\tau_1)) \approx E\varepsilon + \tau_1$).

The next approximations will have to verify both conditions (5.4.3) and (5.4.10). Thus, for the third order approximation, we find that

$$\frac{\partial f_3}{\partial \varepsilon} = E + \frac{3}{2}C_1\varepsilon^{1/2} + C_2\tau + \frac{1}{2}C_3\varepsilon^{-1/2}\tau^2 = E, \quad t \geqslant t_\varepsilon \tag{5.4.11}$$

and

$$\left.\frac{\partial f_3}{\partial \tau}\frac{\partial \tau}{\partial \tau_1}\right|_{\tau=\tau_R^{(3)}(\varepsilon)} = 1 + \left.\frac{C_2\varepsilon + 2C_3\varepsilon^{1/2}\tau + 3C_4\tau^2}{1 + 2E_2\tau}\right|_{\tau=\tau_R^{(3)}(\varepsilon)} = 1. \quad (5.4.12)$$

From (5.4.11) and (5.4.12) we get

$$3C_1\varepsilon + 2C_2\varepsilon^{1/2}\tau + C_3\tau^2 = 0, \quad (5.4.13)$$

$$C_2\varepsilon + 2C_3\varepsilon^{1/2}\tau + 3C_4\tau^2 = 0. \quad (5.4.14)$$

These equations must have at least one common root. But, as our purpose is not to discuss all possibilities, we shall consider here only the case when (5.4.13) and (5.4.14) have two common roots. Such a case arises only when

$$C_2 = 3\lambda C_1, \qquad C_3 = 3\lambda^2 C_1, \qquad C_4 = \lambda^3 C_1, \quad (5.4.15)$$

where λ and C_1 are unknown constants.

Substituting (5.4.15) into (5.4.14) (or (5.4.13)) and taking into account the expression for $f_3(\varepsilon, \tau)$ we obtain

$$\tau = -\frac{1}{\lambda}\varepsilon^{1/2} = \tau_R^{(3)}(\varepsilon), \quad (5.4.16)$$

$$\sigma = \left(E + \frac{1}{\lambda^2}E_2\right)\varepsilon - \frac{1}{\lambda}\varepsilon^{1/2} = \sigma_R^{(3)}(\varepsilon). \quad (5.4.17)$$

The relaxation curve (5.4.17) approximates the parabola $\sigma = \beta\varepsilon^{1/2}$ which has been experimentally found by BELL [1961], if one takes

$$\lambda = -\frac{1}{\beta} \quad (5.4.18)$$

and

$$E + \beta^2 E_2 \approx 0. \quad (5.4.19)$$

The choice (5.4.18) and (5.4.19) leads to $E_2 < 0$ and $\tau > 0$ while (5.4.8) gives the one-to-one correspondence between τ and τ_1.

$$\tau = \frac{-1 - (1 + 4E_2\tau_1)^{1/2}}{2E_2}, \quad (5.4.8')$$

if τ_1 is negative. Relations (5.4.8), (5.4.16), (5.4.18) and (5.4.19) imply that $\beta\sqrt{\varepsilon} - E\varepsilon = \tau_1$ and that τ_1 is negative for $\varepsilon > (\beta/E)^2 = \varepsilon_Y$. The value ε_Y of the strain can be obtained by intersecting the elastic line $\sigma = E\varepsilon$ with the curve $\sigma = \beta\varepsilon^{1/2}$, and it is denoted by ε_{Y_1} in CRISTESCU's and BELL's paper [1970]. The numerical results obtained in this paper, where ε_{Y_1} has been considered as the strain at the flow point, are not in a very good agreement with the experimental results; but one has to observe that these results have been obtained by using $\sigma = \beta\varepsilon^{1/2}$ as the constitutive equation and not as a relaxation curve.

The curves $\sigma = \beta\sqrt{\varepsilon}$ and $\sigma = \sigma_R^{(3)}(\varepsilon)$ of (5.4.17) may therefore be identified only for $\varepsilon > \varepsilon_Y$. τ_1 is negative because ψ is negative for $\varepsilon \geqslant 0$ and $\sigma \geqslant 0$, and since $\varepsilon_0 = 0$, $\sigma_0 = 0$ imply $f_3(\varepsilon, \tau) \leqslant i_3(\varepsilon)$ for all histories τ. To sum up, one has the following results for the third approximation

$$
\begin{aligned}
\sigma &= \sigma_R^{(3)}(\varepsilon) = \beta\sqrt{\varepsilon} + 0(\varepsilon),\\
\tau &= \tau_R^{(3)}(\varepsilon) = \beta\sqrt{\varepsilon},\\
\tau_1 &= \tau - \frac{E}{\beta^2}\tau^2 = \beta\sqrt{\varepsilon} - E\varepsilon, \qquad (5.4.20)\\
\sigma &= f_3(\varepsilon, 0) = i_3(\varepsilon) = E\varepsilon + C_1\varepsilon^{3/2},\\
\sigma &= f_3(\varepsilon, \tau) = E\varepsilon + \tau - \frac{E}{\beta^2}\tau^2 + \frac{C_1}{\beta^3}(\beta\sqrt{\varepsilon} - \tau)^3.
\end{aligned}
$$

The term $0(\varepsilon)$ in the equation of the relaxation curve depends on the approximation (5.4.19). Equation $(5.4.20)_4$ gives the instantaneous curve while $(5.4.20)_5$ describes the whole process.

Let us consider now the fourth approximation

$$
\begin{aligned}
\sigma = f_4(\varepsilon, \tau) = {}& E\varepsilon + \tau + E_2\tau^2 + C_1\varepsilon^{3/2} + C_2\varepsilon\tau + C_3\varepsilon^{1/2}\tau^2 + C_4\tau^3 + \\
&+ D_1\varepsilon^2 + D_2\varepsilon^{3/2}\tau + D_3\varepsilon\tau^2 + D_4\varepsilon^{1/2}\tau^3 + D_5\tau^4. \qquad (5.4.21)
\end{aligned}
$$

For $\tau = \tau_R^{(4)}(\varepsilon)$ we get the equation

$$
\begin{aligned}
&D_4\tau^3 + (C_3 + 2D_3\varepsilon^{1/2})\tau^2 + (2C_3\varepsilon^{1/2} + 3D_2\varepsilon)\tau + \\
&+ (3C_1\varepsilon + 4D_1\varepsilon^{3/2}) = 0.
\end{aligned} \qquad (5.4.22)
$$

Condition (5.4.10) leads to the equation

$$
\begin{aligned}
&4D_5\tau^3 + (3C_4 + 3D_4\varepsilon^{1/2})\tau^2 + (2C_3\varepsilon^{1/2} + 2D_3\varepsilon)\tau + \\
&+ C_2\varepsilon + D_2\varepsilon^{3/2} = 0,
\end{aligned} \qquad (5.4.23)
$$

which has to be satisfied along the relaxation curve.

Again, equations (5.4.22) and (5.4.23) must have at least one root in common. If we require that they have the same roots, we obtain $\sigma_R^{(4)}(\varepsilon) = \sigma_R^{(3)}(\varepsilon)$ and $\tau_R^{(4)}(\varepsilon) = \tau_R^{(3)}(\varepsilon)$, and if the identifications (5.4.18) and (5.4.19) are performed, the only equations that are modified are that of the instantaneous curve and the one describing the whole process. They become

$$\sigma = f_4(\varepsilon, 0) = i_4(\varepsilon) = E\varepsilon + C_1\varepsilon^{3/2} + D_1\varepsilon^2,$$

$$\sigma = f_4(\varepsilon, \tau) = E\varepsilon + \tau - \frac{E}{\beta^2}\tau^2 + \frac{C_1}{\beta^3}(\beta\sqrt{\varepsilon} - \tau)^3 + \tag{5.4.24}$$

$$+ \frac{D_1}{\beta^4}(\beta\sqrt{\varepsilon} - \tau)^4.$$

Roughly speaking, the instantaneous curve $\sigma = f(\varepsilon, 0) = i(\varepsilon)$ controls the first part of the propagation of a wave through an undeformed material. Here the fourth approximation could give a better description of the instantaneous curve for the same relaxation boundary.

5. REMARKS ON THE EXISTENCE OF A "PLATEAU" IN DYNAMIC PLASTICITY

In this section we shall present the results obtained by SULICIU, MALVERN and CRISTESCU [1972, 1973]; we shall discuss the possibility of the existence of an absolute or asymptotic plateau for stresses, strains and velocities in the case of a semi-infinite bar suddenly loaded at one end, provided a rate-type constitutive equation is used to describe the material behaviour.

In the first papers on plastic wave propagation (v. KÁRMÁN-DUWEZ [1950], TAYLOR [1946], RAKHMATULIN [1945]) it has been noticed that for a semi-infinite bar loaded at one end with a constant velocity the maximum strain near the impacted end is constant (see Chapter III, Section 1). This result has been theoretically obtained with the aid of a finite constitutive equation and for a sudden impact at the bar end. The first experiments have also shown a kind of strain plateau (see LEE [1956]).

However, an extensive discussion on the possibility of the existence of such a plateau started only after MALVERN published his two papers [1951 a, b]. After introducing a semilinear rate-type constitutive equation in order to describe plastic wave propagation, MALVERN noticed that the numerical examples given by him did not show any plateau. For many years the presence or absence of such a plateau has been considered as the principal argument in favour of one or another of the two at that time existing theories on the plastic wave propagation: the rate-type theory (depending on the strain rate) and the finite theory (independent of the strain rate).

In general, the results obtained in the experiments with constant velocity have always shown a kind of plateau, though in certain cases there appeared quite close to the impacted end a somewhat larger strain than that of the plateau (see Chapter III and for additional references, see BELL [1968, 1973], CRISTESCU [1967]).

As we mentioned before, MALVERN examples [1951a, b] using a semilinear rate-type theory presented no plateau at all. Later RUBIN [1954] has shown that one can obtain an asymptotic plateau when $t \to \infty$. Again, CRISTESCU's numerical examples [1965], this time for a finite bar, have given no plateau at all. However, WOOD and PHILLIPS [1967], EFRON and MALVERN [1969] have obtained a kind of asymptotic plateaus by using semilinear rate-type theories.

It is already well established that a finite stress-strain law will predict a plateau for stress as well as for strain and velocity. We shall discuss in this section the possibility of finding such a plateau for one or more of the above mentioned functions, when a quasilinear or semilinear rate type constitutive equation is used.

The complete system of equations which describes the mechanical motion of a bar, is (see Chapter I, Section 1)

$$\rho_0 \frac{\partial v}{\partial t} = -\frac{\partial \sigma}{\partial X},$$
$$\frac{\partial v}{\partial X} = -\frac{\partial \varepsilon}{\partial t}, \tag{5.5.1}$$
$$\frac{\partial \sigma}{\partial t} = \varphi(\varepsilon, \sigma)\frac{\partial \varepsilon}{\partial t} + \psi(\varepsilon, \sigma).$$

Unlike in the other sections, we take here both stress and strain to be positive in compression. In particular, when $\varphi = E =$ const., $(5.5.1)_3$ reduces to MALVERN's semilinear model. A particular form of this model which will be used here is

$$\dot{\sigma} = E\dot{\varepsilon} - k[\sigma - \sigma_R(\varepsilon)], \tag{5.5.2}$$

where k is a positive constant and $\sigma - \sigma_R(\varepsilon)$ is the overstress relative to a curve $\sigma = \sigma_R(\varepsilon)$, typical for each material and called the relaxation curve or relaxation boundary (CRISTESCU [1972 b]).

We can apply here to equation $(5.5.1)_3$ the integration procedure with respect to time developed in Section 4 (see also SULICIU [1973], [1974 c]).

By this we eliminate the stress from (5.5.1), so that the system governing the bar motion contains only kinematic quantities that can be measured in a dynamic experiment. Since equation (5.5.2) allows an explicit computation which constitutes a good example for the procedure described in Section 4, we shall give it below.

Let us consider equation (5.5.2) with only the first term at the right-hand side; by integrating it from an initial moment t_0 (for a fixed but arbitrary X), we get

$$\sigma(X, t) = E[\varepsilon(X, t) - \varepsilon(X, t_0)] + \sigma(X, t_0),$$

where $\varepsilon(X, t_0) = \varepsilon_0$ is the initial strain, while $\sigma(X, t_0) = \sigma_0$ is the initial stress. In order to find a solution for the whole equation (5.5.2), we replace $\sigma(X, t_0)$ in the above relation by a function $\tau(X, t)$ to be determined by means of Lagrange's method of variation of parameters.

Thus, the last formula becomes

$$\sigma = f(\varepsilon, \varepsilon_0, \tau) = E(\varepsilon - \varepsilon_0) + \tau. \tag{5.5.3}$$

By substituting (5.5.3) into (5.5.2) we find the following equation for τ:

$$\dot{\tau} = -k[\tau + E(\varepsilon - \varepsilon_0) - \sigma_R(\varepsilon)] \tag{5.5.4}$$

with $\tau(t_0) = \sigma_0$. A direct integration of (5.5.4) gives τ as a functional of ε on an interval $[t_0, t]$.

Thus the integrated form of (5.5.2) can be written as

$$\sigma(t) = E\varepsilon(t) + e^{-k(t-t_0)}\left\{\sigma_0 - E\varepsilon_0 - k\int_{t_0}^{t} [E\varepsilon(s) - f(\varepsilon(s))]\, e^{k(s-s_0)}\, \mathrm{d}s\right\}. \tag{5.5.5}$$

One can obviously see from (5.5.3) and (5.5.5) that τ is determined by the explicit form of φ and ψ and depends on the strain history. This is the reason why τ is called the history parameter. The structure of the function f in (5.5.3) is determined by φ only, while for fixed ε_0 and τ this function describes the instantaneous response.

Now, for the semilinear model (5.5.2) we obtain from (5.5.5) and $(5.5.1)_1$ that

$$\rho_0 \frac{\partial v}{\partial t} + E \frac{\partial \varepsilon}{\partial X} = k e^{-k(t-t_0)} \left\{ \frac{\partial \sigma_0}{\partial X} - E \frac{\partial \varepsilon_0}{\partial X} + \int_{t_0}^{t} [E - \sigma_R'(\varepsilon(s))] \times \right.$$
$$\left. \times \frac{\partial \varepsilon(s)}{\partial X}\, e^{k(s-t_0)}\, \mathrm{d}s \right\}; \tag{5.5.6}$$

equation (5.5.6) together with $(5.5.1)_2$ form a system of equations describing the motion by means of kinematic variables only.

In the general case of the constitutive equation $(5.5.1)_3$, if the functions φ and ψ verify the conditions of Theorem 5.3.3 and the strain $\varepsilon(X, t)$ is of class $\mathbf{R}^1$ with respect to (X, t), then we may apply (5.3.22) with $l(s) = \frac{\partial \varepsilon}{\partial X}(X, s)$, thus obtaining instead of (5.5.6), the integro-differential equation

$$\rho_0 \frac{\partial v}{\partial t} + \varphi(\varepsilon, f(\varepsilon, \tau)) \frac{\partial \varepsilon}{\partial X} =$$

$$= - \frac{\partial f}{\partial \tau} \exp\left[\int_0^t \frac{\partial \mu}{\partial \tau}(\varepsilon(s), \tau(s))\, \mathrm{d}s\right] \int_0^t \left[\exp\left(-\int_0^s \frac{\partial \mu}{\partial \tau}(\varepsilon(s_1), \tau(s_1)\, \mathrm{d}s_1\right)\right] \times$$

$$\times \frac{\partial \mu}{\partial \varepsilon} \frac{\partial \varepsilon}{\partial X}\, \mathrm{d}s, \tag{5.5.7}$$

where μ is given by (5.3.12), i.e.

$$\mu(\varepsilon, \tau) = \frac{\psi(\varepsilon, f(\varepsilon, \tau))}{\frac{\partial f}{\partial \tau}(\varepsilon, \tau)}.$$

Therefore

$$\frac{\partial \mu}{\partial \varepsilon} = \frac{\frac{\partial \psi}{\partial \varepsilon} + \frac{\partial \psi}{\partial \sigma}\varphi - \psi \frac{\partial \varphi}{\partial \sigma}}{\frac{\partial f}{\partial \tau}}, \quad \frac{\partial \mu}{\partial \tau} = \frac{\frac{\partial \psi}{\partial \tau}\left(\frac{\partial f}{\partial \tau}\right)^2 - \psi \frac{\partial^2 f}{\partial \sigma^2}}{\left(\frac{\partial f}{\partial \tau}\right)^2}. \tag{5.5.8}$$

For the sake of simplicity we have taken $t_0 = 0$, $\sigma_0 = 0$ and $\varepsilon_0 = 0$ in (5.5.7). This assumption will be retained in the following if not mentioned otherwise.

Let us establish now the necessary condition for the existence of a plateau.

Formulas (5.5.6) and (5.5.7) contain the term $\frac{\partial \varepsilon}{\partial X}$ which is related to quantities that may be experimentally measured. The parameter we can measure is the "surface angle α", i.e. the angle between the position of the normal to the lateral surface during the bar deformation and the initial position of the same normal. The variation of this angle during the whole process of the dynamic deformation can be determined experimentally (see Chapter III, Section 1 and Bell [1968]).

If one uses the balance of mass, one may obtain the relation

$$\operatorname{tg} \alpha = R_0 \sqrt{\rho_0} \frac{1}{2\rho^{3/2}(1-\varepsilon)^{5/2}} \left[\frac{\partial \rho}{\partial X} - \varepsilon \frac{\partial \rho}{\partial X} - \rho_0 \frac{\partial \varepsilon}{\partial X} \right], \quad (5.5.9)$$

where R_0 is the initial radius of the bar, ρ_0 the initial mass density and ρ the actual mass density.

Some simplified versions of this formula have been given by different authors. Thus the formula

$$\alpha = -\frac{R_0}{2} \frac{\partial \varepsilon}{\partial X} \quad (5.5.10)$$

has been given by BELL, the formula

$$\alpha = -\frac{R_0}{2} \frac{\partial \varepsilon}{\partial X} \left(1 + \frac{5}{2} \varepsilon\right) \quad (5.5.11)$$

has been given by FILBEY and

$$\alpha = -\frac{R_0}{2} \frac{1}{(1-\varepsilon)^{5/2}} \frac{\partial \varepsilon}{\partial X} \quad (5.5.12)$$

by CRISTESCU (for references, see BELL [1969]). All these relations may be obtained from (5.5.9) by means of some additional simplifying assumptions (e.g. by assuming incompressibility or by neglecting the higher powers of ε).

Observing the variation of α during a loading experiment (see BELL [1968]), we conclude by means of formula (5.5.12), that $\frac{\partial \varepsilon}{\partial X}$ in (5.5.6) and (5.5.7) preserves its sign.

We say there exists a plateau when there is a domain in the characteristic XOt-plane where one or both derivatives $\frac{\partial \omega}{\partial t}, \frac{\partial \omega}{\partial X}$ are equal to zero. Here ω denotes one of the functions v, ε, σ. One can find a plateau either for one of these functions or for two or even all three of them. We call it a *plateau in time* if $\frac{\partial \omega}{\partial t} = 0$, and a *plateau in space* if $\frac{\partial \omega}{\partial X} = 0$. We shall also consider the case when both derivatives vanish simultaneously.

An analysis of formula (5.5.7) shows that $\frac{\partial f}{\partial \tau} > 0$ (see $(5.3.6)_1$). Among the terms that enter the right-hand side of equation (5.5.7) only

$\frac{\partial \mu}{\partial \varepsilon}$, given by $(5.5.8)_1$, may change its sign. If the term $\frac{\partial \mu}{\partial \varepsilon}$ changes its sign, then there is a t_* in the interval of integration where

$$\frac{\partial \psi}{\partial \varepsilon} + \frac{\partial \psi}{\partial \sigma} \varphi - \psi \frac{\partial \varphi}{\partial \sigma} = 0; \tag{5.5.13}$$

all quantities of the above relation being computed at the point $(\varepsilon(X, t_*), f(\varepsilon(X, t_*), \tau(X, t_*)))$. Therefore relation (5.5.13) is a necessary condition for the existence of a simultaneous plateau for ε and σ.

The existence of a plateau has been generally discussed in the case when a semi-infinite bar is impacted with a constant velocity. However, one can find such plateaus even for more practical problems such as the symmetric impact of two identical finite bars. There have been obtained a number of numerical solutions for this problem with the use of constitutive equations of the form $(5.5.1)_3$ for an impact velocity $V = 1600$ inches/s $= 4064$ cm/s at one end, while the other is free.

For one of these cases, called case XII by CRISTESCU [1972 *b*] (see also Chapter IV, Section 4), the three images in fig. 5.5.1 describe the time variation of σ, ε, v at different cross-sections of the bar. They show when one of these functions is constant in time and space or in time only. We have given here only the curves for those cross-sections where at least one plateau shows up. Among the three functions, strain is the one which has time-plateaus over the larger part of the bar.

Now, let us write the necessary condition (5.5.13) for the existence of a plateau in the particular case considered here. CRISTESCU [1972 *b*] used the following explicit form of equation $(5.5.1)_3$

$$\dot{\sigma} = \frac{E}{1 + E\Phi} \dot{\varepsilon} - \frac{E\psi}{1 + E\Phi}, \tag{5.5.14}$$

where

$$\psi(\varepsilon, \sigma) = \begin{cases} \frac{k(\varepsilon)}{E} (\sigma - \sigma_R(\varepsilon)) & \text{if } \sigma > \sigma_R(\varepsilon) \text{ and } \varepsilon \geqslant \varepsilon_Y, \\ 0 & \text{if } \sigma \leqslant \sigma_R(\varepsilon) \text{ or } \varepsilon < \varepsilon_Y, \end{cases} \tag{5.5.15}$$

$\sigma = \sigma_R(\varepsilon)$ is the relaxation curve (or boundary) and

$$k(\varepsilon) = k_0 \left[1 - \exp\left(-\frac{\varepsilon}{\hat{\varepsilon}}\right)\right] \tag{5.5.16}$$

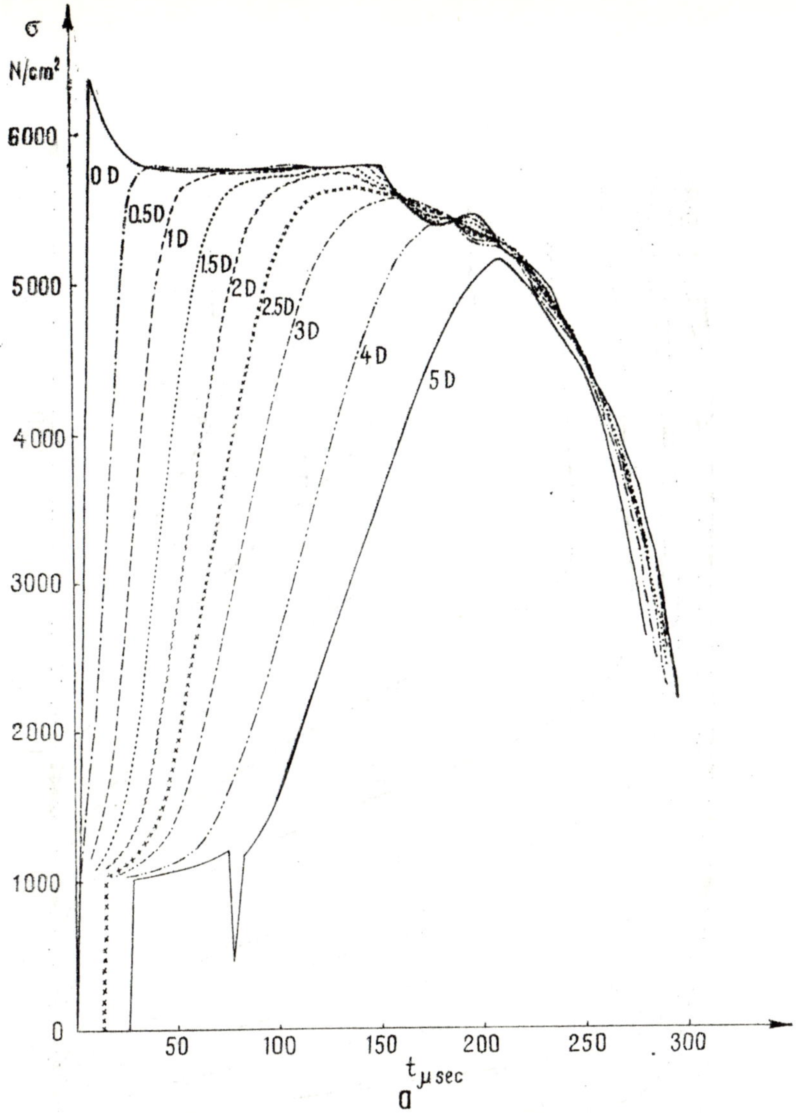

Fig. 5.5.1 a. The "plateau" in time of the stress between $0.5D-1D$

with constant k_0 and $\hat{\varepsilon}$. For those parts of a process where the stress increases the function $\Phi(\varepsilon)$ has been defined as

$$\Phi(\varepsilon) = \frac{3\left[\varepsilon - \varepsilon_Y - \varepsilon^* + \left(\dfrac{a}{3E}\right)^{3/2}\right]^{2/3}}{a} - \frac{1}{E}, \qquad (5.5.17)$$

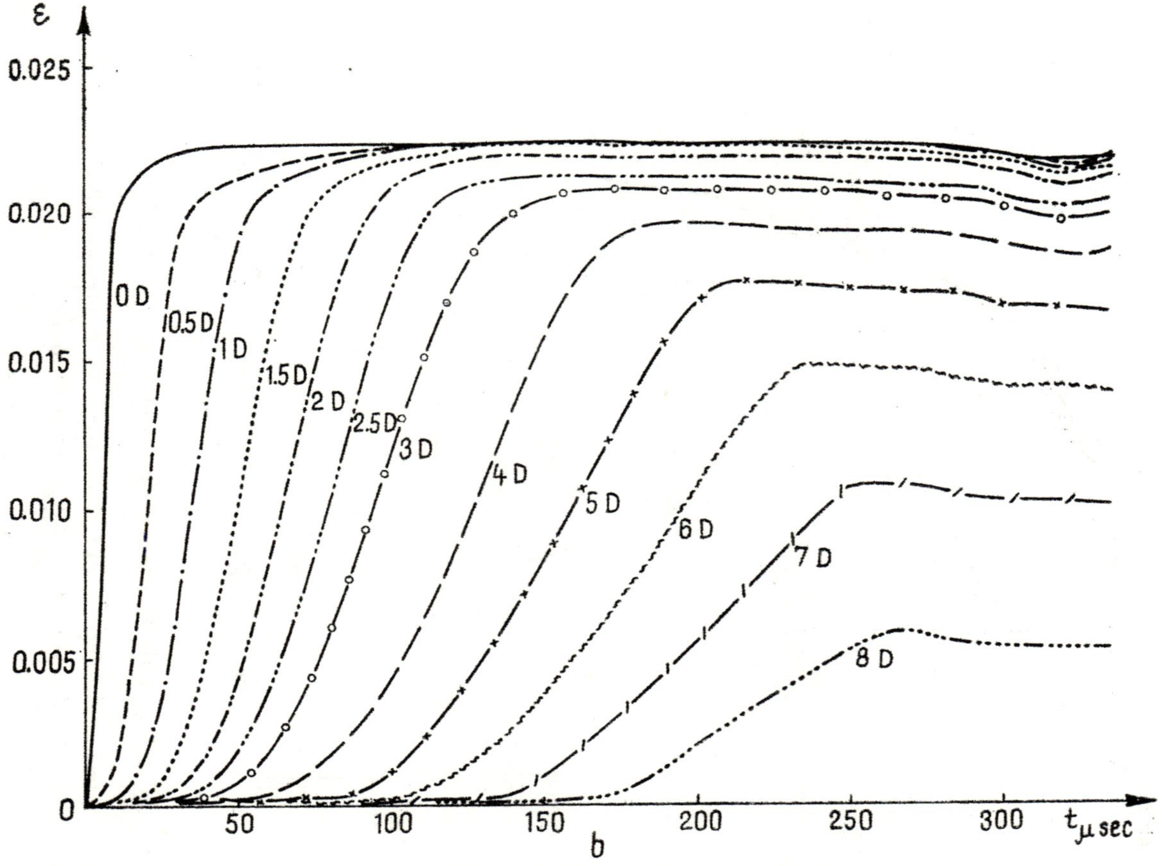

Fig. 5.5.1 b. The "plateau" in time of the strain

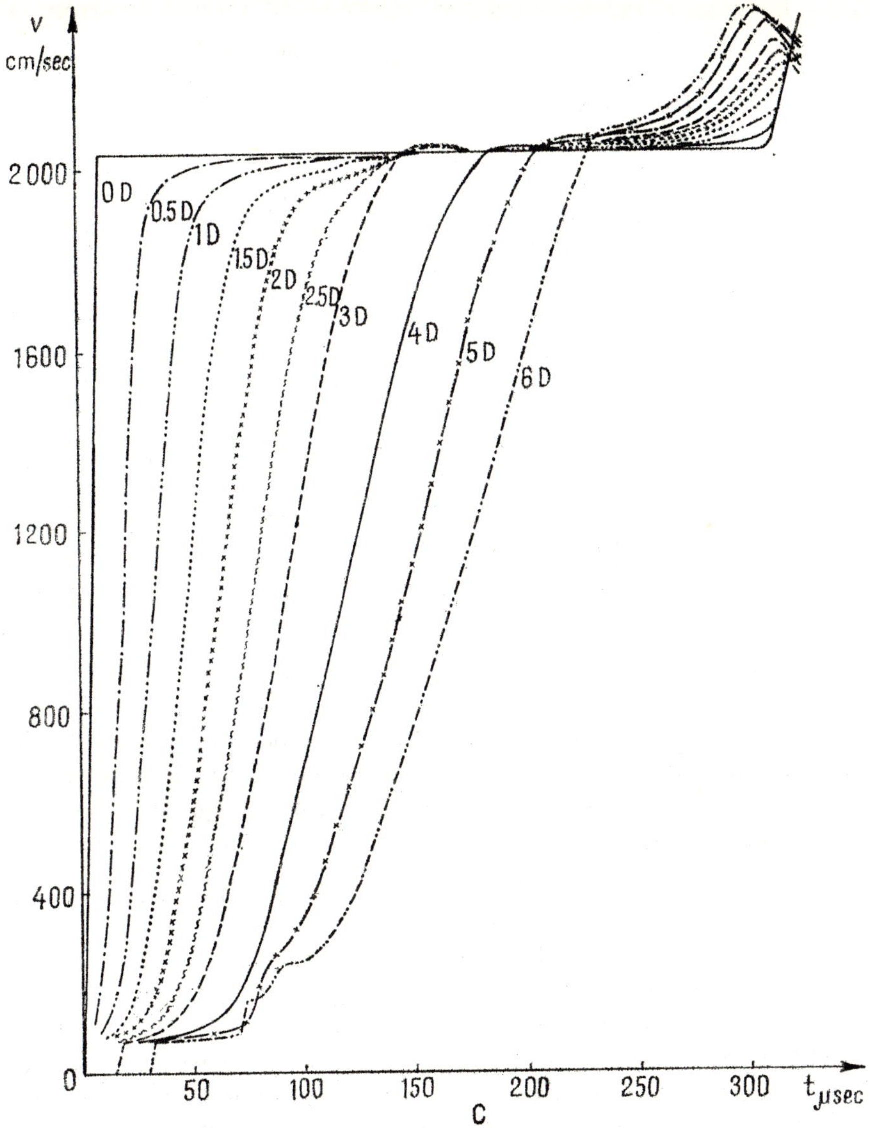

Fig. 5.5.1 c. The "plateau" in time of the particle velocity

where

$$a = m + n \sqrt{\varepsilon} \tag{5.5.18}$$

and m, n, ε_Y and ε^* are constants.

By means of the above notation, condition (5.5.13) may be written as

$$\frac{1}{\left(\frac{1}{E}+\Phi\right)^2}\left[\frac{\partial\psi}{\partial\varepsilon}\left(\frac{1}{E}+\Phi\right)-\psi\frac{\partial\Phi}{\partial\varepsilon}+\frac{\partial\psi}{\partial\sigma}\right]\approx$$

$$\approx\frac{k_0\left[1-\exp\left(-\frac{\varepsilon}{\hat{\varepsilon}}\right)\right]}{(1+E\Phi)^2}\left[1-\left(\frac{1}{E}+\Phi\right)\sigma_R'(\varepsilon)-\right.$$

$$\left.-\frac{2(\sigma-\sigma_R(\varepsilon))}{a\left[\varepsilon-\varepsilon_Y-\varepsilon^*+\left(\frac{a}{3E}\right)^{3/2}\right]^{1/3}}\right]+ \tag{5.5.19}$$

$$+\frac{k_0}{1+E\Phi}\left[\left(\frac{1}{E}+\Phi\right)\frac{1}{\hat{\varepsilon}}\{\sigma-\sigma_R(\varepsilon)\}\exp\left(-\frac{\varepsilon}{\hat{\varepsilon}}\right)-\right.$$

$$\left.-\left\{1-\exp\left(-\frac{\varepsilon}{\hat{\varepsilon}}\right)\right\}\{\sigma-\sigma_R(\varepsilon)\}\frac{3n}{2\sqrt{3Ea}}\frac{1}{\left[\varepsilon-\varepsilon_Y-\varepsilon^*+\left(\frac{a}{3E}\right)^{3/2}\right]^{1/3}}\right].$$

At the right-hand side of (5.5.19) we have neglected certain terms of a much smaller magnitude.

For $\varepsilon=\varepsilon_Y$ (the beginning of plastic deformation) the right-hand side of (5.5.19) is positive while the same expression is negative in the neighborhood of the maximum value of Φ, i.e. ε changes sign. This result is quite obvious, since for $\varepsilon=\bar{\varepsilon}$ (the strain at the inflexion point of the $\varepsilon-t$ curve for $X=0$) we have $\left(\frac{1}{E}+\Phi\right)\sigma_R'(\varepsilon)\approx 1.37$. The same condition is verified for $X=1D$ while all the other terms of (5.5.19) are for $\varepsilon=\bar{\varepsilon}$ either negative or negligible.

Thus the main conclusion is that for a quasilinear constitutive equation (5.5.14), the right-hand side of equation (5.5.7) may vanish for a $t_* > 0$ and therefore an absolute plateau in space and time is possible. In other words, for a quasilinear constitutive equation there may be a domain in the characteristic plane where $v(X, t)=$ const., $\varepsilon(X, t)=$ const. and $\sigma(X, t)=$ const. In the examples illustrated by fig. 5.5.1 this plateau actually exists.

Let us show now that the semilinear model (5.5.2), mostly used in the literature, does not admit an absolute plateau for any of the functions v, ε, σ.

Indeed, let us assume that

$$v = \text{const.}, \qquad \varepsilon = \text{const.} \tag{5.5.20}$$

above a curve Γ of the characteristic plane.

Then, for any point above Γ (with $\varepsilon_0 = 0$, $\sigma_0 = 0$), one obtains from (5.5.20) and (5.5.6) that

$$e^{-k(t-t_0)} \int_{t_0}^{t_*} [E - \sigma_R'(\varepsilon(s))] \frac{\partial \varepsilon(s)}{\partial X} e^{k(s-t_0)} \, ds = 0. \tag{5.5.21}$$

Since the experimental results show that α cannot change its sign, one concludes according to (5.5.12) that (5.5.21) cannot hold. Thus a linear dependence of ψ on the overstress, as in (5.5.2), is incompatible with the existence of an absolute plateau.

An absolute plateau cannot occur also for many other forms proposed for ψ in the literature, as for instance for

$$\psi = -k\left[\exp\left(\frac{\sigma - \sigma_R(\varepsilon)}{a} - 1\right)\right] \quad \text{or} \quad \psi = -k(\sigma - \sigma_R(\varepsilon))^\alpha$$

with $\alpha \geqslant 1$, as one can easily see from condition (5.5.13) with $\varphi = E$. However, it is possible that for other choices of the function ψ which verify the conditions of Theorem 5.3.3, condition (5.5.13) could hold even for a semilinear model.

The assertion of Suliciu, Malvern and Cristescu [1972] that the semilinear constitutive equation

$$\dot{\sigma} = E\dot{\varepsilon} + \psi(\varepsilon, \sigma) \tag{5.5.22}$$

with

$$\psi(\varepsilon, \sigma) = \begin{cases} -k(\sigma - \sigma_R(\varepsilon))^\alpha & \text{if } \sigma > \sigma_R(\varepsilon), \\ 0 & \text{if } \sigma \leqslant \sigma_R(\varepsilon), \end{cases} \tag{5.5.23}$$

for $\sigma \geqslant 0$ and $\varepsilon \geqslant 0$ and $\alpha = \dfrac{1}{2}$ does not admit an absolute plateau is false (see Section 1 and also Kukudjanov [1967, 1977]). This conclusion has been reached by applying (5.5.7) and (5.5.13). Relation (5.5.13) can be applied in this case only when ψ is a function of class C^1 on a domain $\mathscr{D}$ and has bounded partial derivatives on $\bar{\mathscr{D}}$ (see Theorem 5.3.2). Let us show (for a special class of strains, namely for strains such as occur when a bar is impacted with a finite velocity) that formula (5.5.7) can be applied, and in the neighbourhood of the impacted end one has (see SULICIU [1975 a])

$$0 \leqslant \int_{t_0}^{t} \left[\exp\left(-\int_{t_0}^{s} \frac{\partial \mu}{\partial \tau} (\varepsilon(s_1), \tau(s_1))\, ds_1 \right) \right] \frac{\partial \mu}{\partial \varepsilon} \frac{\partial \varepsilon}{\partial X}\, ds \leqslant \text{const.} \quad (5.5.24)$$

for any t, while

$$\exp\left[\int_{t_0}^{t} \frac{\partial \mu}{\partial \tau} (\varepsilon(s), \tau(s))\, ds \right] \to 0 \quad \text{when} \quad t \to \bar{t}_* < \infty; \quad (5.5.25)$$

Therefore we conclude that an absolute plateau in all three variables ε, v, σ is possible.

We choose $\mathscr{D} = \{(\varepsilon, \sigma);\ \sigma > f(\varepsilon),\ \varepsilon > 0,\ \sigma > 0\}$. Since close to the impacted end one registers a jump in strain $\varepsilon_i > \varepsilon_Y$ and the corresponding stress lies on the instantaneous curve (which is Hooke's curve in this case), we have $\sigma_i = E\varepsilon_i > \sigma_R(\varepsilon_i)$. Then we compute the value of ψ in (5.3.10), at the point $(\varepsilon_i, \sigma_i) \in \mathscr{D}$, at $t = t_0$, with $\tau(t_0) = 0$ $\left(\text{and obviously } \frac{\partial f}{\partial \tau} = 1\right)$, and therefore formula (5.5.7) remains valid as long as $(\varepsilon(t), \sigma(t) = E\varepsilon(t) + \tau(t)) \in \mathscr{D}$.

We shall show now that the existence of a plateau for ε ,above a curve Γ of the characteristic plane, implies the existence of a curve Γ^* lying above Γ, where v and σ also reach a plateau.

The proof is based on the following proposition.

Proposition 5.5.1. The initial value problem

$$\dot{\tau} = -k(\tau + \Omega)^{\alpha}, \qquad \tau(t_0) = 0 \quad (5.5.26)$$

with $\alpha \in (0, 1)$ and $\Omega\colon [t_0, \infty) \to R$, $\Omega \in C^1$, $\Omega(t_0) = \Omega_0 > 0$, $\Omega(t) > 0$ for $t \in [t_0, \infty)$, $\dot{\Omega}(t) > 0$ for $t \in (t_0, t_*)$, $\dot{\Omega}(t) = 0$ for $t \in [t_*, \infty)$ admits a unique solution of class $C^1[t_0, \infty)$ (see for instance HARTMAN [1964], Chapter III, Section 6). This solution has the following properties: If

$$t_0 < t_* < \frac{\Omega_*^{1-\alpha}}{k(1-\alpha)}, \quad (5.5.27)$$

where $\Omega_* = \Omega(t_*)$, then there exists a $\bar{t}_* > t_*$ such that

$$\tau(t) + \Omega(t) > 0 \text{ for all } t \in [t_0, \bar{t}_*], \quad (5.5.28)$$

while on $[t_*, \infty)$

$$\tau(t) = \begin{cases} -\Omega_* + [(\tau_* + \Omega_*)^{1-\alpha} - k(1-\alpha)(t - t_*)]^{\frac{1}{1-\alpha}} & \text{if } t \in [t_*, \bar{t}_*], \\ -\Omega_* & \text{if } \bar{t}_* \leqslant t; \end{cases} \quad (5.5.29)$$

which follows by a direct integration of equation (5.5.26) on $[t_*, \infty)$, with $\tau(t_*) = \tau_*$.

In order to prove the assertions (5.5.28) and (5.5.29), let us first remark that

$$\Omega_* = \Omega(t_*) \geqslant \Omega(t), \qquad t \in [t_0, \infty)$$

and

$$- k(\tau + \Omega)^\alpha \geqslant - k(\tau + \Omega_*)^\alpha, \qquad t \in [t_0, \infty). \tag{5.5.30}$$

Then (see for instance HARTMAN [1964], Chapter III, Section 4)

$$\tau(t) \geqslant \tau_0(t), \qquad t \in [t_0, \infty), \tag{5.5.31}$$

where $\tau(t)$ is the solution of the problem (5.5.26) and $\tau_0(t)$ is the solution of the problem

$$\dot{\tau}_0 = - k(\tau_0 + \Omega_*)^\alpha, \qquad \tau_0(t_0) = 0, \tag{5.5.32}$$

given by

$$\tau_0(t) = \begin{cases} - \Omega_* + [\Omega_*^{1-\alpha} - k(1-\alpha)(t - t_0)]^{\frac{1}{1-\alpha}} & \text{if } t \in \left[t_0, \dfrac{\Omega_*^{1-\alpha}}{k(1-\alpha)}\right], \\ - \Omega_* & \text{if } \dfrac{\Omega_*^{1-\alpha}}{k(1-\alpha)} < t. \end{cases} \tag{5.5.33}$$

From (5.5.31) and (5.5.33) we find that

$$\tau(t) > - \Omega_* \qquad \text{for } t \in [t_*, \tilde{t}_*), \tag{5.5.34}$$

where

$$\tilde{t}_* = t_* + \frac{[\tau(t_*) + \Omega_*]^{1-\alpha}}{k(1-\alpha)}, \tag{5.5.35}$$

hence (5.5.28) and (5.5.29) follow. Let us take $\Omega = E\varepsilon - \sigma_R(\varepsilon)$. As $\sigma = f(\varepsilon, \tau) = E\varepsilon + \tau$, we have

$$\mu(\varepsilon, \tau) = \psi(\varepsilon, \tau + E\varepsilon)$$

and, from (5.5.8) and (5.5.23), we obtain

$$\begin{aligned} \frac{\partial \mu}{\partial \varepsilon} &= \frac{\partial \psi}{\partial \varepsilon} = - k\alpha \frac{E - \sigma_R'(\varepsilon)}{(E\varepsilon - \sigma_R(\varepsilon) + \tau)^{1-\alpha}}, \\ \frac{\partial \mu}{\partial \tau} &= \frac{\partial \psi}{\partial \sigma} = - \frac{k\alpha}{(E\varepsilon - \sigma_R(\varepsilon) + \tau)^{1-\alpha}}. \end{aligned} \tag{5.5.36}$$

We have assumed that ε reaches an absolute plateau in time and space above a curve Γ, i.e. for $t > t_* = g(X_*)$, $(X_*, t_*) \in \Gamma$, $\dfrac{\partial \varepsilon}{\partial X}(X_*, t) = 0$, $\dfrac{\partial \varepsilon}{\partial t}(X_*, t) = 0$. Since, from (5.5.28) it follows that $(\varepsilon(t_*), \sigma(t_*)) \in \mathscr{D}$ (i.e. this point does not lie on the relaxation curve), hence $\dfrac{\partial \mu}{\partial \varepsilon}$ and $\dfrac{\partial \mu}{\partial \tau}$ are finite, and we obtain relation (5.5.24) since, for $t \geqslant t_*$, the integral remains constant.

According to (5.5.29) and $(5.5.36)_2$, we have

$$\frac{\partial \mu}{\partial \tau} = -\frac{k\alpha}{(\tau_* + \Omega_*)^{1-\alpha} - k(1-\alpha)(t - t_*)}, \qquad t \in [t_*, \tilde{t}_*)$$

and we can write

$$\exp\left[\int_{t_0}^{t} \frac{\partial \mu}{\partial \tau}\, \mathrm{d}s\right] =$$

$$= \left[\exp\left(\int_0^t \frac{\partial \mu}{\partial \tau}\, \mathrm{d}s\right)\right] \frac{1}{(\tau_* + \Omega_*)^\alpha} [(\tau_* + \Omega_*)^{1-\alpha} - k(1-\alpha)(t - t_*)]^{\frac{\alpha}{1-\alpha}}$$

for all $t \in [t_*, \tilde{t}_*)$. This last relation together with (5.5.35) implies (5.5.25) and therefore, for $t \geqslant \tilde{t}_*$, both stress and velocity also reach a plateau.

Since formula (5.5.21) contains a negative exponential, it may seem that the zero value in (5.5.21) is reached asymptotically for $t \gg t_*$. Let us show now by using a semilinear model (5.5.2) that, the equations of motion cannot be satisfied over the whole X, t plane if one of the functions σ, v, ε is assumed constant in time while the other two are assumed to tend asymptotically to a plateau.

Let us assume that one of the unknown functions, say v, is constant (in time and space) above a curve Γ of the characteristic plane. Let us then examine under which conditions another function, say ε, could reach a plateau asymptotically. If

$$v = \text{const. above } \Gamma \tag{5.5.37}$$

then it follows from $(5.5.1)_2$ that

$$\varepsilon = h(X). \tag{5.5.38}$$

Let us denote by (X_*, t_*) the coordinates of the points lying on Γ; then, by integrating (5.5.6) up to $t > t_*$, we obtain

$$0 = -\sigma_R(\varepsilon(t_*, X_*)) \frac{\partial \varepsilon(t_*, X_*)}{\partial X} +$$

$$+ k \exp(-k(t-t_0)) \left\{ \int_{t_0}^{t} [E - \sigma_R'(\varepsilon(s, X_*))] \frac{\partial \varepsilon(s, X_*)}{\partial X} \exp(k(s-t_0)) \, ds - \right. \tag{5.5.39}$$

$$\left. - \frac{1}{k} [E - \sigma_R'(\varepsilon(t_*, X_*))] \frac{\partial \varepsilon(t_*, X_*)}{\partial X} \exp(k(t_* - t_0)) \right\}.$$

This identity has the form $a + b \exp(-k(t-t_0)) = 0$ with a and b constants. We shall show that, generally, such an identity is neither exactly nor asymptotically satisfied.

Indeed, in order that (5.5.39) be exactly satisfied, we must have

$$\sigma_R' \frac{\partial \varepsilon}{\partial X} = 0$$

$$\int_{t_0}^{t_*} [E - \sigma_R'(\varepsilon(s, X_*))] \frac{\partial \varepsilon(s, X_*)}{\partial X} \exp(k(s-t_0)) \, ds - \tag{5.5.40}$$

$$- \frac{1}{k} [E - \sigma_R'(\varepsilon(t_*, X_*))] \frac{\partial \varepsilon(t_*, X_*)}{\partial X} \exp(k(t_* - t_0)) = 0.$$

But equations (5.5.40) are satisfied in very particular cases only. First of all, $(5.5.40)_1$ is satisfied if $\sigma_R'(\varepsilon(t_*, X_*)) = 0$. But then $(5.5.40)_2$ implies that

$$k \int_{t_0}^{t} \frac{\partial \varepsilon(s, X_*)}{\partial X} \exp(k(s-t_0)) \, ds = \frac{\partial \varepsilon(t_*, X_*)}{\partial X} \exp(k(t_* - t_0)).$$

Thus (5.5.40) is satisfied only if there exists no workhardening and if one imposes a very particular law of strain variation.

A second case when $(5.5.40)_1$ is satisfied, is when $\dfrac{\partial \varepsilon(t_*, X_*)}{\partial X} = 0$, that is

$$\int_{t_0}^{t_*} [E - \sigma_R'(\varepsilon(s, X_*))] \frac{\partial \varepsilon(s, X_*)}{\partial X} \exp(k(s-t_0)) \, ds = 0;$$

but this is impossible (see (5.5.21)).

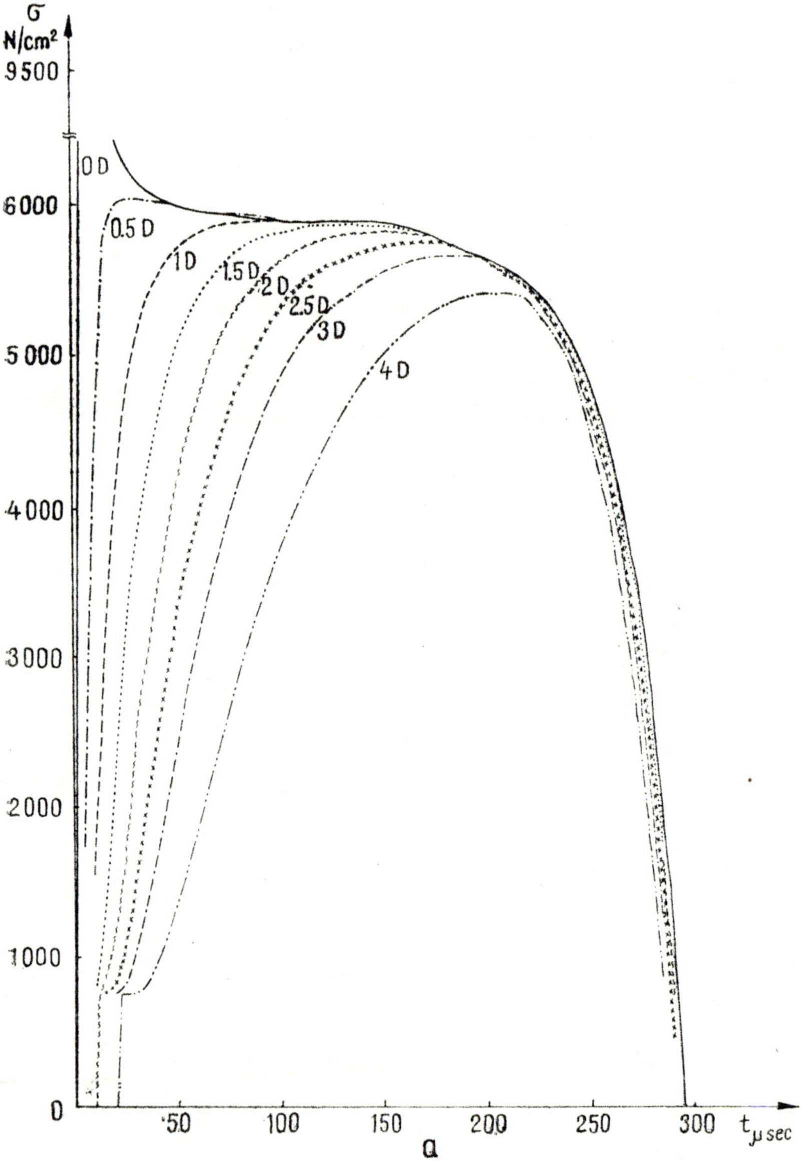

Fig. 5.5.2 a. The asymptotic plateau for the stress

In order to satisfy (5.5.39) asymptotically, $(5.5.40)_1$ has to be satisfied (under the particular conditions already discussed) and at the same time, one has to find a large value of t, say $\tilde{t}_*$, such that (5.5.21) becomes negligible beyond this value (depending on the degree of precision required for determining the unknown functions). But then the equations of motion are no longer satisfied in the strip $(t_*, \tilde{t}_*)$.

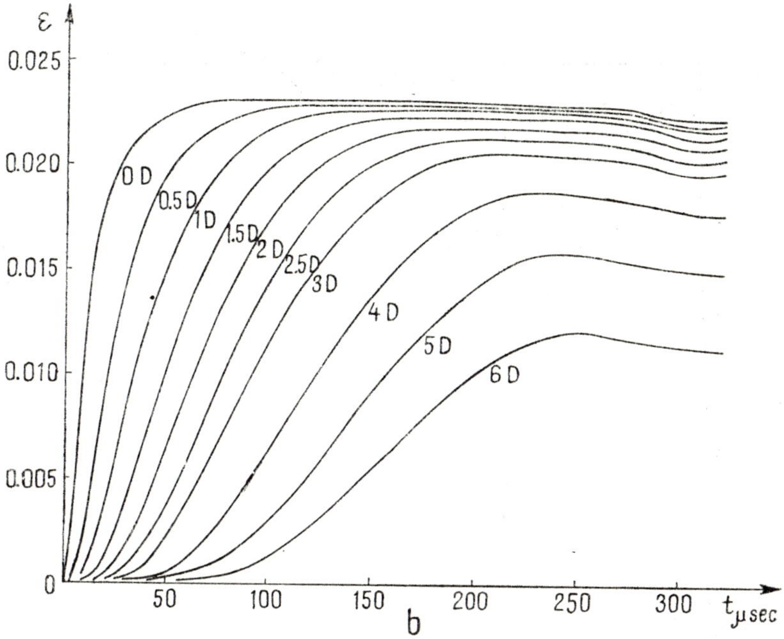

Fig. 5.5.2 b. The asymptotic plateau for the strain

Therefore, if the necessary condition (5.5.13) is not satisfied, then the assumption that there exists an absolute plateau for one of the functions v, ε or σ (the assumption (5.5.37)) is incompatible with a semilinear model (for which ψ is continuous and has bounded partial derivatives on its domain of definition). (If one assumes that ε or σ is constant, the discussion will be similar to the previous one).

Figs. 5.5.2 *a*, *b*, *c* show that, for a semilinear model with ψ given by (5.5.2), even though one can see almost horizontal plateaus, the functions are in fact very slowly increasing or decreasing. Figs 5.5.1 *a*, *b*, *c*, for a quasilinear model show a far better approximation of an absolute plateau along the first two diameters from the impacted end.

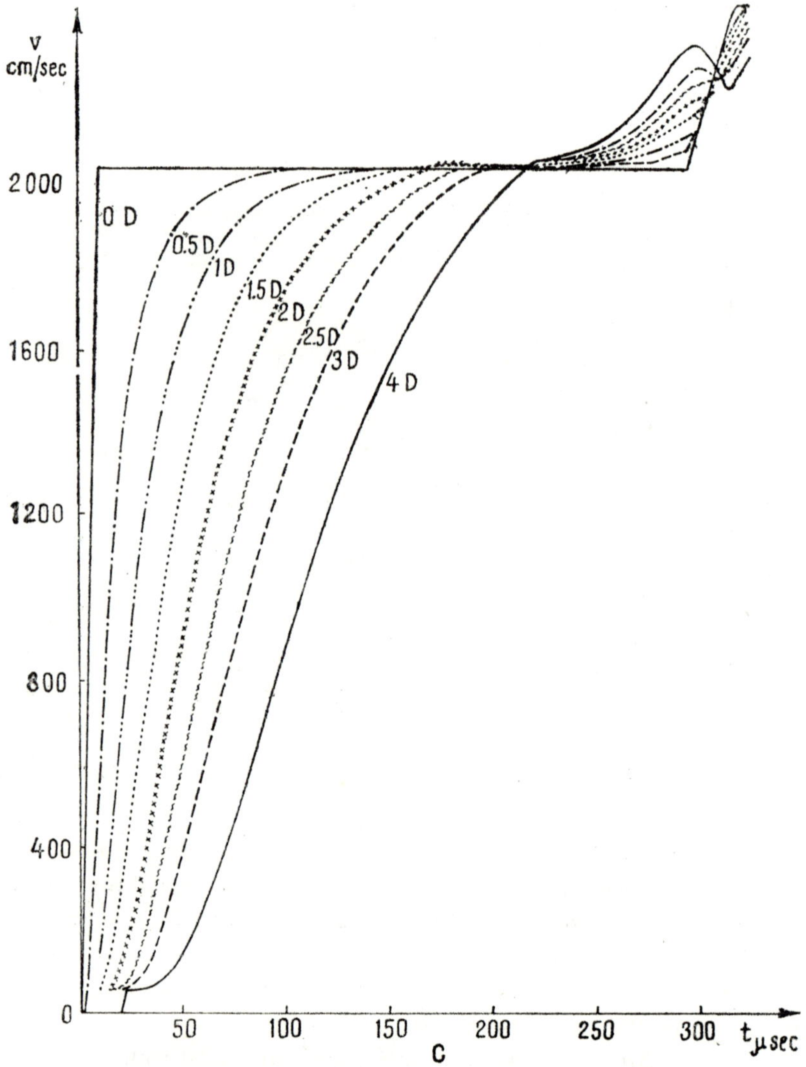

Fig. 5.5.2 c. The asymptotic plateau for the particle velocity

Chapter VI

INELASTIC BEHAVIOUR DESCRIBED BY MEANS OF INTERNAL STATE VARIABLES

1. THERMODYNAMIC RESTRICTIONS ON CONSTITUTIVE EQUATIONS WITH INTERNAL STATE VARIABLES

In this section we present the results obtained by COLEMAN and GURTIN [1967] on the restrictions imposed upon the constitutive equations with internal state variables by the CLAUSIUS-DUHEM inequality. As far as balance equations are concerned, we shall follow here, as in the above mentioned paper, COLEMAN and NOLL's [1963] line of approach. Concerning the constitutive equations and the CLAUSIUS-DUHEM inequality we will rather take the line adopted by MIHĂILESCU-SULICIU and SULICIU [1979] (see also GURTIN, SULICIU and WILLIAMS [1979]). In this way one obtains more than just some restrictions on certain given constitutive equations; indeed, at the same time, the CLAUSIUS-DUHEM inequality defines one of the thermodynamic functions (such as entropy or free energy, etc.) as a function of the thermodynamic state.

1a. **The space of states, constitutive equations.** We consider a material body $\mathscr{B}$, a reference configuration $\mathscr{R}$ of this body and an interval I on the real line. The balance equations of momentum and energy and the CLAUSIUS-DUHEM inequality (see Chapter I, formulas (1.2.9), (1.2.10), (1.2.16)) can be written if for any $(\mathbf{X}, t), \in \mathscr{R} \times I$ the functions

$$\mathbf{x} = \chi(\mathbf{X}, t),\ \mathbf{T} = \mathbf{T}(\mathbf{X}, t),\ e = e(\mathbf{X}, t),\ \mathbf{q} = \mathbf{q}(\mathbf{X}, t),$$
$$\eta = \eta(\mathbf{X}, t),\ \theta = \theta(\mathbf{X}, t),\ \mathbf{b} = \mathbf{b}(\mathbf{X}, t),\ r = r(\mathbf{X}, t) \tag{6.1.1.}$$

are defined and sufficiently smooth (see Chapter I, Section 2 for the physical interpretation of these quantities). Besides the functions (6.1.1), we also assume that the function

$$\boldsymbol{\alpha} = \boldsymbol{\alpha}(\mathbf{X}, t) = (\alpha^1, \alpha^2, \ldots, \alpha^N), \tag{6.1.2}$$

x = X(X, t) - motion   η = η(X, t) - entropy
T = T(X, t) - Cauchy stress tensor   θ = θ(X, t) - temperature
e = e(X, t) - internal energy   b = b(X, t) - body force
q = q(X, t) - heat flux   r = r(X, t) - heat supply

called the *vector of internal state variables,* is defined on $\mathscr{R} \times I$. The physical interpretation of its components depends on the domain of continuum mechanics for which this theory is used. For viscoelasticity and plasticity, for instance, this interpretation will be discussed in the next sections and also in Chapter VII.

The functions (6.1.1) and (6.1.2) form a thermodynamic process for the body $\mathscr{B}$ if they are compatible with the balance equations (1.2.9) and (1.2.10).

In order to introduce the notion of the space of states as well as the constitutive equations, we need to specify some further notations. Let $\mathscr{L}$ be the set of all linear mappings from R^3 into R^3, let $\mathscr{L}^+ \subset \mathscr{L}$ be the set of all nonsingular linear mappings, let $\mathscr{S} \subset \mathscr{L}$ denote the set of all symmetric linear mappings and let R^+ be the set of all positive real numbers.

In this chapter, the *space of states* will be an open subset $\mathscr{D}$ of $\mathscr{L}^+ \times \times R^+ \times R^3 \times R^N$ and its points

$$(\mathbf{F}, \theta, \mathbf{g}, \boldsymbol{\alpha}) \in \mathscr{D} \subset \mathscr{L}^+ \times R^+ \times R^3 \times R^N \tag{6.1.3}$$

will be called *states.* We assume that $\mathscr{D}$ is such that if $(\bar{\mathbf{F}}, \bar{\theta}, \bar{\mathbf{g}}, \bar{\boldsymbol{\alpha}}) \in \mathscr{D}$, then $(\bar{\mathbf{F}}, \bar{\theta}, \mathbf{g}, \bar{\boldsymbol{\alpha}}) \in \mathscr{D}$ for any $\mathbf{g} \in R^3$. Here $\mathbf{F}$ is the deformation gradient defined by (1.2.6), θ is the absolute temperature and $\mathbf{g}$ is the actual temperature gradient defined by (1.2.17). It will be more convenient in the following to use the PIOLA-KIRCHHOFF stress tensor $\mathbf{S}$ (or $\tilde{\mathbf{S}}$) defined by (1.2.12), instead of the CAUCHY stress tensor $\mathbf{T}$.

In the theory of constitutive equations with internal state variables, the following assumptions are made: There exist four sufficiently smooth functions $\hat{\mathbf{S}}: \mathscr{D} \to \mathscr{L}$, $\hat{q}: \mathscr{D} \to R^3$, $\hat{e}: \mathscr{D} \to R$ and $\mathbf{f}: \mathscr{D} \to R^N$ such that χ,$(\mathbf{F} = \operatorname{Grad} \chi)$, θ, η, $\mathbf{b}$, r together with $\mathbf{S}$, $\mathbf{q}$, e, $\boldsymbol{\alpha}$ determined by

$$\left\{\begin{aligned} \mathbf{S} &= \hat{\mathbf{S}}(\mathbf{F}, \theta, \mathbf{g}, \boldsymbol{\alpha}), \\ \mathbf{q} &= \hat{\mathbf{q}}(\mathbf{F}, \theta, \mathbf{g}, \boldsymbol{\alpha}), \\ e &= \hat{e}(\mathbf{F}, \theta, \mathbf{g}, \boldsymbol{\alpha}), \\ \dot{\boldsymbol{\alpha}} &= \mathbf{f}(\mathbf{F}, \theta, \mathbf{g}, \boldsymbol{\alpha}), \end{aligned}\right. \tag{6.1.4}$$

form a thermodynamic process for $\mathscr{B}$, where $\dot{\boldsymbol{\alpha}} = \dfrac{\partial \boldsymbol{\alpha}(\mathbf{X}, t)}{\partial t}$ for any $(\mathbf{X}, t) \in \in \mathscr{R} \times I$.

The above constitutive equations assert that if at a point $\mathbf{X}$ of the reference configuration $\mathscr{R}$ at a moment $t \in I$ there are known the deformation gradient **F**, the absolute temperature θ, its gradient **g** and the internal state variables $\boldsymbol{\alpha}$ and they are such that $(\mathbf{F}, \theta, \mathbf{g}, \boldsymbol{\alpha}) \in \mathscr{D}$, then the stress **S**, the heat flux **q**, the internal energy e and the time rate of the internal state variables are determined by (6.1.4). Equation $(6.1.4)_4$ is called the *evolution equation* of the internal state variables. According to this equation, if (for instance) **f** is smooth with respect to its arguments and if, for a fixed $\mathbf{X} \in \mathscr{R}$, the functions $\mathbf{F}, \theta, \mathbf{g}$ are given as (for instance) smooth functions of $t \in I$, then the knowledge of the values of the internal state variables $\boldsymbol{\alpha}$ at a moment $t_0 \in I$ uniquely determines all the values $\boldsymbol{\alpha}(t)$ for t in an interval $I_1 \subseteq I$.

A thermodynamic process is called an *admissible thermodynamic process*, for the body $\mathscr{B}$, if it is compatible with the constitutive equations (6.1.4).

The constitutive equations (6.1.4) obviously depend on the chosen reference configuration $\mathscr{R}$. For instance, if the body is nonhomogeneous with respect to this configuration, then **X** has to enter these equations explicitly. Since the thermodynamic restrictions will be given for a fixed **X**, this dependence has not been mentioned.

1b. **The implications of the CLAUSIUS-DUHEM inequality on the constitutive equations.** The balance equations (1.2.9) and (1.2.10) together with equations (6.1.4) form a complete system of partial differential equations on $\mathscr{R} \times I$. If the body forces **b** and the heat supply r are given functions on $\mathscr{R} \times I$, this system may be considered as a second order system for the motion χ, the temperature θ and the internal state variables $\boldsymbol{\alpha}$. One can formulate initial and boundary value problems for this system. If such a problem has solutions, then these solutions will determine admissible thermodynamic processes for the body $\mathscr{B}$. These processes have to be compatible with the second law of thermodynamics expressed in the form of the CLAUSIUS-DUHEM inequality (1.2.16); this inequality, under certain conditions, may impose some restrictions on the form of the functions (6.1.4). This way to take into account the second law of thermodynamics is used by MÜLLER [1971] for the constitutive equations of heat conducting fluids and rigid heat conducting solids. He has also used a modified form of the second law of thermodynamics.

COLEMAN and NOLL [1963] assumed that the functions χ, **T**, e and **q** may be arbitrarily prescribed on $\mathscr{R} \times I$, while the functions **b** and r can be chosen so that equations (1.2.9) and (1.2.10) are automatically satisfied on $\mathscr{R} \times I$. This assumption will be adopted in what follows. Another important point is the way one understands the CLAUSIUS-DUHEM inequality. For the constitutive equations with internal state variables COLEMAN and GURTIN [1967] give the following postulate on the entropy production: Any admissible thermodynamic process has to satisfy the inequality (1.2.16), that is

$$\dot{\eta} - \frac{\dot{e}}{\theta} + \theta^{-1}\,\mathbf{S}\cdot\dot{\mathbf{F}} - \frac{1}{\rho\theta^2}\,\mathbf{q}\cdot\mathbf{g} \geqslant 0, \tag{6.1.5}$$

where $\dot{\eta} = \dfrac{\partial\eta(\mathbf{X}, t)}{\partial t}$.

The entropy is said to be *a function of state* if there exists a smooth function $\hat{\eta}:\mathscr{D} \to R$ such that

$$\eta = \hat{\eta}(\mathbf{F}, \theta, \mathbf{g}, \boldsymbol{\alpha}). \tag{6.1.6}$$

Before starting to study the implications of the inequality (6.1.5) on the constitutive equations, let us note several facts (see COLEMAN and NOLL [1963] and COLEMAN and GURTIN [1967]). Let $(\mathbf{F}^0, \theta^0, \mathbf{g}^0, \boldsymbol{\alpha}^0)$ be an arbitrary point in $\mathscr{D}$ and let $(\mathbf{X}^0, t^0) \in \mathscr{R} \times I$. If one chooses an arbitrary tensor $\mathbf{A}$, an arbitrary vector $\mathbf{a}$ and an arbitrary scalar a, then

$$\begin{aligned} \chi(\mathbf{X}, t) &= \mathbf{X}^0 + [\mathbf{F}^0 + (t - t^0)\,\mathbf{A}]\,(\mathbf{X} - \mathbf{X}^0), \\ \theta(\mathbf{X}, t) &= \theta^0 + (t - t^0)\,a + [\mathbf{g}_0 + (t - t^0)\,\mathbf{a}]\cdot\mathbf{F}^0(\mathbf{X} - \mathbf{X}^0) \end{aligned} \tag{6.1.7}$$

represents a motion and a temperature field for $\mathbf{X} \in \mathscr{R}$, $t \in I$, $t \geqslant t^0$ and t close enough to t^0. Now, if $\boldsymbol{\alpha}(\mathbf{X}, t^0) = \boldsymbol{\alpha}^0$, equation $(6.1.4)_4$ together with (6.1.7) determine $\boldsymbol{\alpha}(\mathbf{X}, \tau)$ for any $\mathbf{X} \in \mathscr{R}$ and any $\tau \in [t_0, t]$ (for t close enough to t^0). The function $\boldsymbol{\alpha}$ being thus determined, $\mathbf{S}$, $\mathbf{q}$ and e can be obtained from (6.1.4) for all $\mathbf{X} \in \mathscr{R}$ and $\tau \in [t^0, t]$; finally, the balance equations (1.2.20) determine $\mathbf{b}(\mathbf{X}, t)$ and $r(\mathbf{X}, t)$.

We have thus obtained an admissible thermodynamic process where $(\eta(\mathbf{X}, \tau), (\mathbf{X}, \tau)) \in \mathscr{R} \times [t^0, t]$ is an arbitrary smooth function. It is obvious that if no other constitutive assumption on the entropy function is made, then there are infinitely many possibilities to choose the function $\eta(\mathbf{X}, \tau)$ so that inequality (6.1.5) is satisfied.

An older interpretation of inequality (6.1.5) is the following: One postulates the constitutive equations (6.1.4) together with a constitutive equation for the entropy, and the inequality is considered as a restriction on the time rates that are contained in it (see for instance ONAT [1968]). In the framework of COLEMAN and NOLL [1963] (COLEMAN and GURTIN [1967]), the CLAUSIUS-DUHEM inequality appears as a general statement on the constitutive equations, as one can see from the following theorem.

Theorem 6.1.1. There exists at least one entropy which is a function of state compatible with the CLAUSIUS-DUHEM inequality if and only if (i) $\hat{\mathbf{S}}(\mathbf{F}, \theta, \mathbf{g}, \boldsymbol{\alpha})$ and $\hat{e}(\mathbf{F}, \theta, \mathbf{g}, \alpha)$ are constant with respect to $\mathbf{g}$ on $\mathscr{D}$, i.e.

$$\hat{\mathbf{S}}_{\mathbf{g}}(\mathbf{F}, \theta, \mathbf{g}, \boldsymbol{\alpha}) = \mathbf{0}, \ \hat{e}_{\mathbf{g}}(\mathbf{F}, \theta, \mathbf{g}, \boldsymbol{\alpha}) = \mathbf{0} \text{ on } \mathscr{D}; \tag{6.1.8}$$

(ii) the functions $\frac{1}{\theta}(\hat{e}_{\mathbf{F}} - \hat{\mathbf{S}}), \frac{1}{\theta}\hat{e}_{\theta}, \mathbf{0}$ are the components of the gradient with respect to $(\mathbf{F}, \theta, \mathbf{g})$ of a scalar function $\hat{\eta}: \mathscr{D} \to R$, i.e.

$$\frac{1}{\theta}(\hat{e}_{\mathbf{F}} - \hat{\mathbf{S}}) = \hat{\eta}_{\mathbf{F}}, \ \frac{1}{\theta}\hat{e}_{\theta} = \hat{\eta}_{\theta}, \ \mathbf{0} = \hat{\eta}_{\mathbf{g}} \text{ on } \mathscr{D}; \tag{6.1.9}$$

(iii) the partial derivative of $\hat{\eta}$ (defined by (6.1.9)) with respect to $\boldsymbol{\alpha}$, satisfies the inequality

$$\rho\theta\,(\theta\hat{\eta}_{\boldsymbol{\alpha}} - \hat{e}_{\boldsymbol{\alpha}})\cdot\mathbf{f} - \hat{\mathbf{q}}\cdot\mathbf{g} \geqslant 0. \tag{6.1.10}$$

Proof. Let us assume that the entropy is a function of state, that is there exists $\hat{\eta}: \mathscr{D} \to R$ such that (6.1.6) is true. Let us take an arbitrary state $(\mathbf{F}^0, \theta^0, \mathbf{g}^0, \boldsymbol{\alpha}^0) \in \mathscr{D}$ and the quantities $\mathbf{A}, \mathbf{a}, a$ and introduce them in formulas (6.1.7). Then, at the point $(\mathbf{X}^0, t^0) \in \mathscr{R} \times I$, one has

$$\begin{gathered}\mathbf{F}(\mathbf{X}^0, t^0) = \mathbf{F}^0, \ \theta(\mathbf{X}^0, t^0) = \theta^0, \\ \mathbf{g}(\mathbf{X}^0, t^0) = \mathbf{g}^0, \ \boldsymbol{\alpha}(\mathbf{X}^0, t^0) = \boldsymbol{\alpha}^0\end{gathered} \tag{6.1.11}$$

and

$$\dot{\mathbf{F}}(\mathbf{X}^0, t^0) = \mathbf{A}, \ \dot{\theta}(\mathbf{X}^0, t^0) = a, \ \dot{\mathbf{g}}(\mathbf{X}^0, t^0) = \mathbf{a}. \tag{6.1.12}$$

Therefore, for this process, the inequality (6.1.5) can be written at point $(\mathbf{X}^0, t^0)$ as

$$\begin{gathered}\left[\hat{\eta}^0_{\mathbf{F}} - \frac{1}{\theta^0}(\hat{e}^0_{\mathbf{F}} - \hat{\mathbf{S}}^0)\right]\cdot\mathbf{A} + \left(\hat{\eta}^0_{\theta} - \frac{1}{\theta^0}\hat{e}^0_{\theta}\right)a + \left(\hat{\eta}^0_{\mathbf{g}} - \frac{1}{\theta^0}\hat{e}^0_{\mathbf{g}}\right)\cdot\mathbf{a} + \\ + \left(\hat{\eta}^0_{\alpha} - \frac{1}{\theta^0}\hat{e}^0_{\alpha}\right)\cdot\mathbf{f}^0 - \frac{1}{\rho^0(\theta^0)^2}\hat{\mathbf{q}}^0\cdot\mathbf{g}^0 \geqslant 0,\end{gathered} \tag{6.1.13}$$

where $\eta^0_F = \hat{\eta}_F(\mathbf{F}^0, \theta^0, \mathbf{g}^0, \boldsymbol{\alpha}^0)$, etc. Since $\mathbf{A}$, $\mathbf{a}$ and a are arbitrary quantities the inequality (6.1.13) will be satisfied if and only if

$$\begin{aligned}\hat{\eta}^0_{\mathbf{g}} &= \frac{1}{\theta^0}\hat{e}^0_{\mathbf{g}}, \\ \hat{\eta}^0_{\theta} &= \frac{1}{\theta^0}\hat{e}^0_{\theta}, \\ \hat{\eta}^0_{\mathbf{F}} &= \frac{1}{\theta^0}(\hat{e}^0_{\mathbf{F}} - \hat{\mathbf{S}}^0)\end{aligned} \tag{6.1.14}$$

and

$$\rho^0\, \theta^0(\theta^0\, \hat{\eta}^0_\alpha - \hat{e}^0_\alpha)\cdot \mathbf{f}^0 - \hat{\mathbf{q}}^0\cdot \mathbf{g}^0 \geqslant 0. \tag{6.1.15}$$

Since $(\theta\hat{\eta} - \hat{e})_\theta = \hat{\eta} + (\theta\hat{\eta}_\theta - \hat{e}_\theta)$ and $(\theta\hat{\eta}_\mathrm{g} - \hat{e}_\mathrm{g})_\theta = \hat{\eta}_\mathrm{g} + (\theta\hat{\eta}_\theta - \hat{e}_\theta)_\mathrm{g}$, $(6.1.14)_{1-2}$ imply $\hat{\eta}^0_\mathrm{g} = \mathbf{0}$ and $\hat{e}^0_\mathrm{g} = \mathbf{0}$ while $(6.1.14)_3$ gives $\hat{\mathbf{S}}^0_\mathrm{g} = \mathbf{0}$. Thus one obtains condition (i). Conditions (ii) and (iii) are implicit in formulas $(6.1.14)_{2-3}$ (since $\hat{\mathbf{S}}$ and $\hat{e}$ have to be constant with respect to $\mathbf{g}$) and (6.1.15). Since $(\mathbf{F}^0, \theta^0, \mathbf{g}^0, \boldsymbol{\alpha}^0)$ is an arbitrary state in $\mathscr{D}$, it follows that the first part of the theorem is now proved.

Conversely, let us define the constitutive equation (6.1.4) so that (6.1.8) holds and $\frac{1}{\theta}(\hat{e}_\mathrm{F} - \hat{\mathbf{S}})$, $\frac{1}{\theta}\hat{e}_\theta$, $\mathbf{0}$ are the components of the gradient with respect to $(\mathbf{F}, \theta, \mathbf{g})$ of a scalar function $\hat{\eta}: \mathscr{D} \to R$ whose partial derivative with respect to $\boldsymbol{\alpha}$ verifies the inequality (6.1.10). Then this function may be taken as a constitutive equation for the entropy and the CLAUSIUS-DUHEM inequality is satisfied.

We shall assume in the following that the constitutive equations (6.1.4) are such that conditions (i), (ii) and (iii) of the above theorem are satisfied. Then the free energy (1.2.18) is a function of state and it is constant with respect to $\mathbf{g}$, i.e.

$$\psi = \hat{e}(\mathbf{F}, \theta, \boldsymbol{\alpha}) - \theta\hat{\eta}(\mathbf{F}, \theta, \boldsymbol{\alpha}) = \hat{\psi}(\mathbf{F}, \theta, \boldsymbol{\alpha}) \text{ on } \mathscr{D} \tag{6.1.16}$$

and

$$\begin{aligned} \hat{\psi}_\mathrm{F}(\mathbf{F}, \theta, \boldsymbol{\alpha}) &= \hat{\mathbf{S}}(\mathbf{F}, \theta, \boldsymbol{\alpha}), \\ \hat{\psi}_\theta(\mathbf{F}, \theta, \boldsymbol{\alpha}) &= -\,\hat{\eta}(\mathbf{F}, \theta, \boldsymbol{\alpha}), \end{aligned} \tag{6.1.17}$$

according to (6.1.9). From (6.1.10) and (6.1.16), one has

$$\rho\theta\, \hat{\psi}_\alpha(\mathbf{F}, \theta, \boldsymbol{\alpha})\cdot \mathbf{f}(\mathbf{F}, \theta, \mathbf{g}, \boldsymbol{\alpha}) + \hat{\mathbf{q}}(\mathbf{F}, \theta, \mathbf{g}, \boldsymbol{\alpha})\cdot \mathbf{g} \leqslant 0. \tag{6.1.18}$$

Relations (6.1.17) and (6.1.18) have been obtained by COLEMAN and GURTIN [1967].

Let us observe that inequality (6.1.18) containes two types of dissipations: $\hat{\psi}_\alpha(\mathbf{F}, \theta, \boldsymbol{\alpha})\cdot \mathbf{f}(\mathbf{F}, \theta, \mathbf{g}, \boldsymbol{\alpha}) = \sigma$ called the *internal dissipation* and $(\rho\theta)^{-1}\, \hat{\mathbf{q}}(\mathbf{F}, \theta, \mathbf{g}, \boldsymbol{\alpha})\cdot \mathbf{g} = \omega$ called *dissipation by heat conduction*. Relation (6.1.18) implies that $\sigma + \omega \leqslant 0$ but we do not have , in general, $\sigma \leqslant 0$ and $\omega \leqslant 0$.

According to $(1.1.12)_1$ and (6.1.7), the symmetric Cauchy stress tensor is given by

$$\mathbf{T} = \rho\hat{\psi}_F(\mathbf{F}, \theta, \boldsymbol{\alpha})\, \mathbf{F}^T. \tag{6.1.19}$$

2. THERMODYNAMIC RESTRICTIONS ON THE DISSIPATIVE PART OF THE CONSTITUTIVE EQUATIONS

To simplify notation we set

$$\mathbf{M} = (\mathbf{F}, \theta) \in \mathscr{L}^+ \times R^+, \quad \mathbf{G} = (\mathbf{M}, \mathbf{g}) \in \mathscr{L}^+ \times R^+ \times R^3,$$
$$\mathbf{H} = (\mathbf{G}, \boldsymbol{\alpha}) \in \mathscr{L}^+ \times R^+ \times R^3 \times R^N. \tag{6.2.1}$$

A thermodynamic state $\mathbf{H}^* = (\mathbf{F}^*, \theta^*, \mathbf{0}, \boldsymbol{\alpha}^*)$ is called an *equilibrium state* if

$$\mathbf{f}^* = \mathbf{f}(\mathbf{H}^*) = \mathbf{0}, \tag{6.2.2}$$

and it is called a *strong equilibrium state* if, moreover,

$$\psi^*_{\boldsymbol{\alpha}} = \hat{\psi}_{\boldsymbol{\alpha}}(\mathbf{H}^*) = \mathbf{0} \tag{6.2.3}$$

(see Bowen [1968] and Truesdell [1969]).

In the following, the star as upper index to a function will denote the value of that function at an equilibrium state.

Bowen [1968] has noticed that the function Γ defined by

$$\Gamma(\mathbf{H}) = \hat{\psi}_{\boldsymbol{\alpha}}(\mathbf{M}, \boldsymbol{\alpha}) \cdot \mathbf{f}(\mathbf{H}) + (\rho\theta)^{-1}\, \hat{\mathbf{q}}(\mathbf{H}) \cdot \mathbf{g} \leqslant 0, \tag{6.2.4}$$

has a maximum for $\mathbf{H} = \mathbf{H}^*$; thus if $\mathbf{H}^*$ is an interior point of the domain of definition $\mathscr{D}$ of the functions $\hat{\psi}$, $\mathbf{f}$ and $\hat{\mathbf{q}}$ one may write

$$\Gamma^* = \Gamma(\mathbf{H}^*) = 0, \tag{6.2.5}$$

$$\Gamma^*_{\mathbf{H}} = \Gamma_{\mathbf{H}}(\mathbf{H}^*) = \mathbf{0}, \tag{6.2.6}$$

$$\Gamma^*_{\mathbf{H}^2}[\overline{\mathbf{H}}, \overline{\mathbf{H}}] = \Gamma_{\mathbf{H}^2}(\mathbf{H}^*)\, [\overline{\mathbf{H}}, \overline{\mathbf{H}}] \leqslant 0. \tag{6.2.7}$$

We shall assume in this section that $\hat{\psi}$ is of class C^3 and $\mathbf{f}$ and $\hat{\mathbf{q}}$ are of class C^2.

The inequality (6.2.7) must hold for any $\overline{\mathbf{H}} \in \mathscr{L} \times R \times R^3 \times R^N$ which means that the quadratic form $\Gamma^*_{H^2}[\overline{\mathbf{H}}, \overline{\mathbf{H}}]$ has to be negative semidefinite.

If one writes condition (6.2.6) for each component of $\mathbf{H}$, one gets

$$(\psi^*_{\boldsymbol{\alpha}} \cdot \mathbf{f}^*_{\boldsymbol{\alpha}}) = \mathbf{0}, \tag{6.2.8}$$

$$(\psi^*_{\boldsymbol{\alpha}} \cdot \mathbf{f}^*_{\mathbf{M}}) = \mathbf{0}, \tag{6.2.9}$$

$$(\psi^*_{\boldsymbol{\alpha}} \cdot \mathbf{f}^*_{\mathbf{g}}) + (\rho^*\, \theta^*)^{-1}\, \mathbf{q}^* = \mathbf{0}. \tag{6.2.10}$$

But $\mathbf{f}^*_{\boldsymbol{\alpha}}$ is a linear mapping from R^N into R^N. If its determinant is different from zero, then $\psi^*_{\boldsymbol{\alpha}} = \mathbf{0}$, i.e.

$$\det \mathbf{f}^*_{\boldsymbol{\alpha}} \neq 0, \tag{6.2.11}$$

implies relation (6.2.3).

Thus we have obtained the following result (BOWEN [1968]): If, at an equilibrium state $\mathbf{H} = \mathbf{H}^* = (\mathbf{F}^*, \theta^*, \mathbf{0}, \boldsymbol{\alpha}^*)$, the evolution function $\mathbf{f}$ of the internal state variables in $(6.1.4)_4$ is nondegenerate, then that equilibrium state is a strong equilibrium state (see also TRUESDELL [1969]) and (6.2.3) together with (6.2.10) imply

$$\mathbf{q}^* = \mathbf{0}, \tag{6.2.12}$$

i.e. the heat flux vanishes at strong equilibrium states.

Equality (6.2.3) (hence also (6.2.12)) holds even if (6.2.11) is no longer satisfied but the linear hull in R^N, of $\mathbf{f}^*_{\boldsymbol{\alpha}}(R^N)$ and $\mathbf{f}^*_{\mathbf{M}}(\mathscr{L} \times R)$ is equal to R^N. However, without additional assumptions, relations (6.2.8) and (6.2.9) do not imply that an equilibrium state is always a strong equilibrium state. Moreover, one cannot conclude from (6.2.8) and (6.2.9) that the heat flux $\mathbf{q}$ vanishes at an equilibrium state; this happens if and only if $\psi^*_{\boldsymbol{\alpha}} \cdot \mathbf{f}^*_{\mathbf{g}} = \mathbf{0}$.

The equality $\psi^*_{\boldsymbol{\alpha}} = 0$ has also been obtained by COLEMAN and GURTIN [1967] under the assumption that the equilibrium state is an assymptotically stable state. Here, the same result is obtained assuming only that the function $\mathbf{f}$ depends sufficiently effectively on $\mathbf{F}$, θ and $\boldsymbol{\alpha}$.

Let us assume that there exists an equilibrium state $\mathbf{H}^* = (\mathbf{M}^*, \mathbf{0}, \boldsymbol{\alpha}^*) \in \in \mathscr{D}$ and that (6.2.11) is satisfied at this point; then, according to the implicit function theorem, there exists a neighborhood $U^* \subset \mathscr{L}^+ \times R^+ \times R^N$ of $(\mathbf{M}^*, \alpha^*)$ and a continuous and differentiable function $\boldsymbol{\alpha} = \mathbf{h}(\mathbf{M})$ such that

$$\mathbf{f}(\mathbf{M}, \mathbf{0}, \mathbf{h}(\mathbf{M})) = \mathbf{0}, \quad \boldsymbol{\alpha}^* = \mathbf{h}(\mathbf{M}^*), \tag{6.2.13}$$

and moreover the neighborhood U^* is such that condition (6.2.11) is satisfied at every point of U^*. All the points $(\mathbf{M}, \boldsymbol{\alpha} = \mathbf{h}(\mathbf{M}))$ of U^* are therefore strong equilibrium points (i.e. such that (6.2.3) holds). By means of the function $\mathbf{h}$ one may define the *equilibrium free energy* $\tilde{\psi}(\mathbf{M})$, as well as the *equilibrium stress* and the *equilibrium entropy*, as follows:

$$\tilde{\psi}(\mathbf{M}) = \hat{\psi}(\mathbf{M}, \mathbf{h}(\mathbf{M})), \ \tilde{\mathbf{S}}(\mathbf{M}) = \hat{\mathbf{S}}(\mathbf{M}, \mathbf{h}(\mathbf{M})), \\ \tilde{\eta}(\mathbf{M}) = \hat{\eta}(\mathbf{M}, \mathbf{h}(\mathbf{M})). \tag{6.2.14}$$

The function $\tilde{\psi}$ determines both functions $\tilde{\mathbf{S}}(\mathbf{M})$ and $\tilde{\eta}(\mathbf{M})$ (by equations that are similar to (6.1.17)),

$$\tilde{\mathbf{S}}(\mathbf{M}) = \tilde{\psi}_{\mathbf{F}}(\mathbf{M}), \quad \tilde{\eta}(\mathbf{M}) = -\tilde{\psi}_{\theta}(\mathbf{M}), \tag{6.2.15}$$

since

$$\tilde{\psi}_{\mathbf{M}}(\mathbf{M}) = \hat{\psi}_{\mathbf{M}}(\mathbf{M}, \mathbf{h}(\mathbf{M})) + \hat{\psi}_{\boldsymbol{\alpha}}(\mathbf{M}, \mathbf{h}(\mathbf{M}))\, \mathbf{h}_{\mathbf{M}}(\mathbf{M})$$

and, due to (6.2.11),

$$\hat{\psi}_{\boldsymbol{\alpha}}(\mathbf{M}, \mathbf{h}(\mathbf{M})) = \mathbf{0}.$$

Now, let $(\mathbf{M}, \boldsymbol{\alpha}_0)$ be a point in U^* and let $\boldsymbol{\alpha}_0 \neq \mathbf{h}(\mathbf{M})$; then the existence of the limit

$$\lim_{t\to\infty} \tilde{\boldsymbol{\alpha}}(t, \mathbf{M}, \mathbf{0}, \boldsymbol{\alpha}_0) = \boldsymbol{\alpha}_1 \text{ with } (\mathbf{M}, \boldsymbol{\alpha}_1) \in U^*, \tag{6.2.16}$$

is equivalent to the fact that $(\mathbf{M}, \mathbf{0}, \boldsymbol{\alpha}_1)$ is an assymptotically stable state at constant strain and temperature, in the sense of COLEMAN and GURTIN [1967]. Here $\tilde{\boldsymbol{\alpha}}(t, \mathbf{M}, \mathbf{0}, \boldsymbol{\alpha}_0)$ is the solution of the initial value problem $\dot{\boldsymbol{\alpha}} = \mathbf{f}(\mathbf{M}, \mathbf{0}, \boldsymbol{\alpha})$ with $\boldsymbol{\alpha}(0) = \boldsymbol{\alpha}_0, \mathbf{M} = \text{const.}$, as it was discussed in the previous section. The function $\mathbf{h}$ being unique, it follows that $\boldsymbol{\alpha}_1 = \mathbf{h}(\mathbf{M})$.

Let us go back to inequality (6.2.7). Using the definition (6.2.4) of the function Γ, we can write this inequality in terms of $\hat{\psi}$, $\mathbf{f}$ and $\hat{\mathbf{q}}$, at the equilibrium point $\mathbf{H}^* = (\mathbf{F}^*, \theta^*, \mathbf{0}, \boldsymbol{\alpha}^*)$, as

$$\begin{aligned} &2(\psi^*_{\boldsymbol{\alpha}\mathbf{M}}[\overline{\mathbf{M}}] + \psi^*_{\boldsymbol{\alpha}^2}[\overline{\boldsymbol{\alpha}}])\ \mathbf{f}^*_{\mathbf{H}}[\overline{\mathbf{H}}] + \\ &+ \psi^*_{\boldsymbol{\alpha}} \cdot (\mathbf{f}^*_{\boldsymbol{\alpha}^2}[\overline{\boldsymbol{\alpha}}, \overline{\boldsymbol{\alpha}}] + 2\mathbf{f}^*_{\boldsymbol{\alpha}\mathbf{G}}\,[\overline{\boldsymbol{\alpha}}, \overline{\mathbf{G}}] + \mathbf{f}^*_{\mathbf{G}^2}\,[\overline{\mathbf{G}}, \overline{\mathbf{G}}]) + \\ &+ 2\{(\rho^*\,\theta^*)^{-1}\,(\mathbf{q}^*_{\mathbf{G}}[\overline{\mathbf{G}}] + q^*_{\boldsymbol{\alpha}}\,[\overline{\boldsymbol{\alpha}}] + ((\rho\theta)^{-1})^*_{\mathbf{M}}\,[\overline{\mathbf{M}}]\,q^*\}\cdot\overline{\mathbf{g}} \leqslant 0 \end{aligned} \tag{6.2.17}$$

which has to hold for all $\overline{\mathbf{H}} = (\overline{\mathbf{F}}, \overline{\theta}, \overline{\mathbf{g}}, \overline{\boldsymbol{\alpha}}) \in \mathscr{L} \times R \times R^3 \times R^N$.

Now, if one takes $\overline{\mathbf{H}}$ to be consecutively equal to $(\overline{\mathbf{F}}, 0, \mathbf{0}, \mathbf{0})$, $(\mathbf{0}, \overline{\theta}, \mathbf{0}, \mathbf{0})$, $(\mathbf{0}, 0, \overline{\mathbf{g}}, \mathbf{0})$ and $(\mathbf{0}, 0, \mathbf{0}, \overline{\boldsymbol{\alpha}})$, one obtains from this inequality

$$2(\mathbf{S}^*_{\boldsymbol{\alpha}}[\overline{\mathbf{F}}]) \cdot \mathbf{f}^*_{\mathbf{F}}[\overline{\mathbf{F}}] + \psi^*_{\boldsymbol{\alpha}} \cdot \mathbf{f}^*_{\mathbf{F}^2}\,[\overline{\mathbf{F}}, \overline{\mathbf{F}}] \leqslant 0, \tag{6.2.18}$$

$$-\,2\eta^*_{\boldsymbol{\alpha}} \cdot \mathbf{f}^*_{\theta} + \psi^*_{\boldsymbol{\alpha}}\,\mathbf{f}^*_{\theta^2} \leqslant 0, \tag{6.2.19}$$

$$\psi^*_{\boldsymbol{\alpha}} \cdot \mathbf{f}^*_{\mathbf{g}^2}\,[\overline{\mathbf{g}}, \overline{\mathbf{g}}] + 2(\rho^*\,\theta^*)^{-1}\,(q^*_{\mathbf{g}}[\overline{\mathbf{g}}]) \cdot \overline{\mathbf{g}} \leqslant 0, \tag{6.2.20}$$

$$2(\psi^*_{\boldsymbol{\alpha}^2}[\overline{\boldsymbol{\alpha}}]) \cdot \mathbf{f}^*_{\boldsymbol{\alpha}}[\overline{\boldsymbol{\alpha}}] + \psi^*_{\boldsymbol{\alpha}} \cdot \mathbf{f}^*_{\boldsymbol{\alpha}^2}[\overline{\boldsymbol{\alpha}}, \overline{\boldsymbol{\alpha}}] \leqslant 0. \tag{6.2.21}$$

All these inequalities must hold for any $\overline{\mathbf{F}} \in \mathscr{L}$, $\overline{\mathbf{g}} \in R^3$ and $\overline{\boldsymbol{\alpha}} \in R^N$. In deriving (6.2.18) and (6.2.19) we have made use of (6.1.17). To put it more explicitly, we present below, in detail, inequality (6.2.18):

$$\begin{aligned} &\sum_{i=1}^{N} \left\{ \sum_{k,l=1}^{3} (S^{kl}_{\alpha^i})^*\, \overline{F}^{kl} \right\} \left\{ \sum_{p,r=1}^{3} (f^i_{F^{pr}})^*\, \overline{F}^{pr} \right\} + \\ &+ \sum_{i=1}^{N} \psi^*_{\alpha^i} \left\{ \sum_{k,l\,p,r=1}^{3} (f^i_{F^{kl}F^{pr}})^*\, \overline{F}^{kl}\, \overline{F}^{pr} \right\} \leqslant 0. \end{aligned} \tag{6.2.18'}$$

Inequalities (6.2.18)—(6.2.21) have been derived for the first time by BOWEN [1968] under the more general constitutive assumption that the free energy, the stress, the entropy, the heat flux and the evolution function also depend of $\dot{\mathbf{F}}$, the time derivative of the deformation gradient.

If one assumes that the equilibrium state $\mathbf{H}^*$ is a strong equilibrium state, then the terms containing ψ^*_{α} vanish while inequality (6.2.20) reduces to

$$\mathbf{q}^*_{\mathbf{g}}[\mathbf{g}]\cdot\mathbf{g} \leqslant 0; \tag{6.2.22}$$

thus the heat conduction tensor $\mathbf{k}^* = -\mathbf{q}^*_{\mathbf{g}}$ is positive semidefinite at a strong equilibrium state. If the quadratic form in the left side of (6.2.22) is not identically zero, then this conclusion together with (6.2.12) leads to

$$\hat{\mathbf{q}}(\mathbf{F}^*, \theta^*, \mathbf{g}, \boldsymbol{\alpha}^*) = -\mathbf{k}^*\mathbf{g} + \mathbf{0}(|\mathbf{g}|) \tag{6.2.23}$$

in the neighborhood of a strong equilibrium state, i.e. the FOURIER law is true with an error of the order $\mathbf{0}(|\mathbf{g}|)$ (see COLEMAN and GURTIN [1967]).

The inequalities (6.2.18), (6.2.19) and (6.2.21) have important consequences in the theory of wave propagation through materials whose behaviour may be described by means of internal state variables (see, for instance, COLEMAN and GURTIN [1967 a], CHEN and GURTIN [1971], BOWEN [1969], SULICIU [1975 *a*, *b*], MIHĂILESCU and SULICIU [1976]).

3. THE PRINCIPLE OF FRAME INDIFFERENCE. THE ISOTROPY GROUP OF THE MATERIAL

3a. The principle of frame indifference requires an admissible thermodynamic process to remain admissible after a change of frame defined by (see NOLL [1958], TRUESDELL and NOLL [1965], COLEMAN and GURTIN [1967], BOWEN [1968], MISICU [1967], Section 2.3)

$$\bar{\mathbf{x}} = \mathbf{c} + \mathbf{Q}(\mathbf{x} - \mathbf{0}), \tag{6.3.1}$$

where $\mathbf{0}$ is the origin of the Euclidean space $\mathscr{E}$ and $\mathbf{c}$ is the position vector of the new origin which is, in general, time-dependent; and $\mathbf{Q}$ is a time-dependent linear orthogonal mapping ($\mathbf{QQ}^T = \mathbf{Q}^T\mathbf{Q} = \mathbf{1}$ where $\mathbf{1}$ is the identity map).

The scalar quantities θ, η, e and ψ do not change under this mapping while $\mathbf{F}$, $\mathbf{g}$, $\mathbf{q}$ and $\mathbf{T}$ change as follows:

$$\begin{gathered} \mathbf{F} \to \tilde{\mathbf{F}} = \mathbf{QF},\ \mathbf{g} \to \tilde{\mathbf{g}} = \mathbf{Qg}, \\ \mathbf{q} \to \tilde{\mathbf{q}} = \mathbf{Qq},\ \mathbf{T} \to \tilde{\mathbf{T}} = \mathbf{QTQ}^T. \end{gathered} \tag{6.3.2}$$

If $\boldsymbol{\alpha}$ is a row $\boldsymbol{\alpha} = (\alpha^1, \ldots, \alpha^N)$ of N scalars α^i (see for instance COLEMAN and GURTIN [1967], BOWEN [1968]) then

$$\boldsymbol{\alpha} \to \tilde{\boldsymbol{\alpha}} = \boldsymbol{\alpha}. \tag{6.3.3}$$

If $\boldsymbol{\alpha}^i$ are considered as second order tensors in the Euclidean space $\mathscr{E}$ (see for instance KRATOCHVIL [1973]) then

$$\boldsymbol{\alpha}^i \to \tilde{\boldsymbol{\alpha}}^i = \mathbf{Q}\boldsymbol{\alpha}^i\mathbf{Q}^T, \; i = 1, \ldots, N. \tag{6.3.4}$$

Some authors (see for instance PERZYNA [1973]) considered the internal state variables as consisting of two groups of quantities, some of the form (6.3.3) and others of the form (6.3.4).

In the case when $\boldsymbol{\alpha}$ behave as scalars, the principle of frame indifference is satisfied if and only if the functions $\hat{\psi}, \hat{\eta}, \hat{\mathbf{T}}, \hat{\mathbf{q}}$ and $\mathbf{f}$ verify the identities

$$\begin{aligned}
&\hat{\psi}(\mathbf{F}, \theta, \boldsymbol{\alpha}) = \hat{\psi}(\mathbf{QF}, \theta, \boldsymbol{\alpha}),\\
&\hat{\eta}(\mathbf{F}, \theta, \boldsymbol{\alpha}) = \hat{\eta}(\mathbf{QF}, \theta, \boldsymbol{\alpha}),\\
&\mathbf{Q}\hat{\mathbf{T}}(\mathbf{F}, \theta, \boldsymbol{\alpha})\,\mathbf{Q}^T = \hat{\mathbf{T}}(\mathbf{QF}, \theta, \boldsymbol{\alpha}),\\
&\mathbf{Q}\hat{\mathbf{q}}(\mathbf{F}, \theta, \mathbf{g}, \boldsymbol{\alpha}) = \hat{\mathbf{q}}(\mathbf{QF}, \theta, \mathbf{Qg}, \boldsymbol{\alpha}),\\
&\mathbf{f}(\mathbf{F}, \theta, \mathbf{g}, \boldsymbol{\alpha}) = \mathbf{f}(\mathbf{QF}, \theta, \mathbf{Qg}, \boldsymbol{\alpha})
\end{aligned} \tag{6.3.5}$$

at any point $(\mathbf{F}, \theta, \mathbf{g}, \boldsymbol{\alpha}) \in \mathscr{D}$ and for any time-dependent orthogonal mapping $\mathbf{Q}$.

Among the internal state variables $\boldsymbol{\alpha}$ there must be no quantities that change under $\mathbf{Q}$ in the same way as the vectors of the Euclidean space $\mathscr{E}$ (see COLEMAN and GURTIN[1967]). Indeed, if the internal state variables $\boldsymbol{\alpha}$ contain the vectors $\mathbf{a}^i$ and the second order tensors $\mathbf{A}^j$ which, under a change of frame, change according to the rules $\mathbf{a}^i \to \mathbf{Qa}^i$, $\mathbf{A}^j \to \mathbf{QA}^j\mathbf{Q}^T$, then one must have

$$\dot{\overline{(\mathbf{Qa}^i)}} = \mathbf{f}\mathbf{a}^i\,(\mathbf{QF}, \theta, \mathbf{Qg}, \boldsymbol{\beta}, \mathbf{Qa}^i, \mathbf{QA}^j\,\mathbf{Q}^T).$$

Now, one may choose a $\mathbf{Q}(t)$ such that $\mathbf{Q}(\tau) = \mathbf{1}$ for $t = \tau$ and $\dot{\mathbf{Q}}(\tau) = \mathbf{W}$, where $\mathbf{W}$ is an arbitrary skew-symmetric tensor. Writing the above relation for this $\mathbf{Q}(t)$ at $t = \tau$, we get $\mathbf{Wa}^i = \mathbf{0}$, whence $\mathbf{a}^i = \mathbf{0}$, and therefore $\mathbf{f}\mathbf{a}^i(\mathbf{F}, \theta, \mathbf{g}, \boldsymbol{\beta}, \mathbf{0}, \mathbf{A}^j) = \mathbf{0}$.

The case when the internal state variables also contain quantities that behave under a change of frame as second order tensors presents no difficulties and therefore we shall not consider it here.

Let us consider again the identities (6.3.5). The polar decomposition theorem applied to **F** (see for instance TRUESDELL and NOLL [1965], Section 23) determines uniquely the orthogonal mapping **R** and the symmetric and positive definite mappings **U** and **V**, such that

$$\mathbf{F} = \mathbf{R}\mathbf{U} = \mathbf{V}\mathbf{R}. \tag{6.3.6}$$

One has

$$\mathbf{U}^2 = \mathbf{F}^T\,\mathbf{F} = \mathbf{C},\ \mathbf{V}^2 = \mathbf{F}\mathbf{F}^T = \mathbf{B}. \tag{6.3.7}$$

The tensors **C** and **B**, defined by (6.3.7), are called the *right and left* CAUCHY-GREEN *strain tensors*, respectively. Choosing $\mathbf{Q} = \mathbf{R}^T$ in (6.3.5) and taking (6.3.6) and (6.3.7) into account, we get

$$\begin{aligned}
&\psi = \hat{\psi}(\mathbf{U}, \theta, \boldsymbol{\alpha}) = \check{\psi}(\mathbf{C}, \theta, \boldsymbol{\alpha}),\\
&\eta = \hat{\eta}(\mathbf{U}, \theta, \boldsymbol{\alpha}) = \check{\eta}(\mathbf{C}, \theta, \boldsymbol{\alpha}),\\
&\mathbf{R}^T\,\mathbf{T}\mathbf{R} = \hat{\mathbf{T}}(\mathbf{U}, \theta, \boldsymbol{\alpha}) = \mathbf{U}\check{\mathbf{T}}(\mathbf{C}, \theta, \boldsymbol{\alpha})\,\mathbf{U},\\
&\mathbf{R}^T\,\mathbf{q} = \hat{\mathbf{q}}(\mathbf{U}, \theta, \mathbf{U}^{-1}\,\mathbf{g}_0, \boldsymbol{\alpha}) = \mathbf{U}\check{\mathbf{q}}(\mathbf{C}, \theta, \mathbf{g}_0, \boldsymbol{\alpha}),\\
&\dot{\boldsymbol{\alpha}} = \mathbf{f}(\mathbf{U}, \theta, \mathbf{U}^{-1}\,\mathbf{g}_0, \boldsymbol{\alpha}) = \check{\mathbf{f}}(\mathbf{C}, \theta, \mathbf{g}_0, \boldsymbol{\alpha}),
\end{aligned} \tag{6.3.8}$$

where $\mathbf{g}_0 = \mathbf{F}^T\mathbf{g}$.

Relations $(6.3.8)_{1,2}$ assert that the free energy and the entropy are dependent on **F** by means of the non-singular symmetric tensor **C** only and they are not dependent on rotations. Relations $(6.3.8)_5$ state that the variation of internal state parameters during a deformation process is independent of the local rotations of the material particles; finally, from the relations $(6.3.8)_{3\cdot 4}$ one gets

$$\begin{aligned}
\mathbf{T} &= \mathbf{R}\hat{\mathbf{T}}(\mathbf{U}, \theta, \boldsymbol{\alpha})\,\mathbf{R}^T = \mathbf{F}\check{\mathbf{T}}(\mathbf{C}, \theta, \boldsymbol{\alpha})\,\mathbf{F}^T,\\
\mathbf{q} &= \mathbf{R}\hat{\mathbf{q}}(\mathbf{U}, \theta, \mathbf{U}^{-1}\,\mathbf{g}_0, \boldsymbol{\alpha}) = \mathbf{F}\hat{\mathbf{q}}(\mathbf{C}, \theta, \mathbf{g}_0, \boldsymbol{\alpha}),
\end{aligned} \tag{6.3.9}$$

i.e. the stress tensor **T** and the heat flux **q** depend only on the actual rotation of the material particle and this dependence is described by (6.3.9).

Conversely, if the free energy ψ, the entropy η, the stress tensor **T**, the heat flux **q** and the evolution function are determined by the functions $\check{\psi}$, $\check{\eta}$, $\check{\mathbf{T}}$, $\check{\mathbf{q}}$ and $\check{\mathbf{f}}$ by means of the relations (6.3.8), then the principle of frame indifference will be satisfied.

All the results of Section 1 and 2 are valid for the functions $\check{\psi}$, $\check{\mathbf{T}}$, $\check{\eta}$, $\check{\mathbf{q}}$ and $\check{\mathbf{f}}$ instead of $\hat{\psi}$, $\hat{\mathbf{T}}$, $\hat{\eta}$, $\hat{\mathbf{q}}$ and **f**. Thus, relations (6.1.17) may be written as

$$\check{\eta} = -\,\check{\psi}_\theta,\ \check{\mathbf{T}} = 2\rho\,\check{\psi}_{\mathbf{C}}. \tag{6.3.10}$$

Relations (6.2.8)–(6.2.10), inequality (6.2.17) and its consequences (6.2.18)–(6.2.21) may be obtained either by starting from (6.2.4) where $\hat{\psi}$, **f** and $\hat{\mathbf{q}}$ are replaced by $\check{\psi}$, $\check{\mathbf{f}}$ and $\check{\mathbf{q}}$ respectively, and performing again all the computations, or by directly interchanging the functions and differentiating now with respect to the new arguments **C** and $\mathbf{g}_0$.

3b. The isotropy group $G_{\mathscr{R}}$ of a material with respect to a reference configuration $\mathscr{R}$ consists of all those linear unimodular mappings **K** (i.e. with $|\det \mathbf{K}| = 1$) of the reference configuration $\mathscr{R}$ which leave the response functions of the material unchanged (see NOLL [1958] and also TRUESDELL and NOLL [1965], GURTIN and WILLIAMS [1966]).

One faces again the problem of the variation of internal state variables under changes of the reference configuration; the answer to this problem depends on the physical significance one gives to internal state variables. Following COLEMAN and GURTIN [1967] and BOWEN [1968] we shall assume here that the internal state variables are quantities, so chosen as not to depend on unimodular mappings of the reference configuration (for other choices see Chapter VII). The requirement of unimodularity for the mappings of the reference configuration comes from the condition that any mapping of the isotropy group preserves mass density.

For the constitutive equations postulated in this chapter the isotropy group $G_{\mathscr{R}}$ with respect to the reference configuration $\mathscr{R}$ of the body $\mathscr{B}$ consists of all linear unimodular mappings **K** that satisfy the identities

$$\begin{aligned}
\hat{\psi}(\mathbf{F}, \theta, \boldsymbol{\alpha}) &= \hat{\psi}(\mathbf{FK}, \theta, \boldsymbol{\alpha}),\\
\hat{\eta}(\mathbf{F}, \theta, \boldsymbol{\alpha}) &= \hat{\eta}(\mathbf{FK}, \theta, \boldsymbol{\alpha}),\\
\hat{\mathbf{T}}(\mathbf{F}, \theta, \boldsymbol{\alpha}) &= \hat{\mathbf{T}}(\mathbf{FK}, \theta, \boldsymbol{\alpha}),\\
\hat{\mathbf{q}}(\mathbf{F}, \theta, \mathbf{g}, \boldsymbol{\alpha}) &= \hat{\mathbf{q}}(\mathbf{FK}, \theta, \mathbf{g}, \boldsymbol{\alpha}),\\
\mathbf{f}(\mathbf{F}, \theta, \mathbf{g}, \boldsymbol{\alpha}) &= \mathbf{f}(\mathbf{FK}, \theta, \mathbf{g}, \boldsymbol{\alpha})
\end{aligned} \tag{6.3.11}$$

for any $\mathbf{F}, \theta, \mathbf{g}, \boldsymbol{\alpha}$ in their respective domain of definition.

If the isotropy group $G_{\mathscr{R}}$ contains the mapping $-\mathbf{1}$ and if one postulates the material frame indifference to the complete orthogonal group, then as COLEMAN and NOLL [1963] have noticed (see also COLEMAN and GURTIN [1967]),

$$\hat{\mathbf{q}}(\mathbf{F}, \theta, \mathbf{0}, \boldsymbol{\alpha}) = \mathbf{0}, \tag{6.3.12}$$

independently whether the state $(\mathbf{F}, \theta, \mathbf{0}, \boldsymbol{\alpha})$ is a strong equilibrium state or not. In order to prove this statement we write, using $(6.3.5)_4$ and $(6.3.11)_4$,

$$\mathbf{Q}\hat{\mathbf{q}}(\mathbf{F}, \theta, \mathbf{g}, \boldsymbol{\alpha}) = \hat{\mathbf{q}}(\mathbf{QF}, \theta, \mathbf{Qg}, \boldsymbol{\alpha}) = \hat{\mathbf{q}}(\mathbf{QFK}, \theta, \mathbf{Qg}, \boldsymbol{\alpha}). \tag{6.3.13}$$

Now, by taking $\mathbf{Q} = -\mathbf{1}$ and $\mathbf{K} = -\mathbf{1}$, we get

$$-\hat{\mathbf{q}}(\mathbf{F}, \theta, \mathbf{g}, \boldsymbol{\alpha}) = \hat{\mathbf{q}}(\mathbf{F}, \theta, -\mathbf{g}, \boldsymbol{\alpha}) \tag{6.3.14}$$

which leads to (6.3.12).

The result (6.3.12) obviously holds provided there is satisfied the additional condition, already stressed before, that $\boldsymbol{\alpha}$ be chosen so as to remain invariant under the isotropy group and changes of frame.

The condition that a linear unimodular mapping $\mathbf{K}$ belongs to the isotropy group may be also easily expressed in terms of the functions $\check{\psi}$, $\check{\eta}$, $\check{\mathbf{T}}$, $\check{\mathbf{q}}$ and $\check{\mathbf{f}}$.

A material is said to be *solid* if it has a reference configuration $\mathscr{R}_0$ such that its isotropy group with respect to this reference configuration is contained in the orthogonal group. A material is said to be *fluid* if its isotropy group is the whole unimodular group. A material is called *isotropic* if there exists a reference configuration $\mathscr{R}_0$ such that its isotropy group with respect to it contains the orthogonal group.

For an isotropic solid there exists a reference configuration $\mathscr{R}$ such that its isotropy group with respect to it is equal to the orthogonal group. Then, taking $\mathbf{K} = \mathbf{R}^T$ and using (6.3.6) and (6.3.7), we obtain the following reduced forms of the constitutive equations (6.3.11) for isotropic materials:

$$\begin{aligned}
\psi &= \bar{\psi}(\mathbf{B}, \theta, \boldsymbol{\alpha}) = \hat{\psi}(\mathbf{V}, \theta, \boldsymbol{\alpha}),\\
\eta &= \bar{\eta}(\mathbf{B}, \theta, \boldsymbol{\alpha}) = \hat{\eta}(\mathbf{V}, \theta, \boldsymbol{\alpha}),\\
\mathbf{T} &= \bar{\mathbf{T}}(\mathbf{B}, \theta, \boldsymbol{\alpha}) = \hat{\mathbf{T}}(\mathbf{V}, \theta, \boldsymbol{\alpha}),\\
\mathbf{q} &= \bar{\mathbf{q}}(\mathbf{B}, \theta, \mathbf{g}, \boldsymbol{\alpha}) = \hat{\mathbf{q}}(\mathbf{V}, \theta, \mathbf{g}, \boldsymbol{\alpha}),\\
\dot{\boldsymbol{\alpha}} &= \bar{\mathbf{f}}(\mathbf{B}, \theta, \mathbf{g}, \boldsymbol{\alpha}) = \mathbf{f}(\mathbf{V}, \theta, \mathbf{g}, \boldsymbol{\alpha}).
\end{aligned} \tag{6.3.15}$$

If $\mathscr{B}$ is a fluid body, then the functions $\bar{\psi}$, $\bar{\eta}$, $\bar{\mathbf{T}}$, $\bar{\mathbf{q}}$ and $\bar{\mathbf{f}}$ contain $1/\rho$ instead of $\mathbf{B}$, where ρ is the actual mass density. This is so since in this case $\mathbf{K}$ may be taken as an arbitrary mapping in the unimodular group; one may therefore choose $K = \sqrt[3]{\det \mathbf{F}}\, \mathbf{F}^{-1}$. The balance of mass yields $\det \mathbf{F} = \dfrac{\rho_0}{\rho}$, where ρ_0 is the mass density in the reference configuration; hence the above stated property follows if we take $\mathbf{K} = \sqrt[3]{(\rho_0/\rho)}\, \mathbf{F}^{-1}$ in (6.3.11).

4. CONSEQUENCES OF THE RESTRICTIONS IMPOSED ON THE SPECIFIC FREE ENERGY

4a. As we have already seen in the first two sections of this chapter, the second law of thermodynamics (here used in the form of the CLAUSIUS-DUHEM inequality) imposes rather strong restrictions on the constitutive equations. However, these restrictions still allow the constitutive equations to remain too general, and consequently, it is very difficult, if not impossible, to use them in applications. Moreover, this high degree of generality may introduce into the constitutive equations some effects that are not possible for real materials.

This and the subsequent sections are devoted to discussing the additional restrictions imposed upon the constitutive equations by experimental facts of a general character. These restrictions will specify more explicitly the respective model, which is in this case the model of a body with internal state variables which can describe thermo-elastic/viscoplastic effects, yet they will remain sufficiently general to describe the complex behaviour of a real body.

In order to justify the form of the restrictions imposed on the free energy, let us recall shortly certain well known results concerning linear elasticity (see for instance SOLOMON [1969], Chapter III, Section 3, TRUESDELL and NOLL [1965], Section 51).

In the theory of linear and isotropic elasticity, one always accepts the following two groups of inequalities:

$$\mu > 0, \; 3\lambda + 2\mu > 0 \tag{6.4.1}$$

and

$$\mu > 0, \; \lambda + 2\mu > 0. \tag{6.4.2}$$

λ and μ are called LAMÉ's *constants*. The condition $\mu > 0$ is necessary and sufficient for the shear stress and the shear strain to have the same sign. The condition $(6.4.1)_2$ is necessary and sufficient for the mean stress to have the same sign as the variation of the volume. Also, (6.4.1) are necessary and sufficient conditions for the mechanical work in a non-rigid motion to be positive and they are sufficient, but not necessary conditions for stability. The weaker conditions (6.4.2) are necessary and sufficient for stability as well as necessary and sufficient for the velocities of the acceleration waves to be real. (For other discussions see TRUESDELL and NOLL [1965], Section 51).

For non-linear elasticity the discussion is much more complicated and as yet incomplete, even for the isotropic case and without considering thermal effects. The reader interested in such problems may find detailed discussions on the subject in TRUESDELL and NOLL [1965], Sections 51—53, 87, etc. and also in HAYES [1969].

4b. BOWEN [1968], following the ideas of COLEMAN and NOLL [1959] and GIBBS [1928] proposed for materials with internal state variables the following inequality for the specific free energy

$$\hat{\psi}(\mathbf{M}, \boldsymbol{\alpha}) - \hat{\psi}(\tilde{\mathbf{M}}, \tilde{\boldsymbol{\alpha}}) - (\theta - \tilde{\theta})\, \hat{\psi}_\theta(\mathbf{M}, \boldsymbol{\alpha}) - \hat{\psi}_{\mathbf{F}}(\tilde{\mathbf{M}}, \tilde{\boldsymbol{\alpha}})\,(\mathbf{F} - \tilde{\mathbf{F}}) - \tag{6.4.3}$$
$$- \hat{\psi}_{\boldsymbol{\alpha}}(\tilde{\mathbf{M}}, \tilde{\boldsymbol{\alpha}})\,(\boldsymbol{\alpha} - \tilde{\boldsymbol{\alpha}}) > 0,$$

for all $(\mathbf{M}, \boldsymbol{\alpha}), (\tilde{\mathbf{M}}, \tilde{\boldsymbol{\alpha}})$ in the domain of definition of $\hat{\psi}$ such that $(\mathbf{M}, \boldsymbol{\alpha}) \neq \neq (\tilde{\mathbf{M}}, \tilde{\boldsymbol{\alpha}})$ and $\mathbf{F}\tilde{\mathbf{F}}^{-1} = \mathbf{D} \in \mathscr{S}^+ . \mathscr{S}^+$ is the set of all positive definite linear and symmetric mappings while $\mathscr{S}$ is the set of all linear and symmetric mappings.

Obviously, this inequality must be compatible with the second law of thermodynamics and the principle of frame indifference.

Inequality (6.4.3) has important consequences. Thus, if $(\tilde{\mathbf{M}}, \tilde{\boldsymbol{\alpha}})$ is fixed, the left side of (6.4.3) has a minimum, as a function of $(\mathbf{M}, \boldsymbol{\alpha}) = =(\mathbf{D}\tilde{\mathbf{F}}, \theta, \boldsymbol{\alpha})$, for $\mathbf{D} = \mathbf{1}$ ($\mathbf{1}$ — the identity map), $\theta = \tilde{\theta}$ and $\boldsymbol{\alpha} = \tilde{\boldsymbol{\alpha}}$.

Let $\bar{\mathbf{D}} \in \mathscr{S}$, $\bar{\theta} \in R$ and $\tilde{\boldsymbol{\alpha}} \in R^N$; then, for a sufficiently small $\lambda > 0$, one has $\mathbf{1} + \lambda\bar{\mathbf{D}} \in \mathscr{S}^+$, $\tilde{\theta} + \lambda\bar{\theta} \in R^+$ and $(\mathbf{F} = (\mathbf{1} + \lambda\bar{\mathbf{D}})\,\tilde{\mathbf{F}},\ \theta = \tilde{\theta} + \lambda\bar{\theta}, \boldsymbol{\alpha} = \tilde{\boldsymbol{\alpha}} + \lambda\bar{\boldsymbol{\alpha}})$ is a point in the domain of definition of $\hat{\psi}$. Using the left side of (6.4.3) we introduce a function δ of λ, for fixed $\bar{\mathbf{D}}$, $\bar{\theta}$ and $\bar{\boldsymbol{\alpha}}$, as follows:

$$\delta(\lambda) = \hat{\psi}((\mathbf{1} + \lambda\bar{\mathbf{D}})\,\tilde{\mathbf{F}}, \tilde{\theta} + \lambda\bar{\theta}, \tilde{\boldsymbol{\alpha}} + \lambda\bar{\boldsymbol{\alpha}}) - \hat{\psi}(\tilde{\mathbf{M}}, \tilde{\boldsymbol{\alpha}}) -$$
$$-\lambda\bar{\theta}\, \hat{\psi}_\theta((\mathbf{1} + \lambda\bar{\mathbf{D}})\,\tilde{\mathbf{F}}, \tilde{\theta} + \lambda\bar{\theta},\ \tilde{\boldsymbol{\alpha}} + \lambda\bar{\boldsymbol{\alpha}}) - \lambda\hat{\psi}_{\mathbf{F}}(\tilde{\mathbf{M}}, \tilde{\boldsymbol{\alpha}})\,(\bar{\mathbf{D}}\tilde{\mathbf{F}}) - \tag{6.4.4}$$
$$- \lambda\hat{\psi}_{\boldsymbol{\alpha}}(\tilde{\mathbf{M}}, \tilde{\boldsymbol{\alpha}})\,(\bar{\boldsymbol{\alpha}}).$$

This function has the properties

$$\delta(0) = 0,\ \delta'(0) = 0,\ \delta''(0) \geqslant 0. \tag{6.4.5}$$

$(6.4.5)_{1,2}$ are satisfied identically for all $\bar{\mathbf{D}}, \bar{\theta}$ and $\bar{\boldsymbol{\alpha}}$, while $(6.4.5)_3$ is equivalent to

$$\hat{\psi}_{\mathbf{F}^2}(\mathbf{M}, \boldsymbol{\alpha})\,[\bar{\mathbf{D}}\mathbf{F}, \bar{\mathbf{D}}\mathbf{F}] + 2\hat{\psi}_{\mathbf{F}\boldsymbol{\alpha}}(\mathbf{M}, \boldsymbol{\alpha})\,[\bar{\mathbf{D}}\mathbf{F}, \bar{\boldsymbol{\alpha}}] - \bar{\theta}^2\, \hat{\psi}_{\theta^2}(\mathbf{M}, \boldsymbol{\alpha}) + \tag{6.4.6}$$
$$+ \hat{\psi}_{\boldsymbol{\alpha}^2}(\mathbf{M}, \boldsymbol{\alpha})\,[\bar{\boldsymbol{\alpha}}, \bar{\boldsymbol{\alpha}}] \geqslant 0.$$

The above inequality has to be satisfied for all $\overline{\mathbf{D}} \in \mathscr{S}$, $\overline{\theta} \in R$ and $\overline{\boldsymbol{\alpha}} \in R^N$. Since (6.4.6) must hold for every point $(\tilde{\mathbf{M}}, \tilde{\boldsymbol{\alpha}})$, we have omitted the wave symbol above $\mathbf{M}$ and $\boldsymbol{\alpha}$.

If in (6.4.6) we take subsequently $(\overline{\mathbf{D}} = \mathbf{0}, \overline{\theta}, \overline{\boldsymbol{\alpha}} = 0)$, $(\overline{\mathbf{D}} = 0, \overline{\theta} = 0, \overline{\boldsymbol{\alpha}})$ and $(\overline{\mathbf{D}}, \overline{\theta} = 0, \overline{\boldsymbol{\alpha}} = 0)$ we obtain

$$-\hat{\psi}_{\theta^2}(\mathbf{M}, \boldsymbol{\alpha}) = \hat{\eta}_{\theta}(\mathbf{M}, \boldsymbol{\alpha}) \geqslant 0, \tag{6.4.7}$$

$$\hat{\psi}_{\boldsymbol{\alpha}^2}(\mathbf{M}, \boldsymbol{\alpha})\,[\overline{\boldsymbol{\alpha}}, \overline{\boldsymbol{\alpha}}] = \hat{\psi}_{\alpha_i \alpha_j}(\mathbf{M}, \boldsymbol{\alpha})\,\overline{\alpha}_i\,\overline{\alpha}_j \geqslant 0, \tag{6.4.8}$$

$$\hat{\psi}_{\mathbf{F}^2}(\mathbf{M}, \boldsymbol{\alpha})\,[\overline{\mathbf{D}}\mathbf{F}, \overline{\mathbf{D}}\mathbf{F}] = \hat{\psi}_{F_{ij}F_{kl}}(\mathbf{M}, \boldsymbol{\alpha})\,\overline{D}_{ir}\,F_{rj}\,\overline{D}_{kp}\,F_{pl} \geqslant 0. \tag{6.4.9}$$

In order to deduce (6.4.7) we have used (6.1.17). Both inequalities (6.4.8) and (6.4.9) must hold for each $\overline{\boldsymbol{\alpha}} \in R^N$ and $\mathbf{D} \in \mathscr{S}$, respectively. The quantity

$$C = \theta \hat{\eta}_{\theta}(\mathbf{M}, \boldsymbol{\alpha}) \tag{6.4.10}$$

is called *the specific heat at constant strain and constant internal state variables*. Inequality (6.4.7) implies $C \geqslant 0$.

Let us discuss other consequences of inequality (6.4.3). If in (6.4.3) one takes $\tilde{\mathbf{F}} = \mathbf{F}^*$, $\tilde{\theta} = \theta^*$, $\tilde{\boldsymbol{\alpha}} = \boldsymbol{\alpha}^*$ such that $(\mathbf{F}^*, \theta^*, \mathbf{0}, \boldsymbol{\alpha}^*)$ is a strong equilibrium state (see (6.2.2) and (6.2.3)), then (6.4.3) leads to

$$\hat{\psi}(\mathbf{M}^*, \boldsymbol{\alpha}) > \hat{\psi}(\mathbf{M}^*, \boldsymbol{\alpha}^*). \tag{6.4.11}$$

Thus the specific free energy attains a minimum with respect to $\boldsymbol{\alpha}$ for $\mathbf{M} = \mathbf{M}^* = \text{const}$.

The function

$$V(\boldsymbol{\alpha}) = \hat{\psi}(\mathbf{M}^*, \boldsymbol{\alpha}) - \hat{\psi}(\mathbf{M}^*, \boldsymbol{\alpha}^*) \tag{6.4.12}$$

is a LIAPUNOV function (see for instance HALANAY [1963], Chapter I, Section 1, HARTMAN [1964], Chapter III, Section 8) for the system $(6.1.4)_4$ with $\mathbf{F} = \mathbf{F}^*$, $\theta = \theta^*$, $\mathbf{g} = \mathbf{0}$, i.e. V satisfies the conditions

$$V(\boldsymbol{\alpha}^*) = 0, \quad V(\boldsymbol{\alpha}) > 0 \text{ for all } \boldsymbol{\alpha} \in \mathscr{D}_1 - \{\boldsymbol{\alpha}^*\} \tag{6.4.13}$$

and

$$\dot{V}(\boldsymbol{\alpha}) = \hat{\psi}_{\boldsymbol{\alpha}}(\mathbf{M}^*, \boldsymbol{\alpha})\,\mathbf{f}(\mathbf{M}^*, \mathbf{0}, \boldsymbol{\alpha}) \leqslant 0 \text{ for all } \boldsymbol{\alpha} \in \mathscr{D}_1 - \{\boldsymbol{\alpha}^*\}, \tag{6.4.14}$$

where $\mathscr{D}_1$ is the intersection of the definition domain $\mathscr{D}$ of $\hat{\psi}$, with $\mathbf{M} = \mathbf{M}^*$. (6.4.13) follows from (6.4.12) and (6.4.11) while (6.4.14) follows from (6.2.4).

A solution $\tilde{\boldsymbol{\alpha}}(t)$ (defined for $t \geqslant 0$) of the system

$$\dot{\boldsymbol{\alpha}} = \mathbf{f}(\mathbf{M}^*, \mathbf{0}, \boldsymbol{\alpha})$$

is said to be (Liapunov) *stable* if for any $\varepsilon > 0$ there exists $\delta_\varepsilon > 0$ such that $|\boldsymbol{\alpha}_0 - \tilde{\boldsymbol{\alpha}}(0)| < \delta_\varepsilon$ implies

$$|\boldsymbol{\alpha}(t, \boldsymbol{\alpha}_0) - \tilde{\boldsymbol{\alpha}}(t)| < \varepsilon \text{ for all } t \geqslant 0.$$

It is known that the solution $\bar{\boldsymbol{\alpha}}(t) = \boldsymbol{\alpha}^*$ for $t \geqslant 0$ is stable (see for instance HALANAY [1963], Chapter I, Section 1, Theorem 1.1).

If inequality (6.4.14) is strictly satisfied for all $\boldsymbol{\alpha} \neq \boldsymbol{\alpha}^*$ in a ball in $\mathscr{D}_1$ with center at $\boldsymbol{\alpha}^*$, then the limit $(6.2.16)_1$ exists and it satisfies condition $(6.2.16)_2$, i.e. the solution $\boldsymbol{\alpha}(t) = \boldsymbol{\alpha}^*$ for $t \geqslant 0$ is assymptotically stable.

Inequality (6.4.3) and its consequences which we are discussing here are due to BOWEN [1968].

4c. As we have seen in the previous subsection, the additional inequality (6.4.3) has important implications concerning the constitutive equations. Another type of inequality is the so called *ellipticity condition;* it may be written in the same form as (6.4.3), namely

$$\hat{\psi}(\mathbf{M}, \boldsymbol{\alpha}) - \hat{\psi}(\tilde{\mathbf{M}}, \tilde{\boldsymbol{\alpha}}) - (\theta - \tilde{\theta})\, \hat{\psi}_\theta(\mathbf{M}, \boldsymbol{\alpha}) - \hat{\psi}_{\mathbf{F}}(\tilde{\mathbf{M}}, \tilde{\boldsymbol{\alpha}})(\mathbf{F} - \tilde{\mathbf{F}}) - \hat{\psi}_{\boldsymbol{\alpha}}(\tilde{\mathbf{M}}, \tilde{\boldsymbol{\alpha}})(\boldsymbol{\alpha} - \tilde{\boldsymbol{\alpha}}) > 0. \tag{6.4.15}$$

This inequality is required to hold for all points $(\mathbf{M}, \boldsymbol{\alpha})$, $(\tilde{\mathbf{M}}, \tilde{\boldsymbol{\alpha}})$ in the definition domain of $\hat{\psi}$ and $\mathbf{f}_\mathbf{g}$ (of the form $(\tilde{\mathbf{M}}, \mathbf{g} = \mathbf{0}, \tilde{\boldsymbol{\alpha}})$), such that $(\mathbf{M}, \boldsymbol{\alpha}) \neq (\tilde{\mathbf{M}}, \tilde{\boldsymbol{\alpha}})$ and

$$\mathbf{F} = \tilde{\mathbf{F}} + \mathbf{a} \otimes \mathbf{b},\ \theta = \lambda + \tilde{\theta},\ \boldsymbol{\alpha} = \tilde{\boldsymbol{\alpha}} + \mathbf{f}_\mathbf{g}(\tilde{\mathbf{M}}, \mathbf{0}, \tilde{\boldsymbol{\alpha}})[\mathbf{c}] \tag{6.4.16}$$

with $\mathbf{a}, \mathbf{b}, \mathbf{c} \in R^3$ and $\lambda \in R$ satisfying

$$(\det \tilde{\mathbf{F}}) \det(\tilde{\mathbf{F}} + \mathbf{a} \otimes \mathbf{b}) > 0,\ \lambda + \tilde{\theta} > 0. \tag{6.4.17}$$

The linear mapping $\mathbf{a} \otimes \mathbf{b}$ (the tensor product of $\mathbf{a}$ and $\mathbf{b}$) is given with respect to an orthogonal basis in R^3 by the matrix $a_i b_j$.

The formal difference between the inequalities (6.4.3) and (6.4.15) consists in the relation $(6.4.16)_1$ that connects $\mathbf{F}$ and $\boldsymbol{\alpha}$ to $\tilde{\mathbf{F}}$ and $\tilde{\boldsymbol{\alpha}}$ as well as in the requirement that conditions (6.4.17) are to be satisfied by $\mathbf{a}, \mathbf{b}$ and λ. In fact, the inequality (6.4.15) is, in general, less restrictive since it imposes restrictions on 10 parameters only while inequality (6.4.3) involves $N + 7$ parameters.

Following the method used to obtain (6.4.6) one gets, as a consequence of (6.4.15), that

$$\hat{\psi}_{\mathbf{F}^2}(\mathbf{M}, \boldsymbol{\alpha})[\mathbf{a} \otimes \mathbf{b}, \mathbf{a} \otimes \mathbf{b}] + 2\hat{\psi}_{\mathbf{F}\boldsymbol{\alpha}}(\mathbf{M}, \boldsymbol{\alpha})[\mathbf{a} \otimes \mathbf{b}, \mathbf{f}_\mathbf{g}(\mathbf{M}, \mathbf{0}, \boldsymbol{\alpha})[\mathbf{c}]] - \lambda^2 \hat{\psi}_{\theta^2}(\mathbf{M}, \boldsymbol{\alpha}) + \hat{\psi}_{\boldsymbol{\alpha}^2}(\mathbf{M}, \boldsymbol{\alpha})[\mathbf{f}_\mathbf{g}(\mathbf{M}, \mathbf{0}, \boldsymbol{\alpha})[\mathbf{c}], \mathbf{f}_\mathbf{g}(\mathbf{M}, \mathbf{0}, \boldsymbol{\alpha})[\mathbf{c}]] \geqslant 0, \tag{6.4.18}$$

must hold for all $\lambda \in R$ and $\mathbf{a}, \mathbf{b}, \mathbf{c} \in R^3$.

According to (6.4.18), (6.4.7) is satisfied in the whole domain of definition of $\hat{\psi}$ and, instead of (6.4.8) and (6.4.9), one obtains

$$\hat{\psi}_{\boldsymbol{\alpha}^2}(\mathbf{M}, \boldsymbol{\alpha})\, [\mathbf{f}_g(\mathbf{M}, \mathbf{0}, \boldsymbol{\alpha})\, [\mathbf{c}],\ \mathbf{f}_g(\mathbf{M}, \mathbf{0}, \boldsymbol{\alpha})\, [\mathbf{c}]] \geqslant 0, \tag{6.4.19}$$

$$\hat{\psi}_{\mathbf{F}^2}(\mathbf{M}, \boldsymbol{\alpha})\, [\mathbf{a} \otimes \mathbf{b},\ \mathbf{a} \otimes \mathbf{b}] \geqslant 0 \tag{6.4.20}$$

for all **a** and **b** in R^3.

Condition (6.4.20) has been used for the first time by HADAMARD [1903] as a stability condition. When the inequality is strictly satisfied for $\mathbf{a} \neq \mathbf{0}$ and $\mathbf{b} \neq \mathbf{0}$, it is called the strong *ellipticity condition* and, for a purely mechanical problem, it ensures the ellipticity of the system of partial differential equations that governs the static problems. A series of static consequences of the strong ellipticity condition in the non-linear elastic case have been studied by HAYES [1969]. In case of elastodynamical problems the strong ellipticity condition ensures the existence of real acceleration waves with non-zero finite propagation velocity (see for instance TRUESDELL and NOLL [1965] quoted above and also Chapter VII below).

5. ELASTIC-VISCOPLASTIC MODELS WITH INTERNAL STATE VARIABLES

5a. **Preliminary considerations.** In the mechanics of deformable solids there is quite often used, instead of the deformation gradient **F** or the CAUCHY-GREEN strain tensor, the *strain tensor* **E** defined by

$$\mathbf{E} = \mathbf{C} - \mathbf{1}, \quad \left(2E_{ij} = \frac{\partial \chi_k}{\partial X_i} \frac{\partial \chi_k}{\partial X_j} - \delta_{ij}\right). \tag{6.5.1}$$

This strain tensor **E** has the property of vanishing in the reference configuration (when $\mathbf{x} = \chi(\mathbf{X}, t) = \mathbf{X}$).

In plasticity theory one usually assumes that either **E** is a sum of two tensors: the elastic strain tensor $\mathbf{E}^e$ and the plastic (inelastic) strain tensor $\mathbf{E}^p$, i.e.

$$\mathbf{E} = \mathbf{E}^e + \mathbf{E}^p, \tag{6.5.2}$$

or the deformation gradient **F** is the product of two linear mappings: $\mathbf{F}^e$ which is responsible for the elastic strain and $\mathbf{F}^p$ which is responsible for the plastic strain,

$$\mathbf{F} = \mathbf{F}^e\, \mathbf{F}^p. \tag{6.5.3}$$

5b. **Constitutive assumptions.** There exists a large scope in the choice of the internal state variables and the form of the functions of state. Authors have individual preferences, varying even from paper to paper and sometimes even within one paper. We shall present here shortly some points of view.

Starting from the decomposition (6.5.2) or (6.5.3), some authors take as state variables the total strain $\mathbf{E}$ (or $\mathbf{F}$), the absolute temperature θ, the temperature gradient $\mathbf{g}$ and, as internal state variables, the plastic strain $\mathbf{E}^p$ (or $\mathbf{F}^p$) and a number of other internal state variables $\boldsymbol{\omega}$ (i.e. $\boldsymbol{\alpha} = (\mathbf{E}^p, \boldsymbol{\omega})$) that take into account the spatial display of dislocations and the work-hardening of the material. Some authors consider $\boldsymbol{\omega}$ as second order tensors, the dislocations display tensors (see for instance KRATOCHVIL and DILLON [1969]), while other authors consider $\boldsymbol{\omega}$ as also consisting of scalar quantities (see, for instance, PERZYNA [1971], BALTOV [1971]).

As functions of state are chosen the specific free energy ψ, the specific entropy η, the first PIOLA-KIRCHHOFF stress tensor $\mathbf{S}_I$ (or the CAUCHY stress tensor $\mathbf{T}$, $\mathbf{T} = \rho \mathbf{F} \mathbf{S}_I \mathbf{F}^T$) and the heat flux $\mathbf{q}$; for $\mathbf{E}^p$ (or $\mathbf{F}^p$) and $\boldsymbol{\omega}$ one postulates evolution equations, while assuming that the time variation of these functions is determined by functions of the state variables. It is sometimes assumed that the functions of state depend on the state variables $\mathbf{E}$ and $\mathbf{E}^p$ by depending on the difference $\mathbf{E}^e = \mathbf{E} - \mathbf{E}^p$ only (see for instance KRATOCHVIL and DILLON [1970]), and that the function giving $\mathbf{S}_I$ has an inverse with respect to $\mathbf{E}^e$ (see for instance KESTIN and RICE [1970]) which makes it possible to take the stress as state variable instead of the strain.

Another point of view states that the plastic strain is not a state variable; one therefore postulates an evolution equation for it, depending on the variables describing the motion of dislocations, the internal state variables, the temperature and its gradient as well as on the elastic strain $\mathbf{F}^e$ (see for instance ZARKA [1970], KRÖNER and TEODOSIU [1974], TEODOSIU and SIDOROFF [1976]), or on $\mathbf{E}$ (see PERZYNA [1973]).

Independently of the choice of the state variables and the functions of state, one generally accepts the assumption that the specific free energy is determined by a smooth function of the state variables. The same may be assumed about the functions that determine the entropy, the stress and the heat flux. All the results presented in Sections 1, 3 and 4 can be easily written in terms of the state variables and functions one decides to choose.

The major and still remaining open problem in viscoplasticity concerns the way of establishing the evolution equations for the internal state variables. One assumes that the evolution functions are defined on certain domains in the space of state variables (the space of all points of the form $(\mathbf{F}, \theta, \mathbf{g}, \boldsymbol{\alpha})$) and take values in the space R^N (of the internal state variables) while being continuous but not necessarily differentiable on their domain of definition. It is quite often accepted that the derivatives of the evolution functions may have discontinuities across certain hypersurfaces in the space of state variables. These hypersurfaces may be considered as being transferred to the frame of this theory from the classical theory of loading surfaces.

For different choices of evolution functions we refer the reader to the following works: PERZYNA [1971 a, 1971 b, 1973], TEODOSIU and SIDOROFF [1976], KRATOCHVIL and DILLON [1969, 1970], PERZYNA and WOJNO [1968], ZARKA]1970], MANDEL [1972], ONAT and FARDSHISHEH [1972]. In the next section we present the choices according to KRATOCHVIL and DILLON [1970], ZARKA [1970], ONAT and FARDSHISHEH [1972].

6. ONE-DIMENSIONAL PLASTICITY THEORIES WITH INTERNAL STATE VARIABLES

6a. A quasilinear rate-type constitutive equation for the one-dimensional case, when thermal effects are neglected (see Chapter V, Section 1, (5.1.2)), may be viewed as a constitutive equation with only one internal state variable (see Chapter V, Section 3, formulas (5.3.19), or, in a particular case but much more explicitly, Section 5, formulas (5.5.3) and (5.5.4)). In the notation of the present chapter one may write

$$\sigma = \hat{\sigma}(\varepsilon, \alpha_1), \qquad (6.6.1)$$

$$\dot{\alpha}_1 = f_1(\varepsilon, \alpha_1),$$

where σ and ε are the stress (per initial unit area) and the one-dimensional strain respectively, and α_1 is an internal state variable (α_1 may be identified with τ, $\hat{\sigma}$ with f and f_1 with μ of Chapter V). More often, one assumes that $\hat{\sigma}$ is a linear function in both arguments (see (5.5.3)), i.e.,

$$\hat{\sigma}(\varepsilon, \alpha_1) = E\varepsilon + \alpha_1. \qquad (6.6.2)$$

For further discussions on the forms of the functions $\hat{\sigma}$ and f_1, the reader is referred to Chapter 5, Sections 1 to 5.

6 b. KRATOCHVIL and DILLON [1970] discuss the following explicit set of constitutive equations for the one-dimensional case, when thermal effects are also considered:

$$\psi = \frac{E}{2\rho_0}(\varepsilon - \alpha_1)^2 + \frac{1}{\rho_0} \nu\alpha_2 - \frac{\varkappa}{\rho_0} \theta \left[\ln \frac{\theta}{\theta_0} - 1 \right], \tag{6.6.3}$$

$$q = -k \frac{\partial \theta}{\partial X}, \tag{6.6.4}$$

$$\dot{\alpha}_1 = f_1(\varepsilon, \theta, \alpha_1, \alpha_2) = \begin{cases} A[E(\varepsilon - \alpha_1) - K\alpha_2] & \text{if } E(\varepsilon - \alpha_1) - K\alpha_2 \geqslant 0, \\ 0 & \text{if } E(\varepsilon - \alpha_1) - K\alpha_2 < 0, \end{cases} \tag{6.6.5}$$

$$\dot{\alpha}_2 = f_2(\varepsilon, \theta, \alpha_1, \alpha_2) = R f_1(\varepsilon, \theta, \alpha_1, \alpha_2) - (\alpha_2 - \bar{\alpha}) N(\theta). \tag{6.6.6}$$

Here E, ν, $\varkappa$, A, K, R and $\bar{\alpha}$ are material constants and $N(\theta)$ is a function of θ. α_1 is interpreted as plastic strain and α_2 as the number of dislocations, with the restriction $\alpha_2 \geqslant \bar{\alpha}$. By applying (6.1.17) to (6.6.2), one gets

$$\sigma = E(\varepsilon - \alpha_1), \tag{6.6.7}$$

$$\eta = \frac{\varkappa}{\rho_0} \left(\ln \frac{\theta}{\theta_0} + \frac{\theta}{\theta_0} - 1 \right). \tag{6.6.8}$$

As we have noticed in Section 5, one can no longer apply here the results of Section 2, since the evolution functions f_1 and f_2 are not differentiable everywhere in the domain of definition. A direct application of inequality (6.1.18) leads to

$$k \geqslant 0 \text{ and } N(\theta) \geqslant 0, \ K\bar{\alpha} \geqslant R\nu. \tag{6.6.9}$$

If the function ψ is restricted according to (6.4.15), then, by (6.4.6) and (6.4.20), one reaches the generally accepted conclusions

$$E \geqslant 0, \ \varkappa \geqslant 0. \tag{6.6.10}$$

E is the Young modulus and $\varkappa$ is called the *specific heat at constant strain* (constant volume). Inequality (6.4.19) is trivially satisfied since f_1 and f_2 do not depend here of $g = \dfrac{\partial \theta}{\partial X}$.

The results of the previous sections tell us nothing about the sign of the constants ν, A, K, R and $\bar{\alpha}$. KRATOCHVIL and DILLON [1970] assumed them positive.

One may also notice that the states $(\varepsilon, \theta, 0, \alpha_1, \alpha_2)$ (for $N(\theta) \neq 0$), where

$$\begin{aligned} \alpha_2 &= \bar{\alpha}, \\ \alpha_1 &= \varepsilon - \frac{K}{E} \bar{\alpha}, \end{aligned} \tag{6.6.11}$$

are equilibrium states (see Section 2) but they are not strong equilibrium states provided $\nu = 0$ and $\alpha_1 = \varepsilon$.

A direct test shows that a model determined by the relations (6.6.3) to (6.6.8) may describe phenomena such as work-hardening, creep and relaxation. The reader interested in such computations is sent to MANDEL [1972] where there is also discussed a one-dimensional model with two internal state variables proposed by ZARKA [1970]. The two internal state variables used by ZARKA (see also MANDEL [1972]) are the plastic strain and the cubic root of the density of dislocation segments per unit volume.

Interesting one-dimensional examples may also be obtained by applying to this case the results presented by TEODOSIU and SIDOROFF [1976]. Effective particular forms with two internal state variables have also been constructed by ONAT and FARDSHISHEH [1972] in order to describe creep and relaxation in metals. They do not interpret these variables in terms of the dislocation theory and the evolution functions are determined from (phenomenological) experimental results of creep and relaxation.

Chapter VII

ACCELERATION WAVES IN MATERIALS WITH INTERNAL STATE VARIABLES

1. INTRODUCTION

In this chapter we shall discuss the problem of the existence of real acceleration waves for a constitutive equation with internal state variables. The requirement that real acceleration waves exist, i.e. that the system of partial differential equations which describes the motion of the body is a hyperbolic system, is viewed here as a general requirement imposed on the nature of the constitutive equations. We shall investigate the implications of such a requirement (together with the CLAUSIUS-DUHEM inequality) upon the constitutive equations in the general framework of non-linear theories as well as in the framework of linear and semilinear theories.

Wave propagation through materials for which the heat flux is determined by a constitutive equation of CATTANEO type ([1948], [1958]), have been studied by several authors. Some of them studied the structure of the constitutive equations (see GURTIN and PIPKIN [1968], MÜLLER [1971] KOSIŃSKI and PERZYNA [1972], SULICIU [1975 b, c], KOSIŃSKI [1975], MIHĂILESCU and SULICIU [1976, 1978], MCCARTHY [1970, 1972]), others were dealing with some initial and boundary value problems (see LYKOV [1965], LORD and SCHULMAN [1967], POPOV [1967], ACHENBACK [1968], NEYFEH and NEMAT-NASSER [1972], PODSTRIGATCH and KOLIANO [1976], MAZILU [1978], and for additional references see SULICIU [1975 b] and KOSIŃSKI [1975]). In the framework of non-linear theories one generally finds that acceleration waves are not symmetric with respect to the propagation direction unless additional hypotheses are adopted (see SULICIU [1966], GURTIN and PIPKIN [1968], KOSIŃSKI and PERZYNA [1972], and for additional comments see SULICIU [1975 b]). The symmetry of the acceleration waves as well as the hyperbolicity requirement of the system of partial differential equations are thought as hypotheses of a physical nature, governing thermomechanical propagation phenomena.

The main assumptions used in this chapter are the following:

a) the quasilinearity hypotheses (7.2.7) and (7.2.8) which lead to a system of quasilinear partial differential equations for the unknown functions describing the motion of the body. Then CLAUSIUS-DUHEM inequality implies that the heat flux and the evolution function (i.e. the maps $\hat{\mathbf{A}}$ and $\hat{\mathbf{b}}$ of (7.2.8)) must be given so that besides the conditions of Theorem 6.1.1, also the conditions (7.2.9) are satisfied;

b) the symmetry assumption I which requires that if U is a solution of equation (7.3.17), then so is $-U$;

c) the strong ellipticity condition. This condition, introduced by MIHĂILESCU and SULICIU [1976] is a natural extension of the condition used in the theory of mechanical waves; in our context it implies additional restrictions on the free energy function $\hat{\psi}$ and also on the evolution function $\mathbf{f}$. As far as the space of the internal state variables is concerned the strong ellipticity condition is stated in a weak form that still allows to get all the desired consequences;

d) the existence of a strong equilibrium state. Following BOWEN [1968] (cf. also TRUESDELL [1969]), one assumes there exist strong equilibrium states in the state space $(\mathbf{F}, \theta, \mathbf{g}, \boldsymbol{\alpha})$.

These assumptions imply that at strong equilibrium states all acceleration waves are real. If, instead of assumption d) one adopts the assumption (7.5.10), then this hypothesis together with a)—c) imply that at every state all acceleration waves are real;

e) the second symmetry assumption is made at a strong equilibrium state. Together with a)—d) above, this assumption implies that in the neighborhood of a strong equilibrium state the heat flux can be approximated by a linear function of the internal state variables. In the instantaneously coupled case at least two waves carry jumps of the thermal quantities. In the instantaneously uncoupled case, three waves are pure mechanical ones and the fourth is a pure thermal wave.

For the sake of simplicity, throughout this chapter the state space $\mathscr{D}$ (see Chapter VI, Section 1) will be identified with the space $\mathscr{L}^+ \times R^+ \times \times R^3 \times R^N$.

The internal state variables are chosen to be frame indifferent, but not invariant under the mappings of the isotropy group. If they would remain invariant under the isotropy group too, then the above assumptions would imply $\hat{\mathbf{q}} \equiv 0$, i.e. the constitutive equations obeying the above assumptions would describe heat non-conducting materials only.

Concerning the behaviour of the internal state variables under the isotropy group, one assumes the existence of a decomposition of the form

(7.7.2). This means assuming that the internal state variables contain a quantity that behaves as a three dimensional vector under the isotropy group, and another quantity which behaves as a 3×3 symmetric tensor under the same group, while the remaining components behave as scalar quantities.

The linearity assumption, the assumption (7.2.7) that the heat flux is independent of the actual value of the temperature gradient, together with the strong ellipticity condition lead to the conclusion that in such a body there exist four real and symmetric acceleration waves propagating with constant speeds.

The obtained model includes linear thermoelasticity with CATTANEO's heat conduction law (instead of FOURIER's law) as well as most of the linear rheological models.

In the last section we discuss semilinear models which include the SOKOLOVSKIĬ—MALVERN—PERZYNA type of elasto-viscoplastic model. These models are based on hypotheses which lead to the same structure of the acceleration waves as in the linear case. The evolution equations for the internal state variables are however non-linear (in general, they cannot be linearized, see Chapter V, Section 1) and such that the complete system of partial differential equations governing the behaviour of the body is a semilinear one. The section concludes with a justification of the chosen structure for the internal state variables, giving also an interpretation of the internal state variables in terms of the rate-type constitutive equations.

2. QUASILINEARITY ASSUMPTION

We rewrite the balance laws of mass $(1.2.11)_1$, momentum $(1.2.20)_1$, moment of momentum (1.2.8) and energy $(1.2.20)_2$ in the reference configuration by using the coordinate $\mathbf{X}$ as (this can also be done in the actual coordinate $\mathbf{x}$)

$$\rho_0 = J\rho,$$

$$\rho_0 \frac{\partial \mathbf{v}}{\partial t} - \operatorname{Div} \tilde{\mathbf{S}} = \rho_0 \mathbf{b}, \quad \rho_0 \frac{\partial v_i}{\partial t} - \frac{\partial \tilde{S}_{ij}}{\partial X_j} = \rho_0 b_i,$$

$$\tilde{\mathbf{S}}\mathbf{F}^T = \mathbf{F}\tilde{\mathbf{S}}^T, \quad \tilde{S}_{ik} F_{jk} = F_{ik} \tilde{S}_{jk}, \qquad (7.2.1)$$

$$\rho_0 \frac{\partial e}{\partial t} - \tilde{\mathbf{S}} \cdot \frac{\partial \mathbf{F}}{\partial t} + \operatorname{Div} \tilde{\mathbf{q}} = \rho_0 r, \quad \rho_0 \frac{\partial e}{\partial t} - \tilde{S}_{ij} \frac{\partial F_{ij}}{\partial t} + \frac{\partial \tilde{q}_i}{\partial X_i} = \rho_0 r,$$

where the stress tensors $\mathbf{T}$, $\mathbf{S}$ and $\tilde{\mathbf{S}}$ are related by (1.2.12) and

$$\tilde{\mathbf{q}} = J\mathbf{F}^{-1}\mathbf{q}. \tag{7.2.2}$$

Equations (7.2.1) are supplemented by the constitutive equations (6.1.4). Since these equations have to satisfy the CLAUSIUS-DUHEM inequality, they must be defined so that conditions (6.1.8), (6.1.9) and (6.1.10) are fulfilled. Then the free energy is a function of state $\hat{\psi}(\mathbf{F}, \theta, \boldsymbol{\alpha})$, and the stress function $\tilde{\mathbf{S}}(\mathbf{F}, \theta, \boldsymbol{\alpha})$ and the entropy function $\hat{\eta}(\mathbf{F}, \theta, \boldsymbol{\alpha})$ are given by (6.1.17); moreover $\hat{\psi}(\mathbf{F}, \theta, \boldsymbol{\alpha})$, $\mathbf{f}(\mathbf{F}, \theta, \mathbf{g}, \boldsymbol{\alpha})$ and $\hat{\mathbf{q}}(\mathbf{F}, \theta, \mathbf{g}, \boldsymbol{\alpha})$ must satisfy (6.1.18). Thus the constitutive equations which have to be added to (7.2.1) can be written as

$$\begin{aligned}
\tilde{\mathbf{S}} &= \rho_0\hat{\mathbf{S}}(\mathbf{F}, \theta, \boldsymbol{\alpha}) = \rho_0\hat{\psi}_{\mathbf{F}}(\mathbf{F}, \theta, \boldsymbol{\alpha}),\\
e &= \hat{e}(\mathbf{F}, \theta, \boldsymbol{\alpha}) = -\,\theta\hat{\psi}_\theta(\mathbf{F}, \theta, \boldsymbol{\alpha}) + \hat{\psi}(\mathbf{F}, \theta, \boldsymbol{\alpha}),\\
\tilde{\mathbf{q}} &= J\mathbf{F}^{-1}\hat{\mathbf{q}}(\mathbf{F}, \theta, \mathbf{g}, \boldsymbol{\alpha}) = \tilde{\mathbf{q}}(\mathbf{F}, \theta, \mathbf{g}_0, \boldsymbol{\alpha}),\\
\dot{\boldsymbol{\alpha}} &= \mathbf{f}(\mathbf{F}, \theta, \mathbf{g}, \boldsymbol{\alpha}) = \tilde{\mathbf{f}}(\mathbf{F}, \theta, \mathbf{g}_0, \boldsymbol{\alpha}),
\end{aligned} \tag{7.2.3}$$

where $\mathbf{g} = \operatorname{grad}\theta$ and $\mathbf{g}_0 = \operatorname{Grad}\theta$ are related by

$$\mathbf{g}_0 = \mathbf{F}^T\mathbf{g}, \quad \frac{\partial\theta}{\partial X_i} = \frac{\partial\theta}{\partial x_j}\,\frac{\partial\chi_j}{\partial X_i}. \tag{7.2.4}$$

In (7.2.1) the initial mass density ρ_0 is a given function of $\mathbf{X}$ in the reference configuration $\mathscr{R}$, the body forces $\mathbf{b}$ and the heat supply r are given functions of $\mathbf{X} \in \mathscr{R}$ and $t \geqslant 0$; and the particle velocity $\mathbf{v}$ and the deformation gradient $\mathbf{F}$ are determined from the motion $\boldsymbol{\chi}(\mathbf{X}, t)$ by

$$\mathbf{v} = \frac{\partial\boldsymbol{\chi}}{\partial t}, \quad \mathbf{F} = \operatorname{Grad}\boldsymbol{\chi} = \left\{\frac{\partial\chi_i}{\partial X_j}\right\}, \quad \mathbf{g}_0 = \operatorname{Grad}\theta = \left\{\frac{\partial\theta}{\partial X_i}\right\}.$$

By substituting the constitutive equations $(7.2.3)_{1-3}$ into the balance equations $(7.2.1)_{4,2}$, one obtains the complete system of partial differential equations

$$\begin{aligned}
&\rho_0\frac{\partial\mathbf{v}}{\partial t} - \operatorname{Div}\{\rho_0\hat{\mathbf{S}}(\mathbf{F}, \theta, \boldsymbol{\alpha})\} = \rho_0\mathbf{b},\\
&\rho_0\frac{\partial}{\partial t}\hat{e}(\mathbf{F}, \theta, \boldsymbol{\alpha}) - \rho_0\hat{\mathbf{S}}(\mathbf{F}, \theta, \boldsymbol{\alpha})\frac{\partial\mathbf{F}}{\partial t} + \operatorname{Div}\tilde{\mathbf{q}}(\mathbf{F}, \theta, \mathbf{g}_0, \boldsymbol{\alpha}) = \rho_0 r,\\
&\frac{\partial\boldsymbol{\alpha}}{\partial t} = \tilde{\mathbf{f}}(\mathbf{F}, \theta, \mathbf{g}_0, \boldsymbol{\alpha}),
\end{aligned} \tag{7.2.5}$$

for the motion $\chi(\mathbf{X}, t)$, the temperature $\theta(\mathbf{X}, t)$ and the internal state variables $\boldsymbol{\alpha}(\mathbf{X}, t)$. The actual mass density $\rho(\mathbf{X}, t)$ is determined by $(7.2.1)_1$.

As a consequence of the principle of frame indifference (see $(6.3.8)_1$) and of the invariance of $\boldsymbol{\alpha}$ under changes of frame, we may write

$$\frac{\partial\hat{\psi}(\mathbf{F}, \theta, \boldsymbol{\alpha})}{\partial F_{ij}} = \frac{\partial\check{\psi}(\mathbf{C}, \theta, \boldsymbol{\alpha})}{\partial C_{kl}} (\delta_{jk}F_{il} + \delta_{lj}F_{ik}) = 2\frac{\partial\check{\psi}(\mathbf{C}, \theta, \boldsymbol{\alpha})}{\partial C_{jk}} F_{ik}, \quad (7.2.6)$$

where we have taken into account that $\mathbf{C} = \mathbf{F}^T\mathbf{F}$ is a symmetric tensor. This relation, together with $(7.2.3)_1$, show that $(7.2.1)_3$ is satisfied.

The second equation of the system (7.2.5) is the only equation containing the second order derivatives of θ with respect to $\mathbf{X}$, provided that the heat flux function $\tilde{\mathbf{q}}(\mathbf{F}, \theta, \mathbf{g}_0, \boldsymbol{\alpha})$ is not constant with respect to $\mathbf{g}_0$, while the same equation is the only equation of the system which contains only first order time derivatives of θ. This situation generally implies that the system (7.2.5) is of a mixed hyperbolic-parabolic character, and thus some perturbations propagate with infinite speed. In order to avoid this, one assumes that

$$\frac{\partial\tilde{\mathbf{q}}}{\partial\mathbf{g}_0}(\mathbf{F}, \theta, \mathbf{g}_0, \boldsymbol{\alpha}) = \mathbf{0} \quad (7.2.7)$$

for all states $(\mathbf{F}, \theta, \mathbf{g}_0, \boldsymbol{\alpha}) \in \mathscr{D}$. The constitutive equations are further simplified by assuming that the evolution function $\tilde{\mathbf{f}}(\mathbf{F}, \theta, \mathbf{g}_0, \boldsymbol{\alpha})$ is linear in $\mathbf{g}_0$, i.e.

$$\tilde{\mathbf{f}}(\mathbf{F}, \theta, \mathbf{g}_0, \boldsymbol{\alpha}) = \hat{\mathbf{A}}(\mathbf{F}, \theta, \boldsymbol{\alpha})\,\mathbf{g}_0 + \hat{\mathbf{b}}(\mathbf{F}, \theta, \boldsymbol{\alpha}), \quad (7.2.8)$$

where, for each $(\mathbf{F}, \theta, \boldsymbol{\alpha})$, $\hat{\mathbf{A}}(\mathbf{F}, \theta, \boldsymbol{\alpha})$ is a linear map from R^3 to R^N and $\hat{\mathbf{b}}(\mathbf{F}, \theta, \boldsymbol{\alpha})$ is a N-dimensional vector. The assumptions (7.2.7) and (7.2.8) turn the system (7.2.6) into a quasilinear system of equations for the motion $\chi(\mathbf{X}, t)$ and for the temperature θ. As we shall see in the following sections of this chapter, the above constitutive assumptions leave the constitutive equations general enough to include the CATTANEO [1948, 1958] hyperbolic heat conduction constitutive law. These hypotheses, for the one-dimensional case, were stated by KOSIŃSKI and PERZYNA [1972].

Following the same procedure as in Theorem 6.1.1, one shows that inequality (6.1.18) and assumptions (7.2.7), (7.2.8) imply (see KOSIŃSKI [1965], SULICIU [1965 b, c] and MIHĂILESCU and SULICIU [1966]) that

$$\begin{gathered}\tilde{\mathbf{q}}(\mathbf{F}, \theta, \boldsymbol{\alpha}) = -\rho_0\theta\hat{\mathbf{A}}^T(\mathbf{F}, \theta, \boldsymbol{\alpha})\frac{\partial\hat{\psi}}{\partial\boldsymbol{\alpha}}(\mathbf{F}, \theta, \boldsymbol{\alpha}), \\ \frac{\partial\hat{\psi}}{\partial\boldsymbol{\alpha}}(\mathbf{F}, \theta, \boldsymbol{\alpha})\cdot\hat{\mathbf{b}}(\mathbf{F}, \theta, \boldsymbol{\alpha}) \leqslant 0.\end{gathered} \quad (7.2.9)$$

Taking into account the above results, we can write the system (7.2.5) in Cartesian coordinates as follows:

$$\rho_0 \frac{\partial v_i}{\partial t} - \frac{\partial}{\partial X_k} \{\rho_0 \hat{S}_{ik}(\mathbf{F}, \theta, \boldsymbol{\alpha})\} = \rho_0 b_i, \quad i = 1,2,3,$$

$$\rho_0 \frac{\partial}{\partial t} \hat{e}(\mathbf{F}, \theta, \boldsymbol{\alpha}) - \rho_0 \hat{S}_{ij}(\mathbf{F}, \theta, \boldsymbol{\alpha}) \frac{\partial F_{ij}}{\partial t} + \frac{\partial}{\partial X_k} \tilde{q}_k(\mathbf{F}, \theta, \boldsymbol{\alpha}) = \rho_0 r, \quad (7.2.10)$$

$$\frac{\partial \alpha_m}{\partial t} - \hat{A}_{mk}(\mathbf{F}, \theta, \boldsymbol{\alpha}) \frac{\partial \theta}{\partial X_k} - \hat{b}_m(\mathbf{F}, \theta, \boldsymbol{\alpha}) = 0, \quad m = 1, \ldots, N.$$

Here the functions $\hat{S}_{ik}(\mathbf{F}, \theta, \boldsymbol{\alpha})$ and $\hat{e}(\mathbf{F}, \theta, \boldsymbol{\alpha})$ are determined by the free energy function $\hat{\psi}(\mathbf{F}, \theta, \boldsymbol{\alpha})$ from the relations $(7.2.3)_{1,2}$, the functions $\tilde{q}_i(\mathbf{F}, \theta, \boldsymbol{\alpha})$ by $\hat{A}_{mi}(\mathbf{F}, \theta, \boldsymbol{\alpha})$ and $\hat{\psi}(\mathbf{F}, \theta, \boldsymbol{\alpha})$ by $(7.2.9)_1$, while $\hat{\psi}(\mathbf{F}, \theta, \boldsymbol{\alpha})$ and $\hat{\mathbf{b}}(\mathbf{F}, \theta, \boldsymbol{\alpha})$ must be given so that inequality $(7.2.9)_2$ holds.

In the next sections we shall discuss conditions on $\hat{\psi}(\mathbf{F}, \theta, \boldsymbol{\alpha})$, $\hat{\mathbf{A}}(\mathbf{F}, \theta, \boldsymbol{\alpha})$ and $\hat{\mathbf{b}}(\mathbf{F}, \theta, \boldsymbol{\alpha})$ that turn the system (7.2.10) into a hyperbolic system of partial differential equations (at least for some states $(\mathbf{F}, \theta, \boldsymbol{\alpha})$).

3. SYMMETRIC ACCELERATION WAVES

A regular surface $\varphi(\mathbf{X}, t) = 0$, where $\mathbf{X} \in \mathscr{R}$ and $t \in R$ will be called an *acceleration wave* if $\mathbf{v} = \dot{\mathbf{x}}$, $\mathbf{F} = \operatorname{Grad} \boldsymbol{\chi}$, θ and $\boldsymbol{\alpha}$ are continuous across this surface, but their derivatives with respect to $\mathbf{X}$ and t have jump discontinuities when crossing it. It will be assumed that the body forces $\mathbf{b}$, the heat supply r and the initial mass density ρ_0 have no jumps across the wave.

The geometric and kinematic compatibility conditions that have to be satisfied by the derivatives of $\mathbf{v}$, $\mathbf{F}$, θ and $\boldsymbol{\alpha}$ are as follows (the reader unfamiliar with kinematic and dynamic compatibility conditions is referred to TRUESDELL and TOUPIN [1960], WANG and TRUESDELL [1973], TRUESDELL [1975] or, for their applications to different types of materials, to COLEMAN, GURTIN, HERRERA and TRUESDELL [1965]); see also Chapter I, Section 2, Subsection 4):

$$\left[\frac{\partial^2 \chi_k}{\partial t^2}\right] = U^2 a_k, \quad \left[\frac{\partial F_{kl}}{\partial X_j}\right] = a_k n_l n_j, \quad \left[\frac{\partial F_{ij}}{\partial t}\right] = -U a_i n_j, \quad (7.3.1)$$

$$\left[\frac{\partial \theta}{\partial t}\right] = -U v, \quad \left[\frac{\partial \theta}{\partial X_j}\right] = v n_j, \quad (7.3.2)$$

$$\left[\frac{\partial \alpha_i}{\partial t}\right] = -U \gamma_i, \quad \left[\frac{\partial \alpha_i}{\partial X_j}\right] = \gamma_i n_j. \quad (7.3.3)$$

Here n_j are the components of the unit vector **n**, normal to the discontinuity surface for each fixed t, and called the *direction of propagation*, U is the speed of propagation of the discontinuity surface and these satisfy

$$n_i = (\partial\varphi/\partial X_i)/|\text{Grad } \varphi|, \quad U = -(\partial\varphi/\partial t)/|\text{Grad } \varphi|. \tag{7.3.4}$$

a_i are the components of the vector **a**, called the *mechanical amplitude* of the wave, the scalar v is called the *thermal amplitude* of the wave and the N-dimensional vector $\gamma = (\gamma_1, \gamma_2, \ldots, \gamma_N)$ is called the *internal state amplitude* of the wave.

Taking into account relations $(7.2.3)_{1,2}$ and $(7.2.9)_1$ as well as the above formulas, we may write the following jump relations

$$\left[\frac{\partial}{\partial X_k}(\rho_0 \hat{S}_{ik})\right] = \rho_0 \left\{\frac{\partial^2\hat{\psi}}{\partial F_{ik}\partial F_{jl}} a_j n_l n_k + \frac{\partial^2\hat{\psi}}{\partial F_{ik}\partial\theta} v n_k + \right.$$

$$\left. + \frac{\partial^2\hat{\psi}}{\partial F_{ik}\partial\alpha_m} \gamma_m n_k\right\}, \tag{7.3.5}$$

$$\left[\frac{\partial e}{\partial t}\right] = \theta U \left(\frac{\partial^2\hat{\psi}}{\partial\theta\partial F_{ij}} a_i n_j + \frac{\partial^2\hat{\psi}}{\partial\theta^2} v + \frac{\partial^2\hat{\psi}}{\partial\theta\partial\alpha_m} \gamma_m\right) -$$

$$- U \left(\frac{\partial\hat{\psi}}{\partial F_{ij}} a_i n_j + \frac{\partial\hat{\psi}}{\partial\alpha_m} \gamma_m\right), \tag{7.3.6}$$

$$\left[\frac{\partial\tilde{q}_k}{\partial X_k}\right] = \frac{\partial\tilde{q}_i}{qF_{kl}} a_k n_i n_l + v \frac{\partial\tilde{q}_i}{\partial\theta} n_i + \frac{\partial\tilde{q}_i}{\partial\alpha_m} \gamma_m n_i. \tag{7.3.7}$$

The dynamical compatibility conditions for the system (7.2.10) are obtained by directly applying to the equations of the system the jump operation [ ] (see also Chapter I, Section 2, Subsection 4). Thus from $(7.2.10)_1$, $(7.3.1)_1$ and (7.3.5) we obtain

$$U^2 a_i = \frac{\partial^2\hat{\psi}}{\partial F_{ij}\partial F_{kl}} a_k n_j n_l + v \frac{\partial^2\hat{\psi}}{\partial\theta\partial F_{ij}} n_j + \frac{\partial^2\hat{\psi}}{\partial F_{ij}\partial\alpha_m} \gamma_m n_j, \quad i = 1,2,3. \tag{7.3.8}$$

From $(7.2.10)_2$, $(7.3.1)_3$, (7.3.6), (7.3.7) and $(7.2.3)_1$ follows

$$\rho_0\theta U \frac{\partial^2\hat{\psi}}{\partial\theta\partial F_{ij}} a_i n_j + \frac{\partial\tilde{q}_i}{\partial F_{kl}} a_k n_i n_l + \rho_0\theta U v \frac{\partial^2\hat{\psi}}{\partial\theta^2} + v \frac{\partial\tilde{q}_i}{\partial\theta} n_i +$$

$$+ \rho_0\theta U \frac{\partial^2\hat{\psi}}{\partial\theta\partial\alpha_m} \gamma_m - \rho_0 U \frac{\partial\hat{\psi}}{\partial\alpha_m} \gamma_m + \frac{\partial\tilde{q}_i}{\partial\alpha_m} \gamma_m n_i = 0. \tag{7.3.9}$$

Finally, $(7.2.10)_3$ and the kinematical compatibility conditions $(7.3.2)_2$ and $(7.3.3)_1$ lead to

$$U\gamma_m + vA_{mj}n_j = 0, \quad m = 1,2,\ldots, N. \tag{7.3.10}$$

For a fixed state $(\mathbf{F}, \theta, \boldsymbol{\alpha})$ and a given direction of propagation $\mathbf{n}$, the system (7.3.8)—(7.3.10) is a homogeneous linear system of $N+4$ equations with the $N+4$ unknowns $\mathbf{a}$, v and $\boldsymbol{\gamma}$. It has a non-trivial solution if and only if its determinant vanishes.

In order to simplify the presentation, we introduce the following notation

$$Q_{ik} = \frac{\partial^2\hat{\psi}}{\partial F_{ij}\partial F_{kl}} n_j n_l, \quad v = -\rho_0\theta\frac{\partial^2\hat{\psi}}{\partial\theta^2}, \tag{7.3.11}$$

$$P_i = \frac{\partial^2\hat{\psi}}{\partial\theta\partial F_{ij}} n_j, \tag{7.3.12}$$

$$D_i = \frac{\partial^2\hat{\psi}}{\partial F_{ij}\partial\alpha_m}\hat{A}_{mk}n_j n_k, \quad E_i = \frac{\partial\tilde{q}_k}{\partial F_{il}} n_k n_l, \tag{7.3.13}$$

$$z = -\frac{\partial\tilde{q}_i}{\partial\alpha_m}\hat{A}_{mk}n_i n_k, \quad w = \rho_0\frac{\partial\hat{\psi}}{\partial\alpha_m}\hat{A}_{mk}n_k - \rho_0\theta\frac{\partial^2\hat{\psi}}{\partial\theta\partial\alpha_m}\hat{A}_{mk}n_k + \frac{\partial\tilde{q}_i}{\partial\theta}n_i, \tag{7.3.14}$$

$$H_{im} = \frac{\partial^2\hat{\psi}}{\partial F_{ij}\partial\alpha_m} n_j, \quad G_m = \rho_0\theta\frac{\partial^2\hat{\psi}}{\partial\theta\partial\alpha_m} - \rho_0\frac{\partial\hat{\psi}}{\partial\alpha_m}, \tag{7.3.15}$$

$$K_m = \frac{\partial\tilde{q}_i}{\partial\alpha_m} n_i, \quad L = \frac{\partial\tilde{q}_i}{\partial\theta} n_i, \quad M_m = \hat{A}_{mj}n_j. \tag{7.3.16}$$

The quantities $\mathbf{Q} = (Q_{ij})$ and v/ρ_0 are called the *homothermal acoustical tensor* and the *heat capacity at constant strain* and *constant internal state variables*, respectively.

Using this notation, we write the determinant of the system (7.3.8)—(7.3.10) as

$$\Delta = \begin{vmatrix} Q_{11}-U^2 & Q_{12} & Q_{13} & P_1 & H_{11}\ldots & H_{1N} \\ Q_{21} & Q_{22}-U^2 & Q_{23} & P_2 & H_{21}\ldots & H_{2N} \\ Q_{31} & Q_{32} & Q_{33}-U^2 & P_3 & H_{31}\ldots & H_{3N} \\ \rho_0\theta UP_1+E_1 & \rho_0\theta UP_2+E_2 & \rho_0\theta UP_3+E_3 & -Uv+L & UG_1+K_1\ldots & UG_N+K_N \\ 0 & 0 & 0 & M_1 & U & 0 \\ \vdots & \vdots & \vdots & \vdots & \vdots & \vdots \\ 0 & 0 & 0 & M_N & 0 & U \end{vmatrix} \tag{7.3.17}$$

$$= U^{N-1}\{- vU^8 + wU^7 + (vI_Q + z + \rho_0\theta\mathbf{P}\cdot\mathbf{P})\, U^6 + [- wI_Q +$$
$$+ \mathbf{P}\cdot(\mathbf{E} - \rho_0\theta\mathbf{D})]\, U^5 - [vII_Q + zI_Q + \mathbf{D}\cdot\mathbf{E} + \rho_0\theta I_Q\mathbf{P}\cdot\mathbf{P} - \rho_0\theta\mathbf{QP}\cdot\mathbf{P}]U^4 +$$
$$+ [wII_Q + I_Q\mathbf{P}\cdot(\rho_0\theta\mathbf{D} - \mathbf{E}) + \mathbf{QP}\cdot(\mathbf{E} - \rho_0\theta\mathbf{D})]\, U^3 +$$
$$+ [vIII_Q + zII_Q + I_Q\mathbf{D}\cdot\mathbf{E} - \mathbf{QD}\cdot\mathbf{E} - \rho_0\theta I_Q\mathbf{QP}\cdot\mathbf{P} + \rho_0\theta II_Q\mathbf{P}\cdot\mathbf{P} +$$
$$+ \rho_0\theta\mathbf{Q}^2\mathbf{P}\cdot\mathbf{P}]\, U^2 + [- wIII_Q + II_Q\mathbf{P}\cdot(\mathbf{E} - \rho_0\theta\mathbf{D}) + I_Q\mathbf{QP}\cdot(\rho_0\theta\mathbf{D} - \mathbf{E}) +$$
$$+ \mathbf{Q}^2\,\mathbf{P}\cdot(\mathbf{E} - \rho_0\theta\mathbf{D})]\, U - zIII_Q + I_Q\mathbf{QD}\cdot\mathbf{E} - II_Q\mathbf{D}\cdot\mathbf{E} - (\mathbf{Q}^2\mathbf{D})\cdot\mathbf{E}\} = 0,$$

where I_Q, II_Q, III_Q are the invariants of $\mathbf{Q}$.

The polynomial equation (7.3.17) which determines the speed of propagation of acceleration waves has the root $U = 0$ of order $N - 1$, due to the constitutive assumptions $(7.2.10)_3$ and (7.2.7). Acceleration waves for which $U = 0$ are called *stationary acceleration waves.*

From the physical point of view one expects to get symmetric acceleration waves, and therefore symmetric roots for equation (7.3.17) (i.e. if $U = U_1$ is a root of equation (7.3.17) then $U = - U_1$ is also a root of the same equation). If in (7.3.17) which is a polynomial equation of degree $N + 7$ in U, the coefficient of U^{N-1} is different from zero, then there are only $N - 1$ roots that are equal to zero. The remaining eight roots are generally neither real nor symmetric. Let us introduce the following constitutive assumption that leads to symmetric speeds of propagation.

Symmetry assumption I. For any fixed state $(\mathbf{F}, \theta, \boldsymbol{\alpha})$ and any fixed direction of propagation $\mathbf{n}$, the equation that gives the speeds of propagation of acceleration waves admits symmetric roots.

Symmetry assumption *I* is satisfied if and only if the conditions

$$\begin{aligned} & w = 0, \\ & \mathbf{P}\cdot(\mathbf{E} - \rho_0\theta\mathbf{D}) = 0, \\ & \mathbf{QP}\cdot(\mathbf{E} - \rho_0\theta\mathbf{D}) = 0, \\ & \mathbf{Q}^2\mathbf{P}\cdot(\mathbf{E} - \rho_0\theta\mathbf{D}) = 0 \end{aligned} \tag{7.3.18}$$

are verified for any state $(\mathbf{F}, \theta, \boldsymbol{\alpha})$ and any direction of propagation $\mathbf{n}$. Under these conditions and with $Z = U^2$, equation (7.3.17) becomes

$$vZ^4 - (vI_Q + z + \rho_0\theta\mathbf{P}\cdot\mathbf{P})\, Z^3 + \{vII_Q + zI_Q + \mathbf{D}\cdot\mathbf{E} + \rho_0\theta(I_Q\mathbf{P}\cdot\mathbf{P} -$$
$$- \mathbf{QP}\cdot\mathbf{P})\}\, Z^2 - \{vIII_Q + zII_Q + I_Q\mathbf{D}\cdot\mathbf{E} - \mathbf{QD}\cdot\mathbf{E} + \rho_0\theta(II_Q\mathbf{P}\cdot\mathbf{P} -$$
$$- I_Q\mathbf{QP}\cdot\mathbf{P} + \mathbf{Q}^2\mathbf{P}\cdot\mathbf{P})\}\, Z + zIII_Q - I_Q\mathbf{QD}\cdot\mathbf{E} + II_Q\mathbf{D}\cdot\mathbf{E} + \mathbf{Q}^2\mathbf{D}\cdot\mathbf{E} = 0. \tag{7.3.19}$$

Conditions $(7.3.18)_{2-4}$ can also be written as

$$\mathbf{P}\cdot\mathbf{W} = 0,$$
$$\mathbf{QP}\cdot\mathbf{W} = \mathbf{P}\cdot\mathbf{QW} = 0, \qquad (7.3.20)$$
$$\mathbf{Q}^2\mathbf{P}.\mathbf{W} = \mathbf{QP}\cdot\mathbf{QW} = \mathbf{P}\cdot\mathbf{Q}^2\mathbf{W} = 0,$$

where $\mathbf{W} = \mathbf{E} - \rho_0\theta\mathbf{D}$. Therefore, if $\mathbf{P}$ and $\mathbf{W}$ are both different from zero then either $\mathbf{P}$ or $\mathbf{W}$ is an eigenvector of $\mathbf{Q}$. Let $(\mathbf{F}, \theta, \boldsymbol{\alpha})$ and $\mathbf{n}$ be fixed and let $U^2 \neq 0$ be a corresponding root of equation (7.3.19). From (7.3.10) and (7.3.16) we get

$$\gamma_k = -\frac{v}{U} M_k. \qquad (7.3.21)$$

Using the notation (7.3.11)–(7.3.16), we obtain from (7.3.9) together with $(7.3.18)_1$ and (7.3.21) that

$$(vU^2 - z)\, v = U(U\rho_0\theta\mathbf{P} + \mathbf{E})\cdot\mathbf{a}, \qquad (7.3.22)$$

and from (7.3.8) and (7.3.21) it follows that

$$U^2\mathbf{a} - \mathbf{Qa} = v\left(\mathbf{P} - \frac{\mathbf{D}}{U}\right). \qquad (7.3.23)$$

4. STRONG EQUILIBRIUM STATES. STRONG ELLIPTICITY CONDITION

4a. Due to the constitutive assumptions (7.2.8) and (7.2.7), one can restate the notions of equilibrium and strong equilibrium state (see formulas (6.2.2) and (6.2.3)) in terms of the functions $\hat{\mathbf{b}}(\mathbf{F}, \theta, \boldsymbol{\alpha})$ and $\hat{\psi}(\mathbf{F}, \theta, \boldsymbol{\alpha})$. A state $(\mathbf{F}^*, \theta^*, \mathbf{g}^* = \mathbf{0}, \boldsymbol{\alpha}^*)$ is called an *equilibrium state* if

$$\mathbf{f}^* = \mathbf{b}^* = \hat{\mathbf{b}}(\mathbf{F}^*, \theta^*, \boldsymbol{\alpha}^*) = \mathbf{0}. \qquad (7.4.1)$$

If, moreover,

$$\frac{\partial\psi^*}{\partial\alpha_m} = \frac{\partial\hat{\psi}}{\partial\alpha_m}(\mathbf{F}^*, \theta^*, \boldsymbol{\alpha}^*) = 0, \qquad (7.4.2)$$

the state $(\mathbf{F}^*, \theta^*, \mathbf{g}^* = \mathbf{0}, \boldsymbol{\alpha}^*)$ is called a *strong equilibrium state*. In the following, a star used as upper index to a quantity, will mean that the quantity refers to a strong equilibrium state.

The consequence $(7.2.9)_1$ of the CLAUSIUS-DUHEM inequality, applied at a strong equilibrium state $(\mathbf{F}^*, \theta^*, \boldsymbol{\alpha}^*)$, leads to the relations

$$\begin{aligned}
&\tilde{\mathbf{q}}^* = \tilde{\mathbf{q}}(\mathbf{F}^*, \theta^*, \boldsymbol{\alpha}^*) = \mathbf{0}, \\
&\frac{\partial \tilde{\mathbf{q}}^*}{\partial \theta} + \rho_0 \theta^* \mathbf{A}^{*T} \frac{\partial^2 \psi^*}{\partial \boldsymbol{\alpha} \partial \theta} = \mathbf{0}, \\
&\frac{\partial \tilde{\mathbf{q}}^*}{\partial \mathbf{F}} + \rho_0 \theta^* \mathbf{A}^{*T} \frac{\partial^2 \psi^*}{\partial \boldsymbol{\alpha} \partial \mathbf{F}} = \mathbf{0}, \\
&\frac{\partial \tilde{\mathbf{q}}^*}{\partial \boldsymbol{\alpha}} + \rho_0 \theta^* \mathbf{A}^{*T} \frac{\partial^2 \psi^*}{\partial \boldsymbol{\alpha}^2} = \mathbf{0}.
\end{aligned} \tag{7.4.3}$$

4b. The CLAUSIUS-DUHEM inequality, interpreted in the sense of COLEMAN and NOLL [1963], imposes important restrictions on the form of the constitutive equations. For the materials with internal state variables these restrictions have been established by COLEMAN and GURTIN [1967] (see also ONAT [1967] and VALANIS [1968]) and presented here in Chapter VI, Sections 1 and 2, and in Chapter VII, Section 2. However, these restrictions still allow the constitutive equations to remain too general, permitting them to include certain effects that are inadmissible from the physical point of view.

In order to remove such undesirable effects in non-linear elasticity, in the literature there are proposed several types of additional inequalities (for a detailed discussion on this subject see TRUESDELL and NOLL [1965], Sections 51, 52, 53, 68, 90, WANG and TRUESDELL [1973], Chapter III). For materials with internal state variables BOWEN [1968] suggests an inequality involving the function $\hat{\psi}$; this inequality is of the same type as that of COLEMAN and NOLL [1959]. He investigates its consequences on equilibrium states and on the second order derivatives of the function $\hat{\psi}$ with respect to its arguments (see Chapter VI, Sections 4a, 4b). Here we shall follow the same idea, while imposing on the functions $\hat{\psi}$ and $\mathbf{f}$ an elliptic type inequality (see Chapter VI, Section 4c, HADAMARD [1903], TRUESDELL [1961, 1965], TRUESDELL and NOLL [1965], WANG and TRUESDELL [1973]). In the next section we shall see the important implications which this inequality bears upon the propagation of waves in the neighborhood of strong equilibrium states.

Instead of the ellipticity condition (6.4.15) or its consequences (6.4.18), we shall impose a strong ellipticity condition which is in fact the strict inequality (6.4.18), under the constitutive assumption (7.2.8), i.e. we shall require that

$$\frac{\partial^2\hat{\psi}}{\partial F_{ij}\partial F_{kl}} a_i a_k b_j b_l + 2\frac{\partial^2\hat{\psi}}{\partial F_{ij}\partial \alpha_m} A_{ml} a_i b_j c_l - \mu^2 \frac{\partial^2\hat{\psi}}{\partial \theta^2} +$$

$$+ \frac{\partial^2\hat{\psi}}{\partial \alpha_m \partial \alpha_p} A_{mk} A_{pl} c_k c_l > 0 \tag{7.4.4}$$

for every $(\mathbf{F}, \theta, \boldsymbol{\alpha})$, for all vectors $\mathbf{a} = (a_1, a_2, a_3)$, $\mathbf{b} = (b_1, b_2, b_3)$, $\mathbf{c} = (c_1, c_2, c_3)$, and for each number $\mu \in R$ such that $\mathbf{a}, \mathbf{b}, \mathbf{c}$ and μ do not vanish simultaneously. Among the immediate consequences of this inequality are the following:

$$\frac{\partial^2\hat{\psi}}{\partial F_{ij}\partial F_{kl}} a_i a_k b_j b_l > 0 \tag{7.4.5}$$

for every state $(\mathbf{F}, \theta, \boldsymbol{\alpha})$ and all vectors $\mathbf{a}, \mathbf{b} \in R^3$ which do not vanish simultaneously;

$$\frac{\partial^2\hat{\psi}}{\partial \theta^2} < 0 \tag{7.4.6}$$

for every state $(\mathbf{F}, \theta, \boldsymbol{\alpha})$;

$$\frac{\partial^2\hat{\psi}}{\partial \alpha_m \partial \alpha_p} A_{mi} A_{pj} c_i c_j > 0 \tag{7.4.7}$$

for every state $(\mathbf{F}, \theta, \boldsymbol{\alpha})$ and each non-zero vector $\mathbf{c} \in R^3$.

According to (7.4.5), the tensor $\mathbf{Q}$ given by $(7.3.11)_1$ is positive definite, while (7.4.6) says that v defined by $(7.3.11)_2$ is positive, i.e.

$$v = -\rho_0 \theta \frac{\partial^2\hat{\psi}}{\partial \theta^2} > 0. \tag{7.4.8}$$

4c. Let us present now several results that hold for strong equilibrium states. From $(7.3.14)_1$, $(7.4.3)_4$ and (7.4.7) we have

$$z^* = -\frac{\partial \tilde{q}_i^*}{\partial \alpha_m} A_{mj} n_i n_j = \rho_0 \theta^* \frac{\partial^2 \psi^*}{\partial \alpha_m \partial \alpha_p} A^*_{mi} A^*_{mj} n_i n_j > 0. \tag{7.4.9}$$

The symmetry condition $(7.3.18)_1$ (see the notation $(7.3.14)_2$) written at a strong equilibrium state, together with $(7.4.3)_2$, yield

$$\frac{\partial \tilde{q}_i^*}{\partial \theta} = 0, \quad \frac{\partial^2 \psi^*}{\partial \alpha_m \partial \theta} A^*_{mi} = 0. \tag{7.4.10}$$

Thus, from $(7.4.3)_3$ and (7.3.13), we get

$$\mathbf{E}^* = -\rho_0\theta^*\mathbf{D}^*, \qquad (7.4.11)$$

while (7.4.11) and $(7.3.20)_1$ imply

$$\mathbf{P}^*\cdot\mathbf{D}^* = 0. \qquad (7.4.12)$$

Thus if $\mathbf{P}^* \neq \mathbf{0}$ and $\mathbf{D}^* \neq \mathbf{0}$, then due to (7.3.20), either $\mathbf{P}^*$ or $\mathbf{D}^*$ are eigenvectors for $\mathbf{Q}^*$.

5. REAL ACCELERATION WAVES

5a. We shall prove that in the neighborhood of a strong equilibrium state, under the above stated hypotheses of symmetry, quasilinearity and strong ellipticity, all acceleration waves are real.

First, let us show that at a strong equilibrium state the symmetry assumption and the strong ellipticity hypothesis imply the existence of a symmetric matrix $\mathbf{R} = (R_{ij})_{i,j=1,\dots,4}$ whose characteristic equation

$$\det(\mathbf{R} - Z\mathbf{1}) = 0 \qquad (7.5.1)$$

coincides with the equation (7.3.19). Hence all the roots of (7.3.19) will be real for any direction of propagation $\mathbf{n}$.

We choose

$$R_{kl} = \lambda_k^*\delta_{kl}, \quad k, l = 1,2,3, \qquad (7.5.2)$$

where λ_k^*, $k = 1,2,3$ are the eigenvalues of $\mathbf{Q}^*$ (at a strong equilibrium state). The remaining components $R_{i4} = R_{4i}$, $i = 1,2,3,4$ are determined so as to make equation (7.5.1) coincide with equation (7.3.19), written at a strong equilibrium state, where relation (7.4.11) holds. We get (see Mihăilescu and Suliciu [1976])

$$R_{44} = \frac{1}{v^*}(z^* + \rho_0\theta^*\mathbf{P}^*\cdot\mathbf{P}^*) \qquad (7.5.3)$$

and

$$\begin{aligned} R_{14}^2 + R_{24}^2 + R_{34}^2 &= \frac{\rho_0\theta^*}{v^*}(\mathbf{Q}^*\mathbf{P}^*\cdot\mathbf{P}^* + \mathbf{D}^*\cdot\mathbf{D}^*), \\ \lambda_1^*R_{14}^2 + \lambda_2^*R_{24}^2 + \lambda_3^*R_{34}^2 &= \frac{\rho_0\theta^*}{v^*}(\mathbf{Q}^{*2}\mathbf{P}^*\cdot\mathbf{P}^* + \mathbf{Q}^*\mathbf{D}^*\cdot\mathbf{D}^*), \qquad (7.5.4) \\ \lambda_1^{*2}R_{14}^2 + \lambda_2^{*2}R_{24}^2 + \lambda_3^{*2}R_{34}^2 &= \frac{\rho_0\theta^*}{v^*}(\mathbf{Q}^{*3}\mathbf{P}^*\cdot\mathbf{P}^* + \mathbf{Q}^{*2}\mathbf{D}^*\cdot\mathbf{D}^*). \end{aligned}$$

The last relation can also be written as

$$\lambda_2^*\lambda_3^*R_{14}^2 + \lambda_1^*\lambda_3^*R_{24}^2 + \lambda_1^*\lambda_2^*R_{34}^2 = \frac{\rho_0\theta^*}{v^*}\lambda_1^*\lambda_2^*\lambda_3^*(\mathbf{P}^*\cdot\mathbf{P}^* + \mathbf{Q}^{*-1}\mathbf{D}^*\cdot\mathbf{D}^*).$$

$$(7.5.5)$$

If both vectors $\mathbf{P}^*$ and $\mathbf{D}^*$ are written in the basis consisting of the orthonormal eigenvectors $\mathbf{e}_i^*$ of $\mathbf{Q}^*$, i.e. $\mathbf{P}^* = \mathbf{P}_k^* \mathbf{e}_k^*$ and $\mathbf{D}^* = D_k^* e_k^*$, then the system (7.5.4) (or $(7.5.4)_{1-2}$ and (7.5.5)) has a solution

$$R_{i4}^2 = \frac{\rho_0 \theta^*}{v^*} (\lambda_i^* P_i^{*2} + D_i^{*2}), \quad i = 1,2,3. \tag{7.5.6}$$

Let us prove now that the symmetric matrix $\mathbf{R}$ determined in this way is positive definite. Indeed, let us apply the Sylvester criterion. Since $\mathbf{Q}$ is a symmetric positive definite matrix, all its eigenvalues are positive, $\lambda_i > 0$, thus

$$\lambda_1^* > 0, \quad \lambda_1^* \lambda_2^* > 0, \quad \lambda_1^* \lambda_2^* \lambda_3^* > 0.$$

It remains to show that $\det \mathbf{R} > 0$; we have

$$\det \mathbf{R} = \lambda_1^* \lambda_2^* \lambda_3^* R_{44} - \lambda_2^* \lambda_3^* R_{14}^2 - \lambda_1^* \lambda_3^* R_{24}^2 - \lambda_1^* \lambda_2^* R_{34}^2.$$

Taking into account (7.5.3) and (7.5.5), we get

$$\det \mathbf{R} = \frac{\lambda_1^* \lambda_2^* \lambda_3^*}{v^*} (z^* - \rho_0 \theta^* \mathbf{Q}^{*-1} \mathbf{D}^* \cdot \mathbf{D}^*). \tag{7.5.7}$$

Thus $\det \mathbf{R} > 0$ if and only if

$$z^* - \rho_0 \theta^* \mathbf{Q}^{*-1} \mathbf{D}^* \cdot \mathbf{D}^* > 0. \tag{7.5.8}$$

But this inequality is a consequence of the strong ellipticity condition (7.4.4) written at a strong equilibrium state. Indeed, if we take $\mathbf{b} = \mathbf{c} = \mathbf{n}$ and $\mu = 0$, in (7.4.4) and we use the notation $(7.3.11)_1$, $(7.3.13)_1$ and relation (7.4.9), we obtain

$$\mathbf{Q}_{ik}^* a_i a_k + 2 D_i^* a_i + \frac{z^*}{\rho_0 \theta^*} > 0$$

for any vector $\mathbf{a} = (a_1, a_2, a_3) \in R^3$. To get inequality (7.5.8), we have to choose now $\mathbf{a} = -\mathbf{Q}^{*-1} \mathbf{D}^*$.

Thus the proposition stated at the beginning of this section is proved, i.e. all the roots of the equation

$$\begin{aligned} & v^* Z^4 - (v^* I_{\mathbf{Q}^*} + z^* + \rho_0 \theta^* \mathbf{P}^* \cdot \mathbf{P}^*) Z^3 + [v^* II_{\mathbf{Q}^*} + z^* I_{\mathbf{Q}^*} - \rho_0 \theta^* \mathbf{D}^* \cdot \mathbf{D}^* + \\ & + \rho_0 \theta^* (I_{\mathbf{Q}^*} \mathbf{P}^* \cdot \mathbf{P}^* - \mathbf{Q}^* \mathbf{P}^* \cdot \mathbf{P}^*)] Z^2 - [v^* III_{\mathbf{Q}^*} + z^* II_{\mathbf{Q}^*} - \\ & - \rho_0 \theta^* I_{\mathbf{Q}^*} \mathbf{D}^* \cdot \mathbf{D}^* + \rho_0 \theta^* \mathbf{Q}^* \mathbf{D}^* \cdot \mathbf{D}^* + \rho_0 \theta^* (II_{\mathbf{Q}^*} \mathbf{P}^* \cdot \mathbf{P}^* - \\ & I_{\mathbf{Q}^*} \mathbf{Q}^* \mathbf{P}^* \cdot \mathbf{P}^* + \mathbf{Q}^{*2} \mathbf{P}^* \cdot \mathbf{P}^*)] Z + z^* III_{\mathbf{Q}^*} + \rho_0 \theta^* I_{\mathbf{Q}^*} \mathbf{Q}^* \mathbf{D}^* \cdot \mathbf{D}^* - \\ & - \rho_0 \theta^* II_{\mathbf{Q}^*} \mathbf{D}^* \cdot \mathbf{D}^* - \rho_0 \theta^* \mathbf{Q}^{*2} \mathbf{D}^* \cdot \mathbf{D}^* = 0 \end{aligned} \tag{7.5.9}$$

are real and positive, and this means that all acceleration waves, at a strong equilibrium state, propagate with real and symmetric wave speeds. One can easily see that the waves remain real and symmetric in a neighborhood of a strong equilibrium state.

5b. The above results hold in the neighborhood of a strong equilibrium state. In proving these results, relations (7.4.9) and (7.4.11) were essential. However, these relations will hold at every state if $\hat{\mathbf{A}}(\mathbf{F}, \theta, \boldsymbol{\alpha})$ is constant with respect to $\mathbf{F}$ and $\boldsymbol{\alpha}$, i.e. if

$$\frac{\partial \hat{\mathbf{A}}}{\partial \mathbf{F}} = \mathbf{0}, \quad \frac{\partial \hat{\mathbf{A}}}{\partial \boldsymbol{\alpha}} = \mathbf{0}. \tag{7.5.10}$$

Indeed, let us take the derivative with respect to $\mathbf{F}$ of the relation $(7.2.9)_1$; then, using the notation (7.3.13) and the hypothesis $(7.5.10)_1$, we get

$$\mathbf{E} = -\rho_0\theta\mathbf{D} \tag{7.5.11}$$

at every state $(\mathbf{F}, \theta, \mathbf{g}, \boldsymbol{\alpha})$ and each direction of propagation $\mathbf{n}$. Similarly, taking the derivative of $(7.2.9)_1$ with respect to $\boldsymbol{\alpha}$ and using the notation $(7.3.14)_1$ together with condition $(7.5.10)_2$, we obtain

$$z = -\frac{\partial \tilde{q}_i}{\partial \alpha_m} A_{mj} n_i n_j = \rho_0\theta \frac{\partial^2 \hat{\psi}}{\partial \alpha_m \partial \alpha_p} A_{mi} A_{mj} n_i n_j > 0. \tag{7.5.12}$$

If $\mathbf{a} = \mathbf{b} = \mathbf{0}$, $v = 0$ and $\mathbf{c} = \mathbf{n}$, then the strong ellipticity condition (7.4.4) implies $z > 0$.

Under the symmetry assumption I, the strong ellipticity condition and the assumptions (7.5.10), the characteristic equation (7.3.17) takes the form (7.5.9) for every state $(\mathbf{F}, \theta, \mathbf{g}, \boldsymbol{\alpha})$ and each direction of propagation $\mathbf{n}$. As it was shown in Subsection 5a, this equation has real and positive roots. Thus, all the acceleration waves are real and symmetric at every state and for each direction $\mathbf{n}$, provided the functions $\hat{\psi}(\mathbf{F}, \theta, \boldsymbol{\alpha})$ and $\hat{\mathbf{A}}(\mathbf{F}, \theta, \boldsymbol{\alpha})$ are such that the conditions (7.5.10), (7.4.4), $(7.3.18)_1$ and

$$\mathbf{P}\cdot\mathbf{D} = 0, \quad \mathbf{QP}\cdot\mathbf{D} = 0, \quad \mathbf{Q}^2\mathbf{P}\cdot\mathbf{D} = 0, \tag{7.5.13}$$

are satisfied at every state and for each direction of propagation $\mathbf{n}$.

6. THE SECOND SYMMETRY ASSUMPTION AND THE HEAT FLUX IN THE NEIGHBORHOOD OF STRONG EQUILIBRIUM STATES

6a. Considering the coupling between the thermal and mechanical behaviour, we have to distinguish two important cases concerning the behaviour of the stress-temperature moduli $\partial^2\hat{\psi}/\partial\theta\partial\mathbf{F}$. The instantaneous coupled case is the case when the vector **P** given by (7.3.12) is different from zero at every state $(\mathbf{F}, \theta, \boldsymbol{\alpha})$ and for each unit vector $\mathbf{n} \in R^3$, i.e.

$$\mathbf{P} = \left(\frac{\partial^2 \hat{\psi}}{\partial\theta \partial F_{1j}} n_j, \frac{\partial^2 \hat{\psi}}{\partial\theta \partial F_{2j}} n_j, \frac{\partial^2 \hat{\psi}}{\partial\theta \partial F_{3j}} n_j \right) \neq \mathbf{0} \tag{7.6.1}$$

for all $\mathbf{n} \in R^3$, $\mathbf{n}\cdot\mathbf{n} = 1$ and every state $(\mathbf{F}, \theta, \boldsymbol{\alpha})$. The instantaneous uncoupled case is the one when **P** is zero at every state and for each unit vector **n**, i.e.

$$\frac{\partial^2 \hat{\psi}}{\partial\theta \partial F_{ij}} (\mathbf{F}, \theta, \boldsymbol{\alpha}) = 0 \text{ for every } (\mathbf{F}, \theta, \boldsymbol{\alpha}), \quad i, j = 1,2,3. \tag{7.6.2}$$

6b. We shall supplement presently our assumption of quasilinearity, the symmetry assumption I and the strong ellipticity condition by another symmetry assumption, stated at a strong equilibrium state. The form of this new symmetry assumption depends on the value of **P** given by $(7.6.1)_1$ at the strong equilibrium state.

The symmetry assumption II. Suppose $\mathbf{P}^* \neq \mathbf{0}$ for every direction of propagation **n**. Then the mechanical amplitude **a** of the wave is the same for both speeds of propagation $U = \sqrt{Z^*}$ and $U = -\sqrt{Z^*}$ (where Z^* is any root of equation (7.5.9)) for every given direction of propagation **n**.

The symmetry assumption II′. Suppose $\mathbf{P}^* = \mathbf{0}$ for every direction of propagation **n**. Then the thermomechanical amplitude of the wave $(\mathbf{a}, \nu)$ is the same for both speeds of propagation $U = \sqrt{Z^*}$ and $U = -\sqrt{Z^*}$ (where Z^* is any root of equation (7.5.9)) for every given direction of propagation **n**.

6c. Let Z_i^* be any root of equation (7.5.9); then we may find the corresponding to this root wave speeds $U = \sqrt{Z_i^*}$ und $\tilde{U} = -\sqrt{Z_i^*}$ and the thermomechanical amplitudes of the waves $(\mathbf{a}_i, \nu_i)$ and $(\tilde{\mathbf{a}}_i, \tilde{\nu}_i)$. In other words, the linear system (7.3.22)–(7.3.23) possesses the solutions

$$\begin{aligned} &(\mathbf{a}_i, \nu_i) \text{ for } U = \sqrt{Z_i^*}, \\ &(\tilde{\mathbf{a}}_i, \tilde{\nu}_i) \text{ for } U = \tilde{U} = -\sqrt{Z_i^*}. \end{aligned} \tag{7.6.3}$$

The second symmetry assumption, in the presence of condition (7.6.1), requires that $\mathbf{a}_i = \tilde{\mathbf{a}}_i$. Then equation (7.3.23) implies

$$(\nu_i - \tilde{\nu}_i)\, \mathbf{P}^* = \frac{1}{\sqrt{Z_i^*}} (\nu_i + \tilde{\nu}_i)\, \mathbf{D}^*. \tag{7.6.4}$$

Now, multiplying this relation by $\mathbf{P}^*$ and taking into account (7.4.12) and (7.6.1), one obtains

$$v_i = \tilde{v}_i, \quad v_i\mathbf{D}^* = \mathbf{0}, \quad i = 1,2,3,4, \tag{7.6.5}$$

i.e. in the instantaneously coupled case the symmetry of the mechanical amplitude of the wave implies the symmetry of the thermal amplitude of the wave. If at least one $v_i \neq 0$ ($i = 1,2,3,4$), then it follows from $(7.6.5)_2$ that

$$\mathbf{D}^* = \mathbf{0}. \tag{7.6.6}$$

Let us assume now that all $v_i = 0$, $i = 1,2,3,4$. Then, (7.3.23) implies that $Z^*\mathbf{a} - \mathbf{Q}^*\mathbf{a} = \mathbf{0}$, thus three of the roots of equation (7.5.9) are eigenvalues of $\mathbf{Q}^*$, i.e.

$$Z_k^* = \lambda_k^*, \quad k = 1,2,3 \text{ and } Z_4^* = \frac{1}{v^*}\,(z^* + \rho_0\theta^*\mathbf{P}^*\cdot\mathbf{P}^*). \tag{7.6.7}$$

But in this case (7.5.7) implies

$$\mathbf{P}^*\cdot\mathbf{P}^* + \mathbf{Q}^{*-1}\mathbf{D}^*\cdot\mathbf{D}^* = 0. \tag{7.6.8}$$

This yields $\mathbf{P}^* = \mathbf{0}$ and $\mathbf{D}^* = \mathbf{0}$, which contradicts (7.6.1). Thus we deduce that at least two of the v_i must be different from zero and the conclusion (7.6.6) follows.

Let us summarize: The condition $\mathbf{P}^* \neq \mathbf{0}$ and the symmetry assumption II' imply that the thermal amplitude of the wave is also symmetric, that $\mathbf{D}^* = \mathbf{0}$ and at least two waves also carry thermal discontinuities.

6d. In the instantaneously uncoupled case, i.e. the case when (7.6.2) holds, the symmetry of the mechanical amplitude of the wave does not imply the symmetry of the thermal amplitude. That means that $\mathbf{a}_i = \tilde{\mathbf{a}}_i$ in (7.6.3) does not imply $v_i = \tilde{v}_i$ and thus (7.6.6) cannot be deduced. However, if one requires in this case the symmetry of both mechanical and thermal amplitudes of the wave, i.e. if symmetry assumption II' is accepted, then from (7.6.4) follows the conclusion $(7.6.5)_2$. By the same kind of reasoning as in the previous case one can conclude that the roots of equation (7.5.9) are

$$Z_k^* = \lambda_k^*, \; k = 1, 2, 3, \; Z_4^* = \frac{z^*}{v^*}. \tag{7.6.9}$$

The corresponding amplitudes of the waves are

$$(\mathbf{a}_k, 0), \quad k = 1, 2, 3, \; (\mathbf{0}, v_4), \tag{7.6.10}$$

and, since $v_4 \neq 0$, (7.6.6) holds.

These conclusions justify why this case is called the *instantaneously uncoupled* case; indeed, there are three mechanical waves propagating with the speeds $U = \pm\sqrt{Z_k^*}$, $k = 1, 2, 3$, whose thermal amplitudes vanish and a thermal wave propagating with speed $U = \pm\sqrt{Z_4^*}$ whose mechanical amplitude is zero.

6e. We shall study next the properties of the heat flux in the neighborhood of a strong equilibrium state under the quasilinearity assumption, the first and second symmetry assumptions and the strong ellipticity condition.

Since the internal state variables were chosen to be frame indifferent and since $\mathbf{q} = \frac{1}{J}\mathbf{F}\tilde{\mathbf{q}}$, it follows from $(6.3.9)_2$ and (7.2.7) that

$$\tilde{\mathbf{q}}(\mathbf{F}, \theta, \boldsymbol{\alpha}) = J\check{\mathbf{q}}(\mathbf{C}, \theta, \boldsymbol{\alpha}) = \tilde{\tilde{\mathbf{q}}}(\mathbf{C}, \theta, \boldsymbol{\alpha}). \tag{7.6.11}$$

Let us show that at a strong equilibrium state

$$\frac{\partial\tilde{\mathbf{q}}^*}{\partial\mathbf{F}} = \mathbf{0}. \tag{7.6.12}$$

Indeed, from (7.6.6), (7.4.11), using the notation (7.3.13) one gets

$$\frac{\partial\tilde{q}_k^*}{\partial F_{ij}} + \frac{\partial\tilde{q}_j^*}{\partial F_{ik}} = 0; \tag{7.6.13}$$

but (7.6.11) implies

$$\frac{\partial\tilde{q}_k}{\partial F_{ij}} = 2\frac{\partial\tilde{\tilde{q}}_k}{\partial C_{lj}} F_{il}. \tag{7.6.14}$$

Substituting (7.6.14) into (7.6.13) we get

$$\frac{\partial\tilde{\tilde{q}}_k^*}{\partial C_{ij}} + \frac{\partial\tilde{\tilde{q}}_j^*}{\partial C_{ik}} = 0.$$

Since $C_{ij} = C_{ji}$, this last relation yields

$$\frac{\partial\tilde{\tilde{q}}_k^*}{\partial C_{ij}} = 0, \quad i, j, k = 1, 2, 3, \tag{7.6.15}$$

and thus (7.6.12) is proved.

From $(7.4.3)_3$ and (7.6.12) follows that also the equality

$$\frac{\partial^2\hat{\psi}^*}{\partial F_{ij}\partial\alpha_m} A_{mk}^* = 0, \; i, j, k = 1, 2, 3 \tag{7.6.16}$$

holds.

Considering $(7.4.10)_1$ and (7.6.12) in a neighborhood of a strong equilibrium state, we can write

$$\tilde{q}_i = \tilde{q}_i(\mathbf{F}, \theta, \boldsymbol{\alpha}) = \frac{\partial \tilde{q}_i^*}{\partial \alpha_m} (\alpha_m - \alpha_m^*) + 0_i^{(2)} (|\mathbf{F} - \mathbf{F}^*|, |\theta - \theta^*|, |\boldsymbol{\alpha} - \boldsymbol{\alpha}^*|). \tag{7.6.17}$$

Therefore, under the condition stated at the beginning of this subsection, the heat flux in the neighborhood of a strong equilibrium state can be approximated up to an error of order $0^{(2)} (|\mathbf{F} - \mathbf{F}^*|, |\theta - \theta^*|, |\boldsymbol{\alpha} - \boldsymbol{\alpha}^*|)$ by a linear function of the internal state variables only.

Conversely, if (7.6.17) holds at a strong equilibrium state, i.e. if $\partial \tilde{\mathbf{q}}^*/\partial \theta = \mathbf{0}$ and $\partial \tilde{\mathbf{q}}^*/\partial \mathbf{F} = \mathbf{0}$, then at that strong equilibrium state the first and second symmetry assumptions are satisfied.

7. THE LINEAR THEORY

We shall apply presently the results of the previous sections to the linearized isotropic theory.

7a. We have originally assumed that some of the internal state variables are not invariant under the isotropy group of the material. In the first case (see MIHĂILESCU and SULICIU [1976]) we assumed that the internal state variables vector $\boldsymbol{\alpha} = (\alpha_1, \ldots, \alpha_N)$ contains two groups of components

$$\boldsymbol{\alpha} = (\boldsymbol{\beta}, \bar{\boldsymbol{\alpha}}), \tag{7.7.1}$$

where $\boldsymbol{\beta} = (\alpha_1, \alpha_2, \alpha_3)$ behaves as $\tilde{\mathbf{q}}$ and $\bar{\boldsymbol{\alpha}} = (\alpha_4, \ldots, \alpha_N)$ is invariant under the isotropy group. In the second case we shall assume that the vector $\boldsymbol{\alpha}$ contains three groups of components differing in their behaviour under the isotropy group (see MIHĂILESCU-SULICIU and SULICIU [1978]): 1) a group $\bar{\boldsymbol{\alpha}} = (\alpha_{10}, \ldots, \alpha_N)$ containing scalar variables, 2) a group $\boldsymbol{\beta} = (\alpha_1, \alpha_2, \alpha_3)$ that behaves as a three-dimensional vector and 3) a group $\tilde{\boldsymbol{\alpha}} = (\alpha_4, \ldots, \alpha_9)$ of six components that behaves as a symmetric second order tensor, i.e.

$$\boldsymbol{\alpha} = (\boldsymbol{\beta}, \tilde{\boldsymbol{\alpha}}, \bar{\boldsymbol{\alpha}}). \tag{7.7.2}$$

7b. Let $(\mathbf{F} = \mathbf{1}, \theta = \theta_0, \mathbf{g} = \mathbf{0}, \boldsymbol{\alpha} = \mathbf{0})$ be an equilibrium state with the property

$$\psi^* = 0, \ \eta^* = 0, \quad \tilde{\mathbf{S}}^* = \mathbf{0}. \tag{7.7.3}$$

We shall follow a standard procedure of linearization (see for instance GURTIN [1972], Section 23). We assume that

$$(F_{ij} - \delta_{ij})(F_{ij} - \delta_{ij}) \ll 1, \; |\theta - \theta_0|/\theta_0 \ll 1$$

and that $|\boldsymbol{\alpha}|$ is sufficiently small. Then the small strain tensor

$$\boldsymbol{\varepsilon} = \frac{1}{2}(\mathbf{F} + \mathbf{F}^T - 2\mathbf{1}) \text{ and } \vartheta = \frac{\theta - \theta_0}{\theta_0} \tag{7.7.5}$$

satisfies the inequalities

$$\varepsilon_{ij}\varepsilon_{ij} \ll 1, \quad |\vartheta| \ll 1. \tag{7.7.6}$$

We consider first the case when $\boldsymbol{\alpha}$ has the form (7.7.2); the other case will then follow directly.

The principle of frame indifference requires that $\hat{\psi}(\mathbf{F}, \theta, \boldsymbol{\alpha}) = \check{\psi}(\mathbf{C}, \theta, \boldsymbol{\alpha})$ (see $(6.3.8)_1$), where $\mathbf{C} = \mathbf{F}^T\mathbf{F}$, and according to the linearity hypothesis, $\mathbf{C}$ is approximated by its linear part, that is $\mathbf{1} + 2\boldsymbol{\varepsilon}$. Further, one assumes that there is a neighborhood of the state $(\mathbf{1}, \theta_0, \mathbf{0})$ in the space $(\mathbf{F}, \theta, \boldsymbol{\alpha})$ or, equivalently, of the state $(\mathbf{0}, 0, \mathbf{0})$ in the space $(\boldsymbol{\varepsilon}, \vartheta, \boldsymbol{\alpha})$, where the free energy can be approximated by its TAYLOR second degree polynomial $\tilde{\psi}(\boldsymbol{\varepsilon}, \vartheta, \boldsymbol{\beta}, \tilde{\boldsymbol{\alpha}}, \boldsymbol{\alpha})$, i.e. $\hat{\psi}(\mathbf{F}, \theta, \boldsymbol{\alpha}) \cong \tilde{\psi}(\boldsymbol{\varepsilon}, \vartheta, \boldsymbol{\beta}, \tilde{\boldsymbol{\alpha}}, \bar{\boldsymbol{\alpha}})$. On the other hand, isotropy will imply that $\tilde{\psi}(\boldsymbol{\varepsilon}, \vartheta, \boldsymbol{\beta}, \tilde{\boldsymbol{\alpha}}, \bar{\boldsymbol{\alpha}})$ is an isotropic function of the two tensors $\boldsymbol{\varepsilon}$ and $\tilde{\boldsymbol{\alpha}}$ and the vector $\boldsymbol{\beta}$. Then (7.7.3) together with the representation theorem for isotropic scalar functions (see for instance WANG [1969, 1970, 1971]) lead to the following expression for the free energy in the neighborhood of the considered strong equilibrium state:

$$\begin{aligned}\rho_0\tilde{\psi}(\varepsilon, \vartheta, \boldsymbol{\beta}, \tilde{\boldsymbol{\alpha}}, \bar{\boldsymbol{\alpha}}) &= \frac{\lambda}{2}(tr\varepsilon)^2 + \mu tr(\varepsilon^2) - \varkappa\vartheta(tr\varepsilon) + \Lambda_i\bar{\alpha}_i tr\varepsilon + \\ &+ \Lambda' tr(\varepsilon\tilde{\boldsymbol{\alpha}}) + \Lambda''(tr\varepsilon)(tr\tilde{\boldsymbol{\alpha}}) + \Delta'(tr\tilde{\boldsymbol{\alpha}})^2 + \Delta'' tr(\tilde{\boldsymbol{\alpha}}^2) + \Gamma\vartheta tr\tilde{\boldsymbol{\alpha}} + \\ &+ \Gamma_i\vartheta\bar{\alpha}_i + \Delta_i\bar{\alpha}_i tr\tilde{\boldsymbol{\alpha}} - \frac{1}{2}v\theta_0\vartheta^2 + \xi\boldsymbol{\beta}\cdot\boldsymbol{\beta} + \Omega_{ij}\bar{\alpha}_i\bar{\alpha}_j.\end{aligned} \tag{7.7.7}$$

In the expression (7.7.7), λ and μ are the LAMÉ constants, $\varkappa/\theta_0$ is the stress-temperature modulus, v/θ_0 is the heat capacity and Λ', Λ'', $\Delta', \Delta'', \Gamma, \xi, \Lambda_i, \Gamma_i, \Delta_i$ and Ω_{ij} are defined as follows:

$$2\xi\delta_{kl} = \rho_0\frac{\partial^2\hat{\psi}^*}{\partial\beta_k\partial\beta_l}, \quad \Gamma\delta_{ij} = \rho_0\frac{\partial^2\hat{\psi}^*}{\partial\tilde{\alpha}_{ij}\partial\theta},$$

$$2\Delta'\delta_{ij}\delta_{kl} + 2\Delta''\delta_{ik}\delta_{jl} = \rho_0\frac{\partial^2\hat{\psi}^*}{\partial\tilde{\alpha}_{ij}\partial\tilde{\alpha}_{kl}}, \tag{7.7.8}$$

$$\frac{1}{2}\Lambda'(\delta_{ik}\delta_{jl}+\delta_{jk}\delta_{il})+\Lambda''\delta_{ij}\delta_{kl}=\rho_0\frac{\partial^2\hat{\psi}^*}{\partial\tilde{\alpha}_{ij}\partial F_{kl}},$$

$$\Gamma_k=\rho_0\theta_0\frac{\partial^2\hat{\psi}^*}{\partial\bar{\alpha}_k\partial\theta},\quad \Lambda_k\delta_{ij}=\rho_0\frac{\partial^2\hat{\psi}^*}{\partial\bar{\alpha}_k\partial\mathbf{F}_{ij}},\tag{7.7.8}$$

$$\Delta_k\delta_{ij}=\rho_0\frac{\partial^2\hat{\psi}^*}{\partial\bar{\alpha}_k\partial\tilde{\alpha}_{ij}},\quad \Omega_{kl}=\rho_0\frac{\partial^2\hat{\psi}^*}{\partial\bar{\alpha}_k\partial\bar{\alpha}_l}.$$

Substituting (7.7.7) into $(7.2.3)_{1-2}$ and $(6.1.17)_2$ we determine the stress, the entropy and the internal energy as follows:

$$\tilde{S}_{ij}=\lambda(tr\varepsilon)\,\delta_{ij}+2\mu\varepsilon_{ij}-\varkappa\vartheta\delta_{ij}+\Lambda_k\bar{\alpha}_k\delta_{ij}+\Lambda'\tilde{\alpha}_{ij}+\Lambda''(tr\tilde{\boldsymbol{\alpha}})\,\delta_{ij}$$

$$\rho_0\theta_0\eta=\varkappa(tr\,\varepsilon)-\Gamma tr\tilde{\boldsymbol{\alpha}}+v\theta_0\vartheta-\Gamma_i\bar{\alpha}_i,$$

$$\rho_0 e=\frac{\lambda}{2}(tr\varepsilon)^2+\mu tr(\varepsilon^2)+\Lambda_i\bar{\alpha}_i tr\varepsilon+\Lambda' tr(\varepsilon\tilde{\boldsymbol{\alpha}})+\Lambda''(tr\varepsilon)(tr\tilde{\boldsymbol{\alpha}})+\tag{7.7.9}$$

$$+\Delta'(tr\tilde{\boldsymbol{\alpha}})^2+\Delta'' tr(\tilde{\boldsymbol{\alpha}}^2)+\Delta_i\bar{\alpha}_i tr\tilde{\boldsymbol{\alpha}}+\xi\boldsymbol{\beta}\cdot\boldsymbol{\beta}+\Omega_{ij}\bar{\alpha}_i\bar{\alpha}_j+$$

$$+\frac{1}{2}v\theta_0\vartheta(2+\vartheta)+\varkappa tr\varepsilon-\Gamma tr\tilde{\boldsymbol{\alpha}}-\Gamma_i\bar{\alpha}_i.$$

7c. Let us consider the evolution function **f** as a linear function of its arguments ε, ϑ, $\mathbf{g}_0$, $\boldsymbol{\beta}$, $\tilde{\boldsymbol{\alpha}}$, $\bar{\boldsymbol{\alpha}}$ in a neighborhood of the equilibrium state ($\varepsilon=\mathbf{0}$, $\vartheta=0$, $\mathbf{g}_0=\mathbf{0}$, $\boldsymbol{\beta}=\mathbf{0}$, $\tilde{\boldsymbol{\alpha}}=\mathbf{0}$, $\bar{\boldsymbol{\alpha}}=\mathbf{0}$). Due to the decomposition (7.7.2) of $\boldsymbol{\alpha}$, the evolution equation $(7.2.3)_4$ can be written as

$$\dot{\bar{\boldsymbol{\alpha}}}=\bar{\mathbf{f}}(\varepsilon,\vartheta,\mathbf{g}_0,\boldsymbol{\beta},\tilde{\boldsymbol{\alpha}},\bar{\boldsymbol{\alpha}}),$$

$$\dot{\boldsymbol{\beta}}=\check{\mathbf{f}}(\varepsilon,\vartheta,\mathbf{g}_0,\boldsymbol{\beta},\tilde{\boldsymbol{\alpha}},\bar{\boldsymbol{\alpha}}),\tag{7.7.10}$$

$$\dot{\tilde{\boldsymbol{\alpha}}}=\tilde{\mathbf{f}}(\varepsilon,\vartheta,\mathbf{g}_0,\boldsymbol{\beta},\tilde{\boldsymbol{\alpha}},\bar{\boldsymbol{\alpha}}),$$

where $\bar{\mathbf{f}}$, $\check{\mathbf{f}}$ and $\tilde{\mathbf{f}}$ have to be isotropic linear scalar, vector and tensor valued functions respectively, of two symmetric second order tensors and two vectors. Then (see for instance WANG [1969])

$$\dot{\bar{\alpha}}_j=\bar{A}^1_j tr\varepsilon+\bar{A}^2_j tr\tilde{\boldsymbol{\alpha}}+\bar{A}^3_j\vartheta+\bar{A}^4_{jk}\bar{\alpha}_k,$$

$$\dot{\boldsymbol{\beta}}=A\mathbf{g}_0-\frac{1}{\tau}\boldsymbol{\beta},\tag{7.7.11}$$

$$\dot{\tilde{\boldsymbol{\alpha}}}=\tilde{A}_1\varepsilon+\tilde{A}_2\tilde{\boldsymbol{\alpha}}+(\tilde{A}_3\vartheta+\tilde{A}_4 tr\varepsilon+\tilde{A}_5 tr\tilde{\boldsymbol{\alpha}}+\tilde{A}_{6k}\bar{\boldsymbol{\alpha}}_k)\,\mathbf{1},$$

where A, τ, $\tilde{A}_i$, $i=1,\dots,5$, $\tilde{A}_{6k}$, $k=10,\dots,N$, $\bar{A}^p_j$, $p=1,2,3$, $j=10,\dots,N$ and $\bar{A}^4_{kj}$, $j,k=10,\dots,N$ are scalar constants.

Observe that the temperature gradient appears now only in the evolution equation for $\boldsymbol{\beta}$, which contains neither the strain nor the temperature and is completely decoupled from the evolution equations for $\tilde{\boldsymbol{\alpha}}$ and $\bar{\boldsymbol{\alpha}}$.

7d. Until now, the linearity and isotropy assumptions together with the decomposition (7.7.2) of $\boldsymbol{\alpha}$ were the only hypotheses used in deriving the above results. We shall discuss next the implications on the linear theory of the assumptions stated in the previous sections.

The quasilinearity assumption (7.2.8) is automatically fulfilled due to the linearity assumption. Moreover, the material isotropy assumption together with the decomposition (7.7.2) determine the map $\hat{\mathbf{A}}(\mathbf{F}, \theta, \boldsymbol{\alpha})$ which reduces to multiplying the temperature gradient $\mathbf{g}_0$ in $(7.7.11)_2$ by a constant A, the other components being equal to zero.

The linearity and the isotropy of the function $\tilde{\mathbf{q}}(\boldsymbol{\varepsilon}, \vartheta, \mathbf{g}_0, \boldsymbol{\beta}, \tilde{\boldsymbol{\alpha}}, \bar{\boldsymbol{\alpha}})$ imply that it must have the form

$$\tilde{\mathbf{q}} = a\mathbf{g}_0 - \gamma\boldsymbol{\beta}, \tag{7.7.12}$$

where a and γ are constants.

The assumption (7.2.7) requires that $a = 0$, hence

$$\tilde{\mathbf{q}} = -\gamma\boldsymbol{\beta}. \tag{7.7.13}$$

Thus the heat flux $\tilde{\mathbf{q}}$ is proportional to the internal state variable $\boldsymbol{\beta}$. (If in (7.7.12) one takes $\gamma = 0$ then the FOURIER constitutive equation is obtained). Substituting (7.7.7) and (7.7.13) into $(7.2.9)_1$, we obtain

$$\gamma = 2\theta_0 A\xi. \tag{7.7.14}$$

From (7.7.13) and $(7.7.11)_2$ one gets

$$-\tau\dot{\tilde{\mathbf{q}}} = A\gamma\tau\mathbf{g}_0 + \tilde{\mathbf{q}}. \tag{7.7.15}$$

This last constitutive equation is just the CATTANEO *hyperbolic heat conduction constitutive equation* [1948, 1958]. The constant τ is called the *relaxation time* and $k = A\gamma\tau$ is called the *heat conduction coefficient*. From (7.7.15) it can be seen that if $\tau\dot{\tilde{\mathbf{q}}}$ is small, then $\tilde{\mathbf{q}} \simeq -k\mathbf{g}_0$, i.e. the FOURIER constitutive equation holds. On the other hand, if $\mathbf{g}_0$ is held constant for all $t \geqslant 0$ and $\tilde{\mathbf{q}}(0) = 0$, the heat flux is $\tilde{\mathbf{q}}(t) = -k\mathbf{g}_0[1 - \exp(-t/\tau)]$, hence $\tilde{\mathbf{q}}(t) \cong -k\mathbf{g}_0$ if $\tau > 0$, for t large enough.

The constants in (7.7.7) and (7.7.11) must be such that inequality $(7.2.9)_2$ holds, i.e.

$$-\frac{2\xi}{\tau}\beta^2 + [\Lambda'\varepsilon_{ij} + 2\Delta'\tilde{\alpha}_{ij} + (\Lambda'' tr\boldsymbol{\varepsilon} + 2\Delta' tr\tilde{\boldsymbol{\alpha}} + \Gamma\vartheta + \Delta_k\bar{\alpha}_k)\delta_{ij}] \times$$
$$\times [\tilde{A}_1\varepsilon_{ij} + \tilde{A}_2\tilde{\alpha}_{ij} + (\tilde{A}_3\vartheta + \tilde{A}_4 tr\boldsymbol{\varepsilon} + \tilde{A}_5 tr\tilde{\boldsymbol{\alpha}} + \tilde{A}_{6k}\bar{\alpha}_k)\,\delta_{ij}] + (\Lambda_i tr\boldsymbol{\varepsilon} +$$
$$+ \Delta_i tr\tilde{\boldsymbol{\alpha}} + \Gamma_i\vartheta + 2\Omega_{ij}\bar{\alpha}_j)(\bar{A}_i^1 tr\boldsymbol{\varepsilon} + \bar{A}_i^2 tr\tilde{\boldsymbol{\alpha}} + \bar{A}_i^3\vartheta + \bar{A}_{ik}^4\bar{\alpha}_k) \leqslant 0 \tag{7.7.16'}$$

holds for all sufficiently small $\boldsymbol{\varepsilon}, \vartheta, \boldsymbol{\beta}, \tilde{\boldsymbol{\alpha}}, \bar{\boldsymbol{\alpha}}$. Among other restrictions, (7.7.16′) implies that

$$\frac{\xi}{\tau} \geqslant 0. \tag{7.7.16}$$

From (7.7.14) follows $\gamma^2 = 2\theta_0 A\gamma\xi \geqslant 0$; since $\theta_0 > 0$ and ξ and τ have the same sign, one concludes

$$k = A\gamma\tau \geqslant 0. \tag{7.7.17}$$

The strong ellipticity condition (7.4.4) applied to the constitutive equations (7.7.7) and (7.7.11) is satisfied if and only if

$$\mu > 0, \ \lambda + 2\mu > 0, \ v > 0, \ \xi > 0, \ A \neq 0. \tag{7.7.18}$$

From $(7.7.18)_{4-5}$, (7.7.16) and (7.7.14) follows that

$$\tau > 0, \quad \gamma \neq 0, \tag{7.7.19}$$

and thus

$$k > 0 \text{ and } \gamma A > 0. \tag{7.7.20}$$

7e. In order to write equation (7.3.17) for the speed of the acceleration waves in this case, one needs to know the form of the quantities defined by (7.3.11)–(7.3.14). These are obtained from the free energy (7.7.7) and the constitutive equation (7.7.11):

$$\begin{gathered} \rho_0 Q_{ij} = (\lambda + \mu) n_i n_j + \mu \delta_{ij}, \ P_i = -\frac{\varkappa}{\rho_0\theta_0} n_i, \\ D_i = 0, \quad E_i = 0, \\ z = \gamma A, \ w = 2\xi A(\beta_i n_i). \end{gathered} \tag{7.7.21}$$

Let us further introduce the following notations:

$$a_l^2 = \frac{\lambda + 2\mu}{\rho_0}, \ a_t^2 = \frac{\mu}{\rho_0}, \ a_c^2 = \frac{z}{v}, \ a_d^2 = \frac{\varkappa^2}{\rho_0\theta_0 v}. \tag{7.7.22}$$

Using (7.7.21), (7.7.22) and the linearity assumption, we can write equation (7.3.17) as

$$U^{N-1}(U^2 - a_t^2)^2\{U^4 - (a_l^2 + a_c^2 + a_d^2)U^2 + a_c^2 a_l^2\} = 0, \tag{7.7.23}$$

while the equations (7.3.22) and (7.3.23) for the amplitudes of the waves become

$$\begin{gathered} (vU^2 - z)v = \rho_0\theta_0 U^2 \mathbf{P}\cdot\mathbf{a}, \\ (U^2\mathbf{1} - \mathbf{Q})\mathbf{a} = v\mathbf{P}. \end{gathered} \tag{7.7.24}$$

An important observation concerning equations (7.7.23) and (7.7.24) is that both symmetry assumptions are automatically satisfied in the linear and isotropic case. Thus the existence of real and symmetric acceleration waves is ensured by the so-called quasilinearity assumption which reduces here to condition (7.2.7) (leading to CATTANEO's heat conduction equation) and by the strong ellipticity condition. Let us also note that the equilibrium state $\boldsymbol{\varepsilon} = \mathbf{0}$, $\vartheta = 0$, $\mathbf{g}_0 = \mathbf{0}$, $\tilde{\boldsymbol{\alpha}} = \mathbf{0}$, $\tilde{\boldsymbol{\alpha}} = \mathbf{0}$, $\bar{\boldsymbol{\alpha}} = \mathbf{0}$ is a strong equilibrium state since at this state $\partial\tilde{\psi}/\partial\boldsymbol{\beta} = \mathbf{0}$, $\partial\tilde{\psi}/\partial\tilde{\boldsymbol{\alpha}} = \mathbf{0}$, $\partial\tilde{\psi}/\partial\bar{\boldsymbol{\alpha}} = \mathbf{0}$. The condition on P_i in the second symmetry assumption reduces presently to the condition that $\varkappa \neq 0$ or $\varkappa = 0$.

In the notation (7.7.22) a_l is the speed of the longitudinal (or dilatational) mechanical wave, a_t is the speed of the transversal (or shear) mechanical wave, a_l^2 is an eigenvalue of $\mathbf{Q}$ given by $(7.7.21)_1$ and a_t^2 is a double eigenvalue of the same $\mathbf{Q}$, while a_c is the CATTANEO speed of the pure thermal wave and a_d is the coupling coefficient.

Evidently $U^2 = a_t^2$ is a double root of equation (7.7.23) and also a double eigenvalue of $\mathbf{Q}$; denoting by $\mathbf{a}_t$ a corresponding eigenvector of $\mathbf{Q}$, we get from $(7.7.21)_1$ that $\mathbf{n}\cdot\mathbf{a}_t = 0$. If $a_c \neq a_t$, then $(7.7.24)_1$ implies $v = 0$. Thus the thermomechanical wave propagating with speed $U^2 = a_t^2$ has amplitude $(\mathbf{a}_t, 0)$ (i.e. it is a pure mechanical wave) and it is a transversal wave (since $\mathbf{n}\cdot\mathbf{a}_t = 0$). This is true for the instantaneously coupled or uncoupled case (i.e. for $\varkappa \neq 0$ or $\varkappa = 0$). The result just obtained is the same as in the isotropic linear theory of elasticity, i.e. for any direction of propagation $\mathbf{n}$ there may exist two transversal acceleration waves; across them only the mechanical amplitudes of the waves do not vanish.

Let $U^2 \neq a_t^2$ be a non-zero root of equation (7.7.23); then, using $(7.7.21)_{1-2}$, we can write the equations (7.7.24) for the corresponding amplitude $(\mathbf{a}, v)$ of the wave as

$$
\begin{aligned}
(U^2 - a_c^2)\, v &= -\frac{\varkappa}{v}\, U^2(\mathbf{a}\cdot\mathbf{n}), \\
U^2\mathbf{a} - \left(\frac{\lambda+\mu}{\rho_0}(\mathbf{a}\cdot\mathbf{n})\mathbf{n} + a_t^2\mathbf{a}\right) &= -\frac{\varkappa}{\rho_0\theta_0}\, v\mathbf{n}.
\end{aligned}
\qquad (7.7.25)
$$

If in $(7.7.25)_2$ we write $\mathbf{a} = (\mathbf{a}\cdot\mathbf{n})\mathbf{n} + \mathbf{a}_1$, where $\mathbf{a}_1\cdot\mathbf{n} = 0$ and we multiply this equation by $\mathbf{a}_1$, we get $(U^2 - a_t^2)(\mathbf{a}_1\cdot\mathbf{a}_1) = 0$; thus, if $U^2 \neq a_t^2$, then $\mathbf{a}_1 = \mathbf{0}$ and the wave is longitudinal, i.e. $\mathbf{a} = (\mathbf{a}\cdot\mathbf{n})\mathbf{n}$.

Therefore, for any given direction n there are always two transversal waves that propagate with speed $U^2 = a_t^2$ which does not depend on the direction $\mathbf{n}$, and they carry no thermal jumps, as $v = 0$.

If

$$(\lambda + \mu)\, z \neq \mu \left\{(\lambda + \mu)\, v + \frac{\varkappa}{\theta_0}\right\}, \tag{7.7.26}$$

i.e. if a_t^2 is not a triple root of (7.7.23), then for any direction **n**, there always exist two longitudinal waves whose speeds of propagation are the roots of the equation

$$U^4 - (a_l^2 + a_c^2 + a_d^2)U^2 + a_c^2 a_l^2 = 0. \tag{7.7.27}$$

If $a_d \neq 0$ (i.e. if $P_i \neq 0$), the longitudinal waves are coupled, i.e. both mechanical and thermal amplitudes are different from zero. If $a_d = 0$, then the waves are uncoupled, the roots of (7.7.27) are $U^2 = a_l^2$ and $U^2 = a_c^2$, and the amplitudes of the waves are $(\mathbf{a}, 0)$ and $(\mathbf{0}, v)$ respectively.

We have seen that the strong ellipticity condition implies $z > 0$. If we assume that (7.4.7) vanishes for any $\mathbf{c} \in R^3$, then $z = 0$ for any $\mathbf{n} \in R^3$ which implies $a_c = 0$. Then equation (7.7.26) has the roots $U^2 = 0$ and

$$U_a^2 = a_l^2 + a_d^2; \tag{7.7.28}$$

the last one is sometimes called the *adiabatic sound speed*. The condition $z = 0$ implies $\tilde{\mathbf{q}} \equiv \mathbf{0}$ (see equation (7.7.15) and $(7.7.21)_5$).

If we denote by U_h^2 and U_l^2 the roots of equation (7.7.27) such that $U_h^2 > U_l^2$, then we can write

$$U_l^2 < a_l^2 < U_a^2 < U_h^2.$$

This result might be experimentally tested in a process with fast heat flux variations.

Even for the non-linear case, if $z^* = 0$ at a strong equilibrium state, we obtain from (7.3.22) and (7.3.23) that

$$U^{*2}\mathbf{a} = \mathbf{Q}^*\mathbf{a} + \frac{\rho_0\theta^*}{v^*}\,(\mathbf{P}^*\cdot\mathbf{a})\,\mathbf{P}^*. \tag{7.7.29}$$

The tensor

$$\bar{\mathbf{Q}}^* = \mathbf{Q}^* + \frac{\rho_0\theta^*}{v^*} P^* \otimes \mathbf{P}^* \tag{7.7.30}$$

is called the *adiabatic acoustic tensor*.

Similar conclusions are reached in the theory of wave propagations in materials with fading memory (see COLEMAN and GURTIN [1975]) or in heat nonconducting materials with internal state variables (see BOWEN and CHEN [1972]).

7f. Taking in (7.7.7) and (7.7.11)

$$\Lambda' = 0,\ \Lambda'' = 0,\quad \Delta' = 0,\quad \Delta'' = 0,\ \Delta_i = 0\ (i = 10, \ldots, N), \Gamma = 0 \tag{7.7.31}$$
$$\tilde{A}_k = 0\ (k = 1, \ldots, 5),\ \tilde{A}_{6j} = 0\ (j = 10, \ldots, N),\ \bar{A}_j^2 = 0\ (j = 10, \ldots, N),$$

we obtain the decomposition (7.7.1). The constitutive equations (7.7.7), (7.7.11) and (7.7.13) reduce to the constitutive equations of thermoelasticity; the volume behaves linearly viscoelastic. If, moreover, we take $N = 3$ or, equivalently, if in (7.7.7) and (7.7.11) we set

$$\Lambda_i = 0,\ \Gamma_i = 0,\ \Omega_{ij} = 0\ (i, j = 10, \ldots, N), \tag{7.7.32}$$

then, substituting (7.7.31), (7.7.32) and (7.7.13) in (7.7.7), (7.7.9) and (7.7.11) and rewriting (7.7.15) (with the help of (7.7.17) and (7.7.20)) we obtain the complete set of constitutive equations of the linear thermoelasticity theory (with the CATTANEO law instead of the FOURIER law), namely

$$\rho_0 \tilde{\psi}(\varepsilon, \tau, \tilde{\mathbf{q}}) = \frac{\lambda}{2}(tr\varepsilon)^2 + \mu tr(\varepsilon^2) - \varkappa\vartheta tr\varepsilon - \frac{1}{2}v\theta_0\vartheta^2 + \left(\frac{\xi}{\gamma^2}\right)\tilde{\mathbf{q}}\cdot\tilde{\mathbf{q}},$$

$$\tilde{S}_{ij} = \lambda(tr\varepsilon)\delta_{ij} + 2\mu\varepsilon_{ij} - \varkappa\vartheta\delta_{ij},$$

$$\rho_0\theta_0\eta = \varkappa(tr\varepsilon) + v\theta_0\vartheta, \tag{7.7.33}$$

$$\rho_0 e = \frac{\lambda}{2}(tr\varepsilon)^2 + \mu tr(\varepsilon^2) + \varkappa tr\varepsilon + \frac{1}{2}v\theta_0\vartheta(2 + \vartheta) + \left(\frac{\xi}{\gamma^2}\right)\tilde{\mathbf{q}}\cdot\tilde{\mathbf{q}},$$

$$\tau\frac{\partial\tilde{q}_i}{\partial t} + k\theta_0\frac{\partial\vartheta}{\partial X_i} + \tilde{q}_i = 0,$$

where $i, j = 1, 2, 3$.

Taking into account $(7.7.33)_{2-4}$ and the linearity hypothesis, we can write the equations $(7.2.10)_{1,2}$ as

$$\rho_0\frac{\partial v_i}{\partial t} - \frac{\partial\tilde{S}_{ik}}{\partial X_k} = \rho_0 b_i,\quad i = 1, 2, 3,$$
$$\varkappa\frac{\partial(tr\varepsilon)}{\partial t} + v\theta_0\frac{\partial\vartheta}{\partial t} + \frac{\partial\tilde{q}_k}{\partial X_k} = \rho_0 r, \tag{7.7.34}$$

where $\tilde{S}_{ij} = \tilde{S}_{ji}$ due to $(7.7.33)_2$. If the displacement vector

$$\mathbf{u}(\chi, t) = \chi(\mathbf{X}, t) - \mathbf{X}$$

is introduced, then

$$\varepsilon_{ij} = \frac{1}{2}\left(\frac{\partial u_i}{\partial X_j} + \frac{\partial u_j}{\partial X_i}\right) \quad \text{and} \quad v_i = \frac{\partial u_i}{\partial t}$$

and $(7.7.33)_5$, (7.7.34), with $\tilde{S}_{ij}$ determined by $(7.7.33)_2$, becomes a linear system of seven equations for the seven unknown functions u_i, $\tilde{q}_i$, $i = 1, 2, 3$ and ϑ. This system, under the strong ellipticity condition, is hyperbolic. Let us also note that for static problems (when $\partial \mathbf{v}/\partial t = \mathbf{0}$, $\partial(tr\boldsymbol{\varepsilon})/\partial t = 0$ and $\partial\vartheta/\partial t = 0$) this system reduces to the elliptic system of thermoelasticity with the FOURIER law of heat conduction (see for instance CARLSON [1972]).

7g. The constitutive equations (7.7.7), (7.7.9) and (7.7.11) of the linear theory are generalizing most of the linear rheological models (see for instance REINER [1958] Chapter B). Let us describe now the way one may obtain, what is called in rheology the *linear standard model*. Suppose that (7.7.32) holds and that $(7.7.9)_1$ can be solved for $\tilde{\boldsymbol{\alpha}}$ i.e.

$$\tilde{\boldsymbol{\alpha}} = \mathbf{L}(\tilde{\mathbf{S}}, \boldsymbol{\varepsilon}, \vartheta),$$

where $\mathbf{L}$ is a linear function of its arguments. If $(7.7.9)_1$ is differentiated with respect to t and $\dot{\tilde{\boldsymbol{\alpha}}}$ and $\tilde{\boldsymbol{\alpha}}$ are replaced by $(7.7.11)_3$ (with $\tilde{A}_{6k} = 0$, $k = 10, \ldots, N$) and $\mathbf{L}(\tilde{\mathbf{S}}, \varepsilon, \vartheta)$ respectively, we obtain the linear standard model.

8. THE SEMILINEAR THEORY

8a. The theory which leads to a semilinear system of partial differential equations is called semilinear. One of the characteristics of such a theory is the presence of constant speeds of propagation for acceleration and shock waves.

The semilinear theory is experimentally justified in the domain of small strains and small temperature variations (i.e. under the hypotheses (7.7.4) which will be maintained throughout this section) for certain materials, as metals for instance. Thus, if a stress pulse is superposed over a constant state of strain and temperature, it propagates with the elastic wave speed even in the plastic domain. This kind of behaviour has been shown to occur for several metals, for tension and compression as well (see BELL [1951, 1973], STERNGLASS and STUART [1953], ALTER and CURTIS [1956], BELL and STEIN [1962], FOKUOKA and MASUI [1970]).

These experimental results have been used by SULICIU, MALVERN and CRISTESCU [1974] and SULICIU [1974] in order to formulate the following constitutive assumption consisting of two parts: The speed of acceleration waves at a given equilibrium state is path-independent and, the time a state with constant strain and temperature needs (from the moment it has been achieved) to reach an equilibrium state, is finite; this time has been called the relaxation time. They have studied the consequences of these constitutive assumptions, for the one-dimensional case and for quasilinear rate-type constitutive equations. There exists a large class of semilinear rate-type constitutive equations that satisfy the finite relaxation time hypothesis (see Chapter V, Section 1, formulas (5.1.12), (5.1.13)). A similar study can be performed for materials with internal state variables.

8b. In order to obtain a semilinear theory for isotropic materials the assumption (7.7.4) will be maintained and, with the same decomposition (7.7.2) of the internal state variables vector, we shall postulate that the free energy can be approximated by (7.7.7). Then the constitutive equations are those in (7.7.9). We shall assume a linear constitutive equation for the heat flux; this will imply (7.7.13). If, moreover, the mapping $\hat{\mathbf{A}}$ of (7.2.8) is supposed constant then, under the strong ellipticity condition, the structure of the acceleration waves will be the same as in the linear theory.

Under the above assumption the constitutive equations for the internal state variables are

$$\begin{aligned} \dot{\tilde{\alpha}}_{ij} &= \tilde{A}_{ij,k} g_{0k} + \tilde{b}_{ij}(\boldsymbol{\varepsilon}, \vartheta, \boldsymbol{\alpha}), \\ \dot{\beta}_i &= \check{A}_{ik} g_{0k} + \check{b}_i(\boldsymbol{\varepsilon}, \vartheta, \boldsymbol{\alpha}), \\ \dot{\bar{\alpha}}_m &= \bar{A}_{mk} g_{0k} + \bar{b}_m(\boldsymbol{\varepsilon}, \vartheta, \boldsymbol{\alpha}), \\ \boldsymbol{\alpha} &= (\boldsymbol{\beta}, \tilde{\boldsymbol{\alpha}}, \bar{\boldsymbol{\alpha}}), \\ \tilde{A}_{ij,k} = \text{const.}, &\ \check{A}_{ij} = \text{const.}, \ \bar{A}_{mk} = \text{const.} \\ i, j, k &= 1, 2, 3, \ m = 10, \ldots, N. \end{aligned} \tag{7.8.1}$$

The functions $\bar{b}_m, \check{b}_i, \tilde{b}_{ij}$, $i, j = 1, 2, 3$, $m = 10, \ldots, N$ are assumed to be continuous with respect to all their arguments in a neighborhood U^0 of the equilibrium state $(\boldsymbol{\varepsilon}^0, \theta^0, \mathbf{g} = \mathbf{0}, \boldsymbol{\alpha}^0)$.

Consider now the process (in the body $\mathscr{B}$ with the reference configuration $\mathscr{R}$), generated by

$$\begin{aligned} \boldsymbol{\varepsilon}(\mathbf{Y}, t) &= \boldsymbol{\varepsilon}^0, \\ \theta(\mathbf{Y}, t) &= \theta^0 + \mathbf{g}_0 \cdot (\mathbf{Y} - \mathbf{X}), \ \mathbf{Y} \in \mathscr{R}, \ t \geqslant 0 \end{aligned} \tag{7.8.2}$$

(where $\mathbf{X} \in \mathscr{R}$ is fixed and $\mathbf{g}_0$ is any fixed constant vector in R^3) and which starts at

$$\boldsymbol{\alpha}(0) = \boldsymbol{\alpha}^0. \tag{7.8.3}$$

Then, at $\mathbf{X} \in \mathscr{R}$ and $t = 0$, one obtains

$$\begin{aligned} \dot{\tilde{\alpha}}_{ij}(0) &= \tilde{A}_{ij,k} g_{0k}, \\ \dot{\beta}_i(0) &= \check{A}_{ik} g_{0k}, \\ \dot{\bar{\alpha}}_m(0) &= \bar{A}_{mk} g_{0k}, \end{aligned} \tag{7.8.4}$$

which means that $\tilde{A}_{ij,k}g_{0k}$, $\check{A}_{ik}g_{0k}$, $\bar{A}_{mk}g_{0k}$, $i, j, k = 1, 2, 3$, $m = 10, \ldots, N$ are functions of a vector variable which are isotropic tensor,vector and scalar valued, respectively. Hence

$$\check{A}_{ij} = A\delta_{ij}, \ \tilde{A}_{ij,k} = 0, \ \bar{A}_{mk} = 0, \ i, j, k = 1, 2, 3 \quad m = 10, \ldots, N \tag{7.8.5}$$

$$A = \text{const.}$$

Thus, under the isotropy assumption, the constitutive equations (7.8.1) can be written in the simpler form

$$\begin{aligned} \dot{\tilde{\boldsymbol{\alpha}}} &= \tilde{\mathbf{b}}(\boldsymbol{\varepsilon}, \vartheta, \boldsymbol{\beta}, \tilde{\boldsymbol{\alpha}}, \bar{\boldsymbol{\alpha}}), \\ \dot{\boldsymbol{\beta}} &= A\mathbf{g}_0 + \check{\mathbf{b}}(\boldsymbol{\varepsilon}, \vartheta, \boldsymbol{\beta}, \tilde{\boldsymbol{\alpha}}, \bar{\boldsymbol{\alpha}}), \\ \dot{\bar{\alpha}}_m &= \bar{\mathbf{b}}_m(\boldsymbol{\varepsilon}, \vartheta, \boldsymbol{\beta}, \tilde{\boldsymbol{\alpha}}, \bar{\boldsymbol{\alpha}}). \end{aligned} \tag{7.8.6}$$

Equations (7.8.6) show that in a semilinear theory the temperature gradient is present only in the evolution equation for $\boldsymbol{\beta}$, while $\boldsymbol{\beta}$ may be present in all the equations (compare with the equations (7.7.11) of the linear theory).

Summarizing, the complete set of equations describing a semilinear theory consists of the constitutive equations (7.7.7), (7.7.9), (7.7.13) and (7.8.6) and the balance equations $(7.7.10)_{1,2}$, where $\rho_0\hat{\mathbf{S}} = \tilde{\mathbf{S}}$ (see $7.2.3)_1$) and $\hat{e} \equiv \tilde{e}$; $\tilde{\mathbf{S}}$ and $\tilde{e}$ being given by $(7.7.9)_{1,3}$. The constants A, ξ and γ are related by (7.7.14). The functions $\tilde{\mathbf{b}}$, $\check{\mathbf{b}}$ and $\bar{\mathbf{b}}$ must satisfy the inequality $(7.2.9)_2$ where $\hat{\psi}$ is replaced by $\tilde{\psi}$ given in (7.7.7) and $\hat{\mathbf{b}}$ is taken from (7.8.6) as $\hat{\mathbf{b}} = (\check{\mathbf{b}}, \tilde{\mathbf{b}}, \bar{\mathbf{b}})$, since $\boldsymbol{\alpha} = (\boldsymbol{\beta}, \tilde{\boldsymbol{\alpha}}, \bar{\boldsymbol{\alpha}})$.

Further to (7.7.4), hence (7.7.6), we shall assume that

$$|\boldsymbol{\beta}\cdot\mathbf{g}_0| \ll |\boldsymbol{\beta}|, \quad \left|\vartheta\frac{\partial\vartheta}{\partial t}\right| \ll \left|\frac{\partial\vartheta}{\partial t}\right|, \quad \left|\vartheta\frac{\partial(tr\boldsymbol{\varepsilon})}{\partial t}\right| \ll \left|\frac{\partial(tr\boldsymbol{\varepsilon})}{\partial t}\right|. \tag{7.8.7}$$

The balance of energy $(7.2.10)_2$ can then be written as follows:

$$v\theta_0 \frac{\partial \vartheta}{\partial t} + \varkappa \frac{\partial (tr\varepsilon)}{\partial t} - \gamma \frac{\partial \beta_i}{\partial X_i} + \{tr\varepsilon\Lambda_m\bar{b}_m + tr\tilde{\alpha}\Lambda_m\bar{b}_m +$$

$$+ tr\tilde{\mathbf{b}}\Delta_m\bar{\alpha}_m + 2\Omega_{mn}\bar{\alpha}_m\bar{b}_n + \Lambda' tr(\varepsilon\tilde{\mathbf{b}}) + \Lambda''(tr\varepsilon)(tr\tilde{\mathbf{b}}) + 2\Delta'(tr\tilde{\alpha})(tr\tilde{\mathbf{b}}) +$$

$$+ 2\Delta'' tr(\tilde{\alpha}\tilde{\mathbf{b}}) + 2\xi\boldsymbol{\beta}\cdot\check{\mathbf{b}} - \Gamma_m\bar{b}_m - \Gamma tr\tilde{\mathbf{b}} = \rho_0 r. \tag{7.8.8}$$

8c. The semilinear theory discussed above includes most of the thermo-viscoplasticity theories based on small strains and small temperature variations hypotheses.

For the sake of simplicity, only the case when thermal effects are neglected will be discussed here. Thus, we consider the semilinear rate-type constitutive equation

$$\dot{\tilde{S}}_{ij} = C_{ijkl}\dot{\varepsilon}_{kl} + b_{ij}(\varepsilon, \tilde{\mathbf{S}}), \tag{7.8.9}$$

where C_{ijkl}, $i, j, k, l = 1, 2, 3$ is a nonsingular constant matrix. This constitutive equation includes the Hohenemser and Prager [1932] and Perzyna [1963 a, b] models (see also Freudenthal[1958] and Cristescu [1967], Chapter X). For any smooth strain history on a time interval $[0, t_1]$, $t_1 > 0$ and any fixed initial condition $\tilde{S}_{ij}(0) = \tilde{S}^0_{ij}$, we obtain from (7.8.9) (using the Lagrange method of variation of parameters)

$$\tilde{S}_{ij}(t) = C_{ijkl}(\varepsilon_{kl}(t) - \tilde{\alpha}_{kl}), \tag{7.8.10}$$

where $\tilde{\alpha}$ remains to be determined so that (7.8.10) is a solution of equation (7.8.9) with the initial condition $\tilde{S}_{ij}(0) = \tilde{S}^0_{ij}$, that is

$$\dot{\tilde{\alpha}}_{ij} = -C^{-1}_{ijkl}b_{kl}(\varepsilon, \mathbf{C}[\varepsilon - \tilde{\alpha}]),$$
$$\tilde{\alpha}_{ij}(0) = -C^{-1}_{ijkl}\tilde{S}^0_{kl} \tag{7.8.11}$$

($\varepsilon(0)$ and $\tilde{\mathbf{S}}^0$ can be chosen equal to zero). (7.8.11) is an evolution equation for the internal state variable $\tilde{\alpha}$. This internal state variable is usually interpreted as the viscoplastic part of the strain and $\varepsilon^e_{kl} = \varepsilon_{kl} - \tilde{\alpha}_{kl}$ as its elastic part. Thus, to the constitutive equation (7.8.9) there corresponds an internal state variable which describes the way the stress moves from a given state towards its equilibrium value. This may be an interpretation of the internal state variables.

In thermomechanical phenomena one needs constitutive equations for stress, energy and heat flux; thus a natural way to choose the internal state variables vector $\boldsymbol{\alpha}$ is as it was done in (7.7.2) with $N = 10$. Then $\tilde{\alpha}$ is a symmetric second order tensor in $R^3 \otimes R^3$ corresponding to the stress, $\boldsymbol{\beta}$ is a vector in R^3 corresponding to the heat flux and $\bar{\alpha} = \alpha_{10}$ is a scalar corresponding to the internal energy. For the one-dimensional case this correspondence was considered by SULICIU [1975 *b*]. This yields an interpretation of the internal state variables in terms of rate-type constitutive equations. At the same time, for $N = 10$, the constitutive equations with internal state variables can be thought of as a subclass of the rate-type constitutive equations if certain invertibility conditions with respect to $\boldsymbol{\alpha}$ are fulfilled by the equations $(6.1.4)_{1-3}$.

Chapter VIII

SOME CONSTITUTIVE EQUATIONS USED IN CLASSICAL VISCOPLASTICITY

Due to the variety of the mechanical properties of the materials and to the variety of experimental conditions, several constitutive equations have been proposed in the literature. Generally the proposed constitutive equations have aimed to describe a certain dominant mechanical property observed by experiment. Most of these constitutive equations describe accurately only a relatively narrow range of observed phenomena. By an accurate description we understand here both a qualitative and quantitative description of the phenomena. For those phenomena which can be described qualitatively by a specific constitutive equation, the quantitative description is realized by a careful determination of the constants and coefficient functions involved in the constitutive equations, based on experimental data.

1. THE RIGID/VISCOPLASTIC BINGHAM MODEL

The BINGHAM [1922] model is characterized by the following property: The material (model) starts to flow only if the applied forces exceed a certain limit, called the *yield limit*.

We shall describe the model in Cartesian coordinates. Let x be the spatial coordinate and let v be the material velocity. Besides the components of the rate of deformation tensor

$$D_{ij}=\frac{1}{2}\left(\frac{\partial v_i}{\partial x_j}+\frac{\partial v_j}{\partial x_i}\right) \tag{8.1.1}$$

we shall also consider its deviator

$$D'_{ij}=D_{ij}-\frac{1}{3}D_{kk}\delta_{ij}. \tag{8.1.2}$$

Here δ_{ij} is the Kronecker delta. T_{ij} are the components of the Cauchy stress tensor and

$$T'_{ij} = T_{ij} - \frac{1}{3} T_{kk}\delta_{ij}, \tag{8.1.3}$$

the components of the stress deviator.

Besides the deviators D'_{ij} and T'_{ij} another deviator tensor K_{ij} will be introduced as the part of the stress which corresponds to the plastic properties of the material.

By a *process* we shall mean a collection of smooth functions $D'_{ij}(t)$, $T'_{ij}(t)$, $K_{ij}(t)$ for $t \in [0, T]$ where $T > 0$ is the duration of the process. The Bingham model for rigid/viscoplastic bodies assumes that for any process we have

$$T'_{ij} = K_{ij} + 2\eta D'_{ij}, \tag{8.1.4}$$

$$f(\mathbf{K}) = K_{ij}K_{ij} - 2k^2 \leqslant 0, \tag{8.1.5}$$

and

$$D'_{ij} = 2\lambda K_{ij}, \tag{8.1.6}$$

as in classical plasticity. $\eta > 0$ is a *viscosity coefficient* and k is the *yield stress in pure shear*, i.e. a constant. We can also consider k as a mean yield stress, if work-hardening is also to be described in a global way; generally k is a function of the equivalent strain $\bar{\varepsilon}$ as it is used in the isotropic work-hardening laws of the classical plasticity theory. Finally $\lambda(t)$ is a scalar function defined by

$$\lambda(t) = \Lambda(\mathbf{K}(t), \ \mathbf{K}(t)) = \begin{cases} > 0 & \text{if} \quad f(\mathbf{K}) = 0, \quad \mathrm{d}f(\mathbf{K}) = 0, \\ 0 & \text{if} \quad \begin{cases} f(\mathbf{K}) < 0, \\ f(\mathbf{K}) = 0 \text{ and } \mathrm{d}f(\mathbf{K}) < 0 \end{cases} \end{cases} \tag{8.1.7}$$

and

$$\mathrm{d}f(\mathbf{K}) = \frac{\partial f(\mathbf{K}(t))}{\partial K_{mn}(t)} \frac{\partial K_{mn}(t)}{\partial t} \mathrm{d}t.$$

The yield condition (8.1.5) is the *Mises condition*. According to it, the intensity of the deviator K_{ij}, i.e. the invariant $\mathrm{II}_K = \frac{1}{2} K_{ij}K_{ij}$ cannot exceed the yield stress in pure shear k. From (8.1.6) and (8.1.4) it follows that the deviator of the rate of deformation can vary only if K_{ij} stays on the surface $f(\mathbf{K}) = 2k^2$, moving along it. For all other processes, D_{ij} is zero. This is the reason why equality (8.1.5) is called the yield (or flow) condition.

Usually for the Bingham model the incompressibility of the volume is also assumed, i.e.

$$D_{ii}(t) = 0 \text{ or } tr\, \mathbf{D} = 0 \tag{8.1.8}$$

for any process of any duration $T \geqslant 0$.

Further the Bingham model will be considered with the deviators T'_{ij} and D'_{ij} only. (8.1.4) and (8.1.6) yield

$$\begin{aligned} T'_{ij} &= (1 + 4\eta\lambda)K_{ij}, \\ II_{T'} &= (1 + 4\eta\lambda)^2 II_K. \end{aligned} \tag{8.1.9}$$

It follows from (8.1.7) and (8.1.9) that for $\lambda = 0$, we have $II_{K'} = II_K \leqslant k^2$ and that for $\lambda > 0$ we have $II_{T'} > k^2$, and conversely, if $II_{T'} = II_K \leqslant k^2$ then $\lambda = 0$ and if $II_{T'} > k^2$ then $\lambda > 0$. Therefore the function $\lambda(t)$ is a function which depends on t through the intermediary of $II_{T'}$ and thus (8.1.7) can be written in the form

$$\lambda(t) = \tilde{\Lambda}(II_{T'}(t)) = \begin{cases} \dfrac{1}{4\eta}\left(\dfrac{\sqrt{II_{T'}}}{k} - 1\right) & \text{if } II_{T'} > k^2, \\ 0 & \text{if } II_{T'} \leqslant k^2. \end{cases} \tag{8.1.10}$$

From (8.1.6), (8.1.9) and (8.1.10) we get

$$D_{ij} = \begin{cases} \dfrac{1}{2\eta}\left(1 - \dfrac{k}{\sqrt{II_{T'}}}\right) T'_{ij} & \text{if } II_{T'} > k^2, \\ 0 & \text{if } II_{T'} \leqslant k^2, \end{cases} \tag{8.1.11}$$

It is readily seen that the stress power can be expressed as

$$T'_{ij}\, D_{ij} = 2k\sqrt{II_D} + 4\eta\, II_D \geqslant 0 \quad \text{if} \quad II_{T'} > k^2 \tag{8.1.12}$$

for any process in a Bingham body.

It is also easy to obtain the relation

$$2\eta + \frac{k}{\sqrt{II_D}} = \sqrt{\frac{II_{T'}}{II_D}} = \frac{2\eta}{1 - \dfrac{k}{\sqrt{II_{T'}}}}\, . \tag{8.1.13}$$

Using (8.1.13) together with (8.1.11) we get the constitutive equation in the inverse form

$$\begin{aligned} &II_{T'} \leqslant k^2 \quad \text{if } II_D = 0, \\ &T'_{ij} = \left(2\eta + \frac{k}{\sqrt{II_D}}\right) D_{ij} \quad \text{if } II_D \neq 0. \end{aligned} \tag{8.1.14}$$

Though the Bingham model is a model of fluid body, it was used to describe the deformation and flow of many solid bodies. In the form presented above, the model can be considered to be a generalization to viscoplasticity of the rigid/perfectly plastic model of classical plasticity theory. A variant of the model which could generalize the rigid/work hardening model can be written in the form

$$D_{ij} = \begin{cases} 0 & \text{if } II_{T'} < f(\bar{\varepsilon}), \\ \dfrac{1}{2\eta}\left(1 - \dfrac{f(\bar{\varepsilon})}{\sqrt{3}\sqrt{II_{T'}}}\right) T'_{ij} & \text{if } \sqrt{3}\sqrt{II_{T'}} > f(\bar{\varepsilon}), \end{cases} \tag{8.1.15}$$

where $\bar{\varepsilon} = \dfrac{2}{\sqrt{3}}\sqrt{II_\varepsilon}$ is the equivalent strain, $\bar{\sigma} = f(\bar{\varepsilon})$ is for any $\bar{\varepsilon}$, $f > 0$, $f' > 0, f'' < 0$ a "work-hardening conditon" or a "relaxation boundary", and $\bar{\sigma} = \sqrt{3}\sqrt{II_{T'}}$ is the equivalent stress. The expression $\sqrt{3}\sqrt{II_{T'}} - f(\bar{\varepsilon})$ is sometimes called *overstress*. Similarly, by a generalization of (8,1.14) we get

$$\begin{aligned} &\sqrt{3}\sqrt{II_{T'}} \leqslant f(\bar{\varepsilon}) \quad \text{for} \quad II_D = 0, \\ &T'_{ij} = \left(2\eta + \frac{f(\bar{\varepsilon})}{\sqrt{3}\sqrt{II_D}}\right) D_{ij} \quad \text{for} \quad II_D \neq 0. \end{aligned} \tag{8.1.16}$$

For the stress power we obtain now

$$T'_{ij}D_{ij} = \frac{2}{\sqrt{3}} f(\bar{\varepsilon})\sqrt{II_D} + 4\eta II_D > 0 \quad \text{for} \quad \sqrt{3}\sqrt{II_{T'}} > f(\bar{\varepsilon}). \tag{8.1.17}$$

The equations of motion or of quilibrium must be supplemented by the constitutive equations (8.1.10) and (8.1.7), and further the boundary and initial conditions must be formulated. Generally it is not possible to solve for a Bingham body those initial and boundary value problems where the deformation process starts from the natural state of equilibrium ($T_{ij} = 0$, $\varepsilon_{ij} = 0$ for $t = 0$). This is so since for the determination of the displacements and stresses we have only three equations of motion for $II_{T'} < k^2$, while the unknown functions are the stresses T'_{ij} and $\sigma = \dfrac{1}{3} T_{ii}$. If in a specific problem the number of unknown stress components does not surpass three, then it is generally possible to determine from the boundary and initial conditions a process which starts from a natural equilibrium state. Therefore generally for a Bingham model it is possible to formulate definite problems (the number of unknown functions is equal to the number of equations) in the case $II_{T'} > k^2$ only.

The reader interested in the description of the Bingham model may find details in REINER [1960] Chapter VIII, EIRICH [1956], OGYBALOV and MIRZADJANZADE [1970].

2. THE ELASTIC-VISCOPLASTIC MODEL

This model is in fact an improvement of the Bingham model (see HOHENEMSER and PRAGER [1932], FREUDENTHAL [1958]).

In the Bingham model during a process which starts from a natural equilibrium state the deformations are zero as long as the stress does not exceed the yield limit. In the Hohenemser-Prager model it is assumed that even below this limit the strains are increasing together with the stresses according to the Hooke law. When the stress point reaches the yield surface, the constitutive equation is replaced by a rate-type constitutive equation.

Let us give a mathematical description of this model. For the Bingham body it is assumed that the state of the body is described by the deviators D'_{ij}, T'_{ij}. K_{ij}, the mean strain and mean stress σ. The concept of process remains the same. The constitutive assumptions are the following: The rate of deformation deviator tensor can be decomposed in two parts

$$D'_{ij} = D'^{E}_{ij} + D'^{VP}_{ij}, \tag{8.2.1}$$

where D'^{E}_{ij} is the elastic rate of deformation deviator and D'^{VP}_{ij} is the viscoplastic rate of deformation deviator. Further it is assumed that the rate of the stress deviator and the elastic component of the rate of deformation deviator are connected by a relationship of the Hooke type

$$\overset{\circ}{T}{}'_{ij} = 2G\, D'^{E}_{ij}, \tag{8.2.2}$$

where G is the shear modulus which is assumed constant, and a small circle denotes the objective derivative, for instance

$$\overset{\circ}{T}{}'_{ij} = \dot{T}'_{ij} - w_{ij}T'_{kj} - T'_{ik}w_{kj},$$

where w_{ij} is the spin tensor

$$w_{ij} = \frac{1}{2}\left(\frac{\partial v_i}{\partial x_j} - \frac{\partial v_j}{\partial x_i}\right).$$

The deviator of the viscoplastic rate of deformation D'^{VP}_{ij} is governed by a Bingham type constitutive equation (8.1.10), i.e.

$$D'^{VP}_{ij} = \begin{cases} 0 & \text{if } II_{T'} \leqslant k^2, \\ \dfrac{1}{2\eta}\left(1 - \dfrac{k}{\sqrt{II_{T'}}}\right) T'_{ij} & \text{if } II_{T'} > k^2, \end{cases} \tag{8.2.3}$$

and thus it is assumed that viscoplastic deformation takes place only if the stresses are exceeding a certain yield limit ($II_{T'} > k^2$).

Concerning the mean strain and the mean stress, sometimes it is assumed that the volume deforms elastically only, i.e.

$$\sigma = 3\,K\varepsilon, \tag{8.2.4}$$

where K is the bulk modulus, however sometimes, in order to simplify the problem, an incompressibility condition is accepted. Thus the model is completely described.

Sometimes the constitutive equation is written in terms of total rate of deformation. Adding (8.2.2) and (8.2.3) and taking into account (8.2.1) we can write

$$D'_{ij} = \frac{\overset{\circ}{T}{}'_{ij}}{2G} + \frac{H(II_{T'} - k^2)}{2\eta}\left(1 - \frac{k}{\sqrt{II_{T'}}}\right) T'_{ij}, \tag{8.2.5}$$

where $H(II_{T'} - k^2)$ is the Heaviside function

$$H(II_{T'} - k^2) = \begin{cases} 0 & \text{if } II_{T'} \leqslant k^2, \\ 1 & \text{if } II_{T'} > k^2. \end{cases} \tag{8.2.6}$$

The above model is one the simplest models of the rate-type with instantaneous response (see CRISTESCU and SULICIU [1976] Chapter VII, Section 2). Though it is more complex than the Bingham model, it still remains much too limited in order to describe accurately such phenomena as work-hardening, the Bauschinger effect, the existence of a dynamic relaxation boundary distinct from the quasistatic one etc. (see Chapter III).

One of the main drawbacks of this model is the fact that for any stress state kept constant at value T'^{0}_{ij} and for $II^{0}_{T'} > k^2$ the strain increases linearly in time. Assuming small displacements and rotations we have

$$\varepsilon'_{ij}(t) = \varepsilon'^{0}_{ij} + \frac{1}{2\eta}\left(1 - \frac{k}{\sqrt{II^{0}_{T'}}}\right)(t - t_0)\, T^{0}_{ij}, \tag{8.2.7}$$

where ε'_{ij} is the strain deviator tensor. According to (8.2.7) the creep described by the model is always linear in time, whichever constant stress state is applied.

However, the constitutive equation (8.2.5) can describe relatively well the stress relaxation process at constant strain. Let ε'^{0}_{ij} be kept constant for $t \geqslant t_0 > 0$ and let $T'^{0}_{ij} = T'_{ij}(t_0)$ be the stress state at moment t_0 and let $II^{0}_{T'} > k^2$. From equation (8.2.5) we get (for small rotations)

$$\frac{1}{2G}\dot{T}'_{ij} + \frac{1}{2\eta}\left(1 - \frac{k^2}{II_{T'}}\right) T'_{ij} = 0, \; T_{ij}(t_0) = T^{0}_{ij}. \tag{8.2.8}$$

Multiplying (8.2.8) by T'_{ij} and summing up for i and j, and using $T'_{ij}T'_{ij} = 2II_{T'}$ and $T'_{ij}\dot{T}'_{ij} = \dot{II}_{T'}$ we obtain the following equation for $II_{T'}$:

$$\dot{II}_{T'} = -\frac{2G}{\eta}(\sqrt{II_{T'}} - k)\sqrt{II_{T'}}, \quad II_{T'}(t_0) = II^0_{T'}. \tag{8.2.9}$$

Its solution is

$$\sqrt{II_{T'}} = k + (\sqrt{II^0_{T'}} - k)\exp\left[-\frac{G}{\eta}(t - t_0)\right], \tag{8.2.10}$$

and therefore when $t \to \infty$, we have $II_{T'} \to k$. Substituting (8.2.10) in (8.2.8) we can write for each component T'_{ij}.

$$\int_{T'^0_{ij}}^{T'_{ij}} \frac{\mathrm{d}T'_{ij}}{T'_{ij}} = -\frac{G}{\eta}\int_{t_0}^{t}\left(1 - \frac{k}{\sqrt{II_{T'}(t)}}\right)\mathrm{d}t, \tag{8.2.11}$$

and this yields

$$T'_{ij}(t) = \frac{k}{\sqrt{II^0_{T'}}}\left\{1 + \left(\frac{\sqrt{II^0_{T'}}}{k} - 1\right)\exp\left[\frac{G}{\eta}(t - t_0)\right]\right\}T^0_{ij}. \tag{8.2.12}$$

When $t \to \infty$ the stress deviator relaxes towards a point on the yield surface $II_{T'} = k^2$; this point is

$$T'^{\infty}_{ij} = \frac{k}{\sqrt{II^0_{T'}}}T'^0_{ij}. \tag{8.2.13}$$

This result is connected with the theorem of Chapter VI, Section 4 in Cristescu and Suliciu [1976], though in the way this theorem is formulated, it is not a direct consequence of this result. This is due to the fact that in the mentioned theorem it was required that the state $(\varepsilon'^0_{ij}, T'^0_{ij})$ should reach in a finite time the state $(\varepsilon'^0_{ij}, T'^{\infty}_{ij})$.

The equation (8.2.5) was generalized in various ways in order to allow the description of various real phenomena. Thus Perzyna [1963 a], in order to increase the number of material functions and constants, introduces the notation

$$F = \frac{\sqrt{II_{T'}}}{k} - 1$$

and then a function $\Phi(F)$ which satisfies the conditions

$$\begin{aligned} &\Phi(F) = 0 \text{ if } F \leqslant 0,\\ &\Phi(F) \neq 0 \text{ if } F > 0, \end{aligned} \tag{8.2.14}$$

and postulates the equation (8.2.4) in the form

$$D'_{ij} = \frac{1}{2G}\dot{T}'_{ij} + \gamma\Phi(F)\frac{T_{ij}}{\sqrt{II_{T'}}}, \tag{8.2.15}$$

where γ is a material constant. Also PERZYNA [1963 *b*], in order to describe the work-hardening, chooses

$$F = \frac{f(T_{ij}, \varepsilon_{ij}^{VP})}{\varkappa} - 1, \tag{8.2.16}$$

where $\varkappa$ is a function of viscoplastic work, i.e. a function of the argument

$$\int_0^{\varepsilon_{ij}^{VP}} T_{ij}\, d\varepsilon_{ij}^{VP}.$$

For further details concerning other properties of the model (8.2.5) and the domain of applicability, and for references see for example FREUDENTHAL [1958], CRISTESCU [1967] Chapter X, Section 1, BANERJEE and MALVERN [1974] , NOWACKI [1978].

3. THE QUASILINEAR MODEL

Starting from the quasilinear model discussed in Chapter IV and following the formal generalization procedure used in classical plasticity theory (see for example HILL [1950] Chapter II) the one-dimensional constitutive equation (4.1.11) can be generalized to take the form

$$D'_{ij} = \frac{\mathring{T}'_{ij}}{2G} + \frac{2k(\bar{\varepsilon})}{2E}\left[1 - \frac{f(\bar{\varepsilon})}{\sqrt{3}\sqrt{II_{T'}}}\right]T'_{ij} + \frac{2\Phi(\bar{\sigma}, \bar{\varepsilon})}{4\, II_{T'}}T'_{kl}\mathring{T}_{kl}T'_{ij}. \tag{8.3.1}$$

Here $\bar{\sigma} = \sqrt{3}\sqrt{II_T}$ and $\bar{\varepsilon} = \frac{2}{\sqrt{3}}\sqrt{II_\varepsilon}$ are the equivalent stress and the equivalent strain respectively, and f and Φ are the functions defined in Chapter IV (where certainly the one-dimensional components ε and σ must be replaced by $\bar{\varepsilon}$ and $\bar{\sigma}$). (8.3.1) must be supplemented by the law of volume compressibility

$$\sigma = 3\, K\varepsilon. \tag{8.3.2}$$

This is only one of the possible generalizations of the one-dimensional constitutive equation (4.1.11). Other aspects as for instance the choosing of the equation of the relaxation surface, or of several such surfaces, or of other generalizations of the coefficient function Φ etc., can be thought of (see also CRISTESCU [1967], [1974 *a*]).

Some generalizations of the quasilinear constitutive equation can describe dilatancy properties for the volume as shown by FRANK and GUELLEC [1974],

$$3K\varepsilon = \sigma + K_d \sqrt{II_{T'}},$$

where K_d is the dilatancy coefficient which can be constant or variable.

Other generalizations for finite strain are known; see e.g. CLIFTON [1971, 1972].

Chapter IX

A VISCOPLASTIC CONSTITUTIVE EQUATION FOR ROCKS

1. INTRODUCTION

Rocks have long been considered in mining practice as being linear elastic though it was well known that mechanical properties of rocks are much more complex (see OBERT and DUVALL [1967], BAKLASHOV and KARTOZYA [1975], ERJANOV et al. [1970], KARTASHOV [1973], GOLDSMITH and SACKMAN [1973], LAMA and VUTUKURI [1978], JUMIKIS [1979]). The rheological properties of rocks are significant not only within geological time intervals but also within much shorter time intervals (days, months) of interest to mining industry.

The aim of the present chapter is an attempt to establish a much more precise than hitherto known constitutive equation for rocks to be used in time intervals ranging from a few minutes (sometimes even shorter) to several years. Rheological models for rocks, mainly linear viscoelastic models, were already proposed by many authors (see the literature already mentioned). It was thought, however, that rocks are more complex and that their mechanical properties would rather be described by elastic visco-plastic non-linear models for both shearing properties as well as for the volume compressibility. We shall propose here a tentative model based on several diagnostic tests. In the future additional test of the same kind or other will be still needed to clarify some of the aspects, mainly quantitatively, concerning the model. In the meantime the model must be considered to be a first approximation, pending further experimental data. We shall consider mainly creep properties and deformation processes during loading, as these appear to be predominantly involved in the mining applications we have in mind. Other mechanical properties will be only marginally discussed. We shall follow here CRISTESCU [1979].

2. STANDARD EXPERIMENTS IN COMPRESSION

We had two reasons for doing this type of experiments. On one hand these were the first and simplest tests which could lead to finding the dominant mechanical properties of various rocks in quasistatic compression. Standard testing machines were used, while the cylindrical specimens were 10 cm long and 5 cm in diameter, with unconfined lateral surface. In order to look for possible time effects, even these experiments were done with various loading rates, controlled by the testing machine.

On the other hand, we were going to use the rock response, as revealed by such experiments, to estimate the deformation of the specimen during the first period of deformation in creep test, i.e. during the period when the testing machine used in creep experiments is loaded. Further compressive stresses and strains were defined as positive; in the experiments discussed here only positive stresses and strains were involved.

In fig. 9.2.1 three stress-strain curves obtained in quasistatic experiments for dry schist are shown (porosity from 3 to 8%). The upper

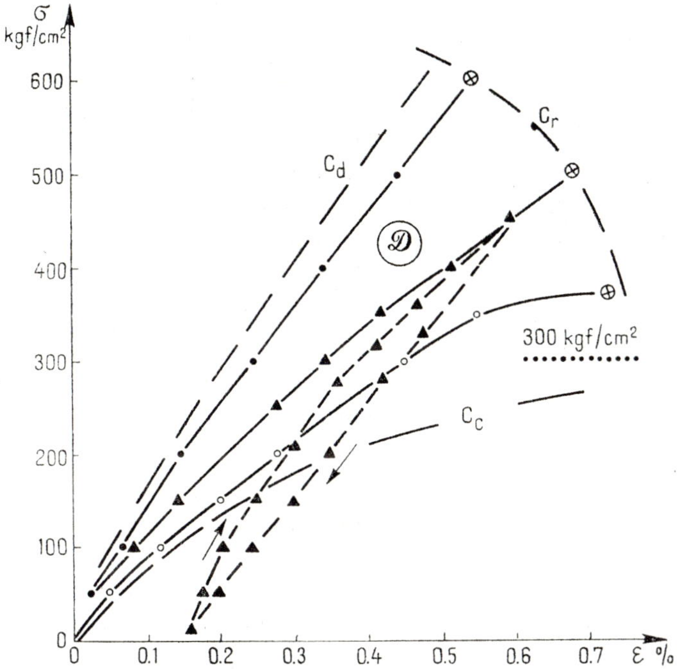

Fig. 9.2.1 Stress-strain curves for dry schist and domains involved in the mathematical model

curve (full dots) corresponds to the loading rate 300 kgf cm^{-2} min^{-1} (2942 N cm^{-2} min^{-1}), the middle one (triangles) to the loading rate 40 kgf cm^{-2} min^{-1} (392.3 N cm^{-2} min^{-1}) and finally the lower one (circles) to a very small loading rate of 1.33 kgf cm^{-2} min^{-1} (13 N cm^{-2} min^{-1}). It is evident from these curves that the stress-strain relations are non-linear and are dependent on the loading rate. This loading rate effect is similar to the one reported by many authors for various rocks (see for instance LAMA and VUTUKURI [1978], PERKINS et al. [1970], KUNTYSH and TEDER [1970]). Successive cycles of unloading followed by reloading have also been done, showing significant hysteresis loops and variation of the tangent modulus during both loading and unloading. In order to keep the figure simple, only one such loop has been shown. The produced unloading, starting from various stages of deformation, indicated significant permanent strains. Generally the order of magnitude of the permanent component of the strain was about one quarter to one third of the total deformation, and therefore it was quite significant.

The ultimate point on various curves shown on fig. 9.2.1 corresponds to the fracture of the specimen. Generally with an increased rate of loading the stress at failure is higher, but the corresponding strain is smaller. The stress at failure with moderate rate of loadings (of the order of 40 kgf cm^{-2} min^{-1} (392.3 N cm^{-2} min^{-1})) is involved in the mathematical model and will be denoted by σ_r. The magnitude of σ_r will be considered to be a typical constant for the particular kind of rock under consideration.

Let us also remark that the stress-strain curve obtained with unconfined lateral surface of the specimen has the concavity directed towards the positive strain axis. Some other rocks (e.g. sandstone) possess even in such kind of experiments stress-strain curves with an opposite curvature (directed towards the positive stress axis), while still some other rocks possess stress-strain curves which are almost linear, though highly sensitive to the loading rate (e.g. limestone).

3. CREEP TESTS

In the creep tests we used specimens of the sizes mentioned before, and the loads were successively increasing (in steps). The specimen was first loaded with a certain constant stress and the strain was recorded for several days or weeks. Generally at the end of a finite well determined time period the strain ceases to change. When it became evident that the strain will increase no more, an additional load was applied to the specimen, and so on. Typical strain-time curves obtained in such kind of creep tests are shown on fig. 9.3.1 and fig. 9.3.2, again for schist. The last points shown on these curves correspond to the failure of the specimen.

Since for the same type of rock there is a broad range of strength characteristics corresponding to various values of the stress at failure, the creep test is better described by recording the variation of the strain as function of the ratio $\sigma_{\text{effective}}/\sigma_r$, rather than of the effective stress $\sigma_{\text{effective}}$. This ratio will be denoted by $\Delta = \sigma_{\text{eff}}/\sigma_r$ and will be called

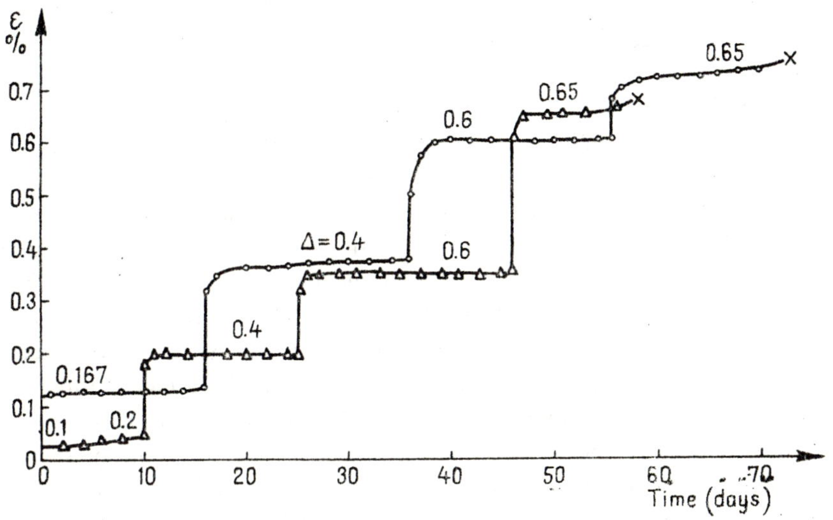

Fig. 9.3.1 Strain-time curves for schist during creep tests

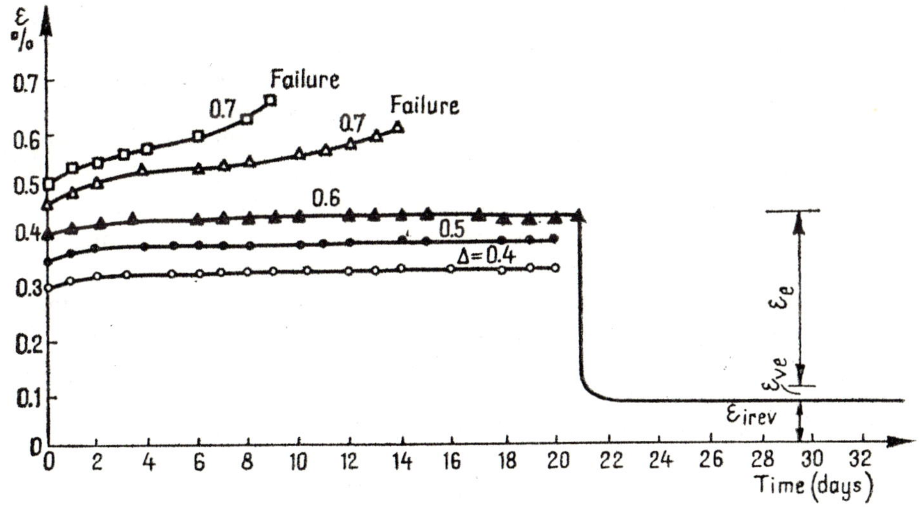

Fig. 9.3.2 Strain-time curves for schist during creep tests and one inverse creep test

loading ratio. Here σ_r is obtained in standard compressive tests with medium loading rate, as described above, and is a typical constant for the rock under consideration. The curve on fig. 9.3.1 is showing that at each increase of the loading ratio we get an instantaneous increase of the strain, followed by a slow increase of the strain due to creep.

A very important aspect in the deformation by creep is the following. If the loading ratio applied to an unstressed and undeformed specimen is not surpassing a certain limit, then after several days or weeks the strain will become constant and will stay constant independently of the duration of the experiment. Let us denote by Δ_s the highest value of Δ for which the deformation by creep will remain stable after a certain finite interval of time. Thus, for any $\Delta > \Delta_s$, the deformation by creep is expected to become unstable in the sense that for such loading ratio the strain is continuously increasing up to the failure of the specimen which occurs after a finite time interval beginning at the moment when the load was applied. For $\Delta \leqslant \Delta_s$ the strain is becoming constant after a certain interval of time. For $\Delta < \Delta_s$ this time interval is finite and can be determined accurately by experiment. For $\Delta \simeq \Delta_s$ it is quite difficult to decide if this time interval is finite or infinite. For the purpose of establishing a model we shall subsequently choose Δ_s so that for any $\Delta \leqslant \Delta_s$ the time of stabilization by creep is finite. Thus Δ_s is depending on σ_{eff} and σ_r but it will be assumed that Δ_s does not depend on the initial strain nor on the loading rate. If for certain rocks this assumption is not acceptable, then in order to uniquely define Δ_s a unique standard loading rate will be used to determine Δ_s in all successive tests starting from the state $\sigma = \varepsilon = 0$.

Inverse creep experiments (unloading) were also performed (fig. 9.3.2) in order to get information concerning the nature of various parts of the strain. When loading was removed practically instantaneously, a part of the strain decreased also instantaneously. Further on, the strain continued to decrease slowly for a certain period of time after which it did not decrease any more. Thus it was found that strain can be decomposed in three parts: an instantaneous reversible one (elastic) ε_e, a noninstantaneous reversible one (viscoelastic) ε_{ve} and finally a permanent one ε_p. It is interesting to mention that generally the deformation of the specimen obtained during the first stage of deformation, i.e. during the time when the load is applied to the machine, contains a significant permanent component. This is obvious if comparison is made between this initial instantaneous component of the strain and ε_e obtained during instantaneous unloading. Recalling the results mentioned in Section 2 we come to the conclusion that this initial permanent component is of viscoplastic nature.

A mathematical model describing the main experimentally found and presented above characteristics, can be obtained as follows. We first consider in the ε, σ plane (fig. 9.2.1) a domain $\mathscr{D}$ bounded by three curves, all three obtained by experiments. Curve C_c is the boundary of stabilization by creep; points on this boundary are the points of maximum strain which can be reached with various stresses (relatively small) during creep tests. The experimental data used to determine C_c suggest that this boundary has a horizontal asymptote $\sigma = a =$ const., and therefore the stresses involved in the experiments determining C_c do not exceed the stress corresponding to this asymptote. If such an asymptote does not exist, then $\sigma = a$ is the ultimate stress still producing a stabilization of deformation by creep. Curve C_d is the curve of instantaneous response of the rock, i.e. this would be the response of the rock if the fastest (within the possibilities of the experiments under consideration) loading is applied to the specimen. For instance, if creep behaviour is to be described, then a fast loading obtained with a standard testing machine would be this kind of fast loading, generally much faster than the one which could be obtained when a creep machine is loaded. Finally, the last curve C_r is the boundary where the fracture of the specimen is produced. These three curves are bounding the domain $\mathscr{D}$ of all possible stress-strain states which can be reached in loading processes by creep tests. We observe here that for stresses in the neighbourhood of the line $\sigma = a$ both curves C_r and C_c are difficult to be found by experiments since very long time intervals are necessary. Therefore it is not certain that the two curves C_r and C_c have a common horizontal asymptote but our experimental data would rather suggest that this is so. The experimental data for concrete obtained by RUSCH [1960] seem also to suggest this hypothesis.

We consider now the general form of the constitutive equation (initially proposed for metals by CRISTESCU [1963])

$$\dot{\varepsilon} = \left(\frac{1}{E} + F(\varepsilon, \sigma, \operatorname{sgn} \dot{\sigma})\right) \dot{\sigma} + \Psi(\varepsilon, \sigma) \tag{9.3.1}$$

with

$$F(\varepsilon, \sigma, \operatorname{sgn} \dot{\sigma}) = \begin{cases} \Phi(\varepsilon, \sigma) & \text{if } \dot{\sigma} \geqslant 0, \\ 0 & \text{if } \dot{\sigma} < 0, \end{cases} \tag{9.3.2}$$

where $\Phi(\varepsilon, \sigma)$ describe the "fast" or "instantaneous" properties of the rock and is a non-negative function of class C^1 for $0 \leqslant \varepsilon$, $0 \leqslant \sigma \leqslant E\varepsilon$. $\Psi(\varepsilon,\sigma)$ describes the "slow" properties of the rock (creep, relaxation). E is an elastic modulus determined in "fast" quasistatic loading tests. We recall that for rocks the modulus E determined in quasistatic loading tests is dependent on the loading rate. Here $E = \text{const.}$ is defined by conventionally choosing a certain "fast" loading rate for machine. But in the model the cases when E depends on the stress state can also be considered, if necessary. The dots above the letters in the formulas above mean time derivatives. It will be assumed that strains and rotations are small. The "instantaneous" response of the rock starting from the state $\sigma = \varepsilon = 0$ will be approximated by a curve

$$\varepsilon = \alpha \left(\frac{\sigma}{E_0} \right)^{\beta}, \tag{9.3.3}$$

where $\alpha > 0$ and $\beta > 0$ are dimensionless characteristic constants of that particular rock and $E_0 = \text{const.}$ is a particular value of E, for instance the value of E obtained in dynamic tests (by low amplitude wave propagation). Then the function $\Phi(\varepsilon, \sigma)$ is defined as

$$\Phi(\varepsilon, \sigma) = \alpha\beta \left(\frac{\sigma}{E_0} \right)^{\beta-1} - \frac{1}{E}, \ \forall\ (\varepsilon, \sigma) \in \mathscr{D}. \tag{9.3.4}$$

The coefficient function $\Psi(\varepsilon, \sigma)$ is found by experiments so that $\Psi(\varepsilon, \sigma) = 0$ becomes the equation of the curve of stabilization by creep, i.e. the equation of the curve C_c. In other words, it will be assumed that there exists a curve $(\varepsilon, \sigma = \sigma_c(\varepsilon)) \in \mathscr{D}$ with the properties

$$\Psi(\varepsilon, \sigma) > 0 \ \text{ if } \ \sigma_c(\varepsilon) < \sigma, \tag{9.3.5}$$
$$\Psi(\varepsilon, \sigma) = 0 \quad \text{if } \ \sigma_c(\varepsilon) \geqslant \sigma.$$

Since the loadings exceeding a certain limit $\sigma = a$ create an unstable deformation by creep (i.e. the deformation finally producing fracture), the equation of the boundary C_c is chosen in the form

$$\sigma = a\,[1 - \exp(-b\varepsilon)], \tag{9.3.6}$$

where $b > 0$ is a dimensionless constant, while $\sigma = a$ is the highest value of the stress still producing stabilization of the deformation by creep. This yields a criterion for choosing the value of a, but a slightly higher value for a can also be used if necessary in specific applications.

Since even for one kind of rock there occur distinct categories of mechanical properties which can be roughly characterized by the conventionally defined fracture resistance σ_r, it is more convenient to introduce in (9.3.6) the maximal loading ratio Δ_s

$$a = \Delta_s \sigma_r \tag{9.3.7}$$

which still produces a stable creep, while σ_r is obtained with standard testing machines. Now (9.3.7) can be written as

$$\frac{\sigma}{\sigma_r} = \Delta_s [1 - \exp(-b\,\varepsilon)]. \tag{9.3.8}$$

Generally the coefficient b is also somewhat dependent on σ_r but this is of lesser importance for the discussion which follows.

Sometimes it is useful to determine the boundary of stabilization by creep using tests intermediate between creep tests and standard tests. These are the tests in which continuously increasing loading is applied but the loading rate is very small. Let us assume that

$$\sigma = \sigma_r f(\varepsilon) \tag{9.3.9}$$

is the stress-strain curve obtained by such a procedure. We found that this curve is close to the creep stabilization curve and sometimes even slightly lower than the later one. Thus (9.3.9) can be used for describing the creep stabilization boundary. For many rocks (9.3.9) can be well approximated by a straight line

$$\sigma/\sigma_r = h\varepsilon. \tag{9.3.10}$$

Thus we may choose the coefficient function $\Psi(\varepsilon, \sigma)$ entering the constitutive equation in the form

$$\Psi(\varepsilon, \sigma) = \begin{cases} \dfrac{k(\sigma)}{E}\{\sigma - \sigma_r \Delta_s [1 - \exp(-b\varepsilon)]\}^m \text{ if} \\ \qquad \sigma > \sigma_r \Delta_s [1 - \exp(-b\,\varepsilon)], \\ 0 \text{ if } \quad 0 \leqslant \sigma \leqslant \sigma_r \Delta_s [1 - \exp(-b\varepsilon)], \end{cases} \tag{9.3.11}$$

with $m > 0$ a constant (further only the value $m = 1$ will be used), or in the form

$$\Psi(\varepsilon, \sigma) = \begin{cases} \dfrac{k(\sigma)}{E}\left\{\exp \dfrac{\sigma - \sigma_r \Delta_s [1 - \exp(-b\varepsilon)]}{n} - 1\right\} \text{ if} \\ \qquad \sigma > \sigma_r \Delta_s [1 - \exp(-b\varepsilon)], \\ 0 \qquad \text{if } 0 \leqslant \sigma \leqslant \sigma_r \Delta_s [1 - \exp(-b\,\varepsilon)], \end{cases} \tag{9.3.12}$$

with $n =$ const., or of various other forms (see Chapter II). Finally if the stabilization boundary is taken of the form (9.3.7) we can write

$$\Psi(\varepsilon, \sigma) = \begin{cases} \dfrac{k(\sigma)}{E}[\sigma - \sigma_r h\varepsilon] \quad \text{if } \sigma > \sigma_r h\varepsilon, \\ 0 \quad \text{if } \ 0 \leqslant \sigma \leqslant \sigma_r h\varepsilon, \end{cases} \tag{9.3.13}$$

or again another variant can be used. In these formulas $E/k(\sigma)$ can be considered to be a "viscosity coefficient" of the rock in creep tests; $k(\sigma)$ is in fact governing the variable value of this coefficient. To keep the model simple here, it will be assumed that k depends on stress alone.

Therefore the mathematical model describing the slow deformation of rocks is fully determined by the equation (9.3.1). The viscosity coefficient can be determined by measuring the strain at various moments during creep tests and using the constitutive equation. It will be given in Poise.

The numerical coefficients were determined for several rocks. For instance for schist several classes of strength characterized by the stress at fracture σ_r were determined. For one of these classes, stresses up $\sigma = 300$ kgf cm^{-2} (2942 N cm^{-2}) are still producing a stable creep while stresses above this value do not. Since for this kind of rocks the mean value for σ_r is $\sigma_r = 500$ kgf cm^{-2} (4903 N cm^{-2}) we get $\Delta_s = 0.6$ and further $a = 300$ kgf cm^{-2} (2942 N cm^{-2}) and $b = 300$. Using formula (9.3.1) with (9.3.11) and $m = 1$ we can find a mean value for k from creep tests. Generally for such rock k is of the order of magnitude $0.2 - 0.07\ d^{-1}$ (d stands for "day") and it is smaller when the applied stress is higher. This implies an order of magnitude of 10^{12} Poise for the "viscosity coefficient" Ek^{-1} in such kind of creep tests with final constant strain. Even for stationary creep we have found a viscosity coefficient of the same order of magnitude. This is smaller than the values 10^{17} to 10^{18} Poise reported by MAXIMOV et al. for stationary creep of argillaceous schist (see VYALOV [1978]). VOLAROVITCH [1977] reports for granite, gabbro and limestone, tested at rates of strains of orders 10^{-4}—10^{-8} sec^{-1}, values for the viscosity coefficient in the range 10^{12} to 10^{14} Poise, i.e. of the same order of magnitude as found in our experiments. If formula (9.3.13) is used in conjunction with (9.3.1), we obtain $h = 69.3$ and further k is somewhere between $0.2\ d^{-1}$ and $0.09\ d^{-1}$. Thus the curve C_c is determined by both approaches. From tests performed with standard testing machines we obtained for the other constants approximately the values $E = 167{,}000$ kgf cm^{-2} (1,637, 769 N cm^{-2}), $\alpha = 4.2$ and $\beta = 1.2$ and this allowed us to establish the curve C_d. Thus the constitutive equation (9.3.1) has been fully determined.

The determination of the numerical coefficients to be used in the constitutive equations of the form (9.3.1) for several other rocks proceeded similarly (rock salt BARONCEA et al. [1977], dolomite, limestone, sandstone, gnaiss etc.). Some of these coefficients varied significantly from one rock to another, and sometimes even within the same kind of rocks. Among all these coefficients the elastic modulus E exhibits the greatest variation from one kind of rock to another. The order of magnitude for the other coefficients is not changing too much when passing from one rock to other.

Finally the third curve C_r, necessary to fully define the boundary of the domain $\mathscr{D}$, is determined experimentally using both standard testing machines (at various loading rates) and creep tests (at various loading stresses). It was found that for higher stresses and/or higher rates of loadings the strain at fracture is smaller. On the other hand, the fracture points determined by standard testing machines are furnishing higher strains (for the same stress) than those furnished by creep tests. The curve C_r in conjunction with the constitutive law (9.3.1) is of great importance for practical applications such as for instance the prediction of the failure of an underground structure, etc.

4. VOLUME COMPRESSIBILITY

It is well known that with respect to volume compressibility, rocks (and soils) exhibit quite different behaviour than other materials, as for instance metals, in the sense that volume compressibility is significant and partially permanent (VOLAROVITCH et al. [1974], LEVYKIN and VAVAKIN [1978], VYALOV [1978], STEPHENS et al. [1970]).

In order to investigate the volume compressibility of various rocks a special device was made. (Variants of this type of device are described for instance by VOLAROVITCH et al. [1974] and STEPHENS et al. [1970]). Small rock samples, 20 mm in diameter and about the same height were compressed with a piston inside a thick walled hard steel cylinder. Thus the lateral surface was confined. The diameter of the cylinder was only slightly larger than that of the specimen. The lateral surface of the specimen was lubricated. The piston was compressed either by a standard testing machine or by a dead-weight loading machine. Generally, due to the very high forces involved, the whole device deformed, though it was made from high strength steel. Probably many experimental errors were involved, and not all of these are fully estimated. Therefore the results obtained until now should be considered as tentative, i.e. indicating only some qualitative properties of the rock. In all experiments only a single continuous loading was followed by a single unloading. It appears that the constitutive equation proposed describes the material response in such experiments.

The stresses and displacement fields involved were estimated assuming linear elasticity for the rock specimen. Consider cylindrical coordinates r, θ, z, the Oz axis coinciding with the specimen symmetry axis. Given a specimen, let us denote by σ_1 the applied axial stress at its upper end $z = l$, by τ the frictional stress at its lateral surface $r = R$ (which will be assumed to be constant, i.e. independent on z and θ), and by l its height. Assuming axial symmetry, that the surface $r = R$ of the steel cylinder is deforming negligibly and that at the bottom of the specimen, at $z = 0$ and $r = R$ the displacement is zero, it is easy to show that the mean stress (pressure) can be written as

$$\sigma = \frac{1}{3}\frac{1+\nu}{1-\nu}\left[\sigma_1 + \frac{2\tau}{R}(l-z)\right], \tag{9.4.1}$$

where ν is the Poisson's ratio. From this formula it follows that if the lateral friction is small (τ small), if the radius of the specimen is relatively big while its length is relatively small, and if ν is not too far from 0.5, then the mean stress does not differ much from σ_1. Formula (9.4.1) was used only to get an indication as to what sizes of specimens should be used. Generally the friction forces are significant since even after unloading, a significant force (of the order of several tons) was necessary to push the specimen out of the cylinder.

Thus in this series of experiments a relationship between the axial stress σ_1 and the axial strain ε_1 was established. Both quantities were measured. The other components of the strain were considered to be small in a first approximation, so that $\varepsilon_1 = 3\varepsilon$, where ε is the mean strain. These relationships suggested, at least qualitatively, the laws of volume compressibility for rocks in confined experiments.

To give an example, on fig. 9.4.1 are represented the relationships between σ_1 and ε_1 obtained for schist with two loading rates (4 ton/min and 1 ton/min); the unloading was done with the same rate. It was always found that the volume compressibility of various rocks is rate sensitive more during loading than during unloading, and that an important component of the volume strain is present. The magnitude of the permanent volume strain was also highly dependent on the loading rate; when the loading rate was higher, then at the same stress the permanent volume strain was smaller. Both loading and unloading curves are non-linear with concavity directed towards the σ_1 axis. The permanent compressibility of the volume was checked both by measuring the size of the specimen after the test and by measuring its density. A few preliminary tests have been performed to check if a creep in volume compressibility is possible. For this purpose the machine was stopped at various levels of the applied stress and further this stress was kept constant or a dead weight machine was used. It was found that volume strain was increasing though the applied stress was kept constant. The process was a transitory one and after a few minutes or tens of minutes the volume strain became constant and remained so (see the small steps on fig. 9.4.1 as well as the small horizontal plateaux at the top of these curves).

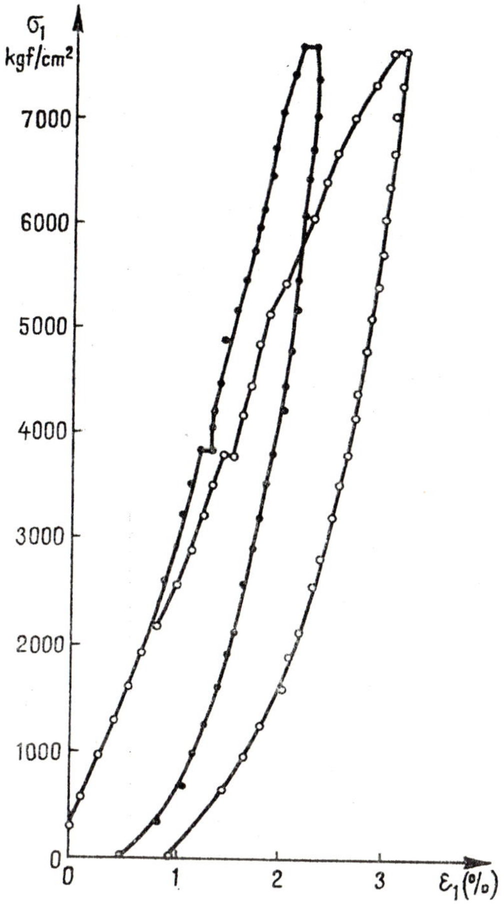

Fig. 9.4.1 Stress-strain curves for volume compressibility for schist for two loading rates

A mathematical model for the volume compressibility can be established by applying the same procedure as above and making the following assumptions (see fig. 9.4.2). It will be assumed that there is a limit volume strain ε_* which cannot be exceeded, whatever the applied pressure may be (locking model). If σ is the pressure and ε the mean strain, then it will be assumed further that by making experiments with the lowest loading rate still relevant for the kind of experiments we have in mind, we find a creep-stabilization boundary for the volume compressibility. The equation of such a boundary can be written for instance as

$$\varepsilon = \varepsilon_* \left[1 - \exp \left(- \frac{\sigma}{\sigma_0} \right) \right], \tag{9.4.2}$$

where $\sigma_0 > 0$ is a constant of the material.

The previous assumption was made also in conjunction with another kind of experiment in which the same device was used but the loading was applied with a dead-weight machine. This time the load was applied in successive steps. The resulting pressure-volumetric strain curve was rising stepwise. A typical curve of this kind, shown on fig. 9.4.2, reminds us of the Masson-Savart (or Portevin-Le Chatelier) phenomenon. Generally at each additional loading we get an "instantaneous" increase of the pressure and of the volumetric strain, followed by a period in which volumetric strain continues to increase under constant pressure. The disadvantage of the experiments done with the deadweight machine is that the maximum

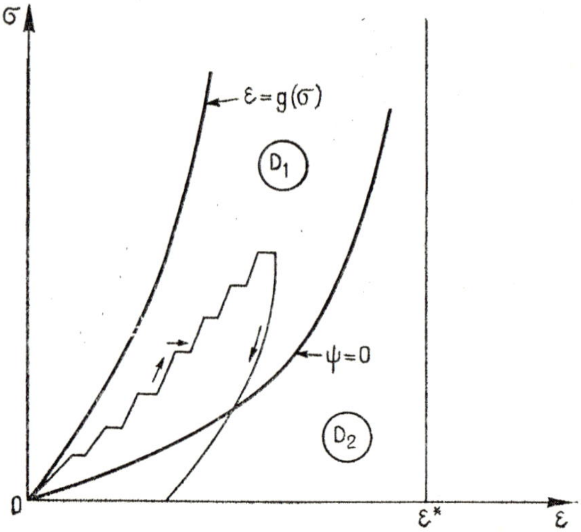

Fig. 9.4.2 Schema of domains and boundaries involved in the constitutive equation for volume compressibility

pressure which can be reached is smaller than in experiments using a standard testing machine (in our experiments we were unable to surpass about 4 kbar with these machines).

The boundary (9.4.2) is thought to have distinct properties than the boundary C_c discussed in the previous paragraph; i.e. starting from any $(\varepsilon, \sigma) \in D_1$ by any process with constant or decreasing pressure the boundary (9.4.2) will be reached in an infinite or finite time interval respectively.

We have also assumed that by making the fastest possible test in the range of experiments we have in mind, we get an "instantaneous" response for volume compressibility which may be expressed for instance as

$$\varepsilon = g(\sigma), \tag{9.4.3}$$

if the experiment starts from zero stress and zero strain states. Here $g(\sigma)$ is a non-negative function of class C^1 for $0 < \sigma$, $0 < \varepsilon < \varepsilon_*$. In fig. 9.4.2, D_2 is the domain

$$\varepsilon_* \left[1 - \exp\left(-\frac{\sigma}{\sigma_0}\right)\right] \leqslant \varepsilon < \varepsilon_*, \qquad \sigma > 0$$

and the domain D_1 is given by

$$g(\sigma) < \varepsilon < \varepsilon_* \left[1 - \exp\left(-\frac{\sigma}{\sigma_0}\right)\right], \qquad \sigma > 0.$$

Points belonging to D_1 are possible strain-stress states which can be reached in loading processes when stress is increasing. Some points in D_2 can be reached during unloading (decreasing stress) while others, as for instance those in the neighbourhood of $\varepsilon = \varepsilon_*$, $\sigma = 0$ cannot be reached by any conceivable experiment. For several rocks power functions seem to be suitable for use in (9.4.3). In particular, (9.4.3) can be a straight line.

In order to describe the volume compressibility a rate-type constitutive equation of the form

$$\dot{\varepsilon} = \left[\frac{1}{3K(\sigma)} + f(\varepsilon, \sigma, \operatorname{sign} \dot{\sigma})\right] \dot{\sigma} + \Psi(\varepsilon, \sigma) \tag{9.4.4}$$

is used. The instantaneous properties in loading are defined by

$$f(\varepsilon, \sigma, \operatorname{sign} \dot{\sigma}) = \begin{cases} \varphi(\varepsilon, \sigma) \text{ if } & \dot{\sigma} \geqslant 0, \\ 0 \quad \text{if} & \dot{\sigma} < 0, \end{cases} \tag{9.4.5}$$

where

$$\varphi(\varepsilon, \sigma) = \frac{dg(\sigma)}{d\sigma} - \frac{1}{3K(\sigma)} \qquad \text{for all } (\varepsilon, \sigma) \in D_1 \tag{9.4.6}$$

and $K(\sigma)$ is the variable modulus in unloading. In a first approximation one can assume $K = \text{const}$.

The function $\psi(\varepsilon, \sigma)$ describing the slow deformation of the volume (creep) is defined by

$$\psi(\varepsilon, \sigma) = \begin{cases} \eta \left\{ \varepsilon_* \left[1 - \exp\left(-\dfrac{\sigma}{\sigma_0} \right) \right] - \varepsilon \right\} & \text{if } \varepsilon < \varepsilon_* \left[1 - \exp\left(-\dfrac{\sigma}{\sigma_0} \right) \right], \\ 0 & \text{if } \quad \varepsilon > \varepsilon_* \left[1 - \exp\left(-\dfrac{\sigma}{\sigma_0} \right) \right], \end{cases} \tag{9.4.7}$$

where η is a volume viscosity coefficient given in $\sec^{-1}$ (or $(\text{day})^{-1}$) which in a first approximation will be considered to be constant.

Making the rough assumption that σ and σ_1 are of the same order of magnitude, the constants involved in (9.4.4) can be determined by the tests mentioned. Thus for schist the approximate values are $\varepsilon_* = 0.0108$, $\sigma_0 = 4000$ kgf cm^{-2} (39228 N cm^{-2}), $\eta = 43$ d^{-1}. For natural chalk which is a soft rock with $\sigma_r = 45$ kgf cm^{-2} (441 N cm^{-2}) and initial density $\rho = 1.67$ g cm^{-3} we get $\varepsilon_* = 0.07$, $\sigma_0 = 550$ kgf cm^{-2} (5394 N cm^{-2}) and $\eta = 3.80 \pm 1.07$ d^{-1}. The value of η was determined from several dozens of tests (of horizontal plateaux at various levels of stresses). Generally it was found that η is quite constant. Since a small number of experiments were used to determine these constants, the values given must be considered only as indicative pending further experimental data.

5. CONCLUSIONS

It was shown how by several diagnostic tests one can determine elastic-viscoplastic constitutive equations for rocks concerning one-dimensional compressive loadings and compressibility of volume. A three-dimensional generalization of the model based on these one-dimensional models is now under consideration taking into account the experiments which would reveal the relationship between shearing properties and volume variation (dilatancy) . Of course, even experiments of the kind discussed are further necessary to make precise more details of the model (mainly to find more exact values of the constants involved).

Chapter X

AN ANALOGY BETWEEN THE CONSTITUTIVE EQUATIONS OF ELECTRIC LINES AND THOSE OF ONE-DIMENSIONAL VISCOPLASTICITY

1. INTRODUCTION

1a. The balance laws of the magnetic flux and the electric charge when applied to a two-wires electric line lead to the following partial differential equations (see for instance RĂDULEŢ, TIMOTIN and ŢUGULEA [1970])

$$\frac{\partial \varphi}{\partial t} + \frac{\partial u}{\partial X} = \tilde{u},$$
$$\frac{\partial q}{\partial t} + \frac{\partial i}{\partial X} = \tilde{i}, \tag{10.1.1}$$

where X is a spatial coordinate along the electric line and t is the time. The quantities $\varphi = \varphi(X, t)$, $q = q(X, t)$, $u = u(X, t)$ and $i = i(X, t)$ are the magnetic flux per unit length, the electric charge per unit length, the voltage and the electric current, respectively. Finally, $\tilde{u}$ is the line voltage drop on unit conductor length and $\tilde{i}$ is the line leakage current on unite conductor length.

Let us denote by D a plane domain and let $(X, t) \in D$ be its points. Let $\mathscr{D}$ denote a bounded domain in the four dimensional space R^4 such that $\mathscr{D}$ contains the origin; all the points (u, i, φ, q) of the domain $\mathscr{D}$ will be called electromagnetic states of the electric line.

In order to turn the system (10.1.1) into a complete system of partial differential equations, two equations relating the four quantities u, i, φ and q are still necessary. It is also necessary to define the functions $\tilde{u}$ and $\tilde{i}$ as functions of $(X, t) \in D$ and of the state $(u, i, \varphi, q) \in \mathscr{D}$. We may assume that

$$\tilde{u} = \overline{u}(X, t) + u^*(u, i, \varphi, q)$$
$$\tilde{i} = \overline{i}(X, t) + i^*(u, i, \varphi, q). \tag{10.1.2}$$

The quantities $\bar{u}(X, t)$ and $\bar{i}(X, t)$ are thought here as given functions of X and t. Thus they have a similar meaning as body forces (or heat supply) in continnum mechanics. It is easy to imagine many situations when such quantities may be nonvanishing along a given electric line. Therefore $\bar{u}$ and $\bar{i}$ represent external influences acting on the electric line whose state is characterized by (u, i, φ, q).

The two equations relating u, i, φ and q and the two functions $u^* = u^*(u, i, \varphi, q)$ and $i^* = i^*(u, i, \varphi, q)$ are called *constitutive equations* and they specify a given line, while equations (10.1.1) are describing the behaviour of any electric line. The form of the constitutive equations depends on the properties of the conductors, on the way they are installed, on the kind of insulation used, etc., and has to be independent of the variations of the quantities $(u, i, \varphi, q) \in \mathscr{D}$, in the time and space where the given electric line is designed to work. Once the constitutive equations are established, one may formulate and solve initial and boundary value problems.

Unfortunately, up to now, there are no general rules to establish constitutive equations in continuum physics. The procedure one generally follows is to choose the constitutive equations in a sufficiently general way, as to include the observed experimental behaviour; then, based on the laws of continuum physics, to restrict the form of these equations and finally, by using experimental results, to determine the remaining unknown functions and constants specific to the problem, e.g. to the given electric line.

1b. The simplest constitutive equations for electric lines are those involved in the telegrapher's equations where it is supposed that

$$\varphi = Li, \; q = Cu, \; u^* = Ri, \; i^* = Gu \tag{10.1.3}$$

and in the absence of external influences,

$$\bar{u} = 0, \quad \bar{i} = 0. \tag{10.1.4}$$

In (10.1.3), L, C, R, G are real positive numbers, their meaning being the usual one (i.e., L is the line inductance, C is the geometric capacitance of the line, R is the resistance of the line and G is the line conductance). If $(10.1.3)_{3-4}$ and (10.1.4) are substituted into (10.1.2) and the obtained result is combined with $(10.1.3)_{1-2}$ and the balance equations (10.1.1), one gets the well known telegrapher's equations. These equations are neglecting important facts such as the skin effect, the relaxation of the electric charge, the corona effect, etc. which may sometimes be important in studying the propagation phenomena on electric lines.

In this chapter the quantities $\tilde{u}$ and $\tilde{i}$ (and thus $\bar{u}, u^*, \bar{i}, i^*$) will be neglected since the main purpose here is to build a model (based on experimental data) which will be able to describe the behaviour of a given electric line when corona discharges are present. At the same time we shall observe that whenever the corona effect is present, the effect of electric relaxation is also present.

Section 2 deals with general quasilinear rate-type constitutive equations for the electric charge and the magnetic flux. We formulate the restrictions imposed on the functions characterizing a given line by the requirement that symmetric and real waves should exist.

The general linear rate-type model is discussed in Section 3. It is shown that these constitutive equations can be written in an integral form which makes this model closer to the one proposed by RĂDULEŢ, TIMOTIN and ŢUGULEA [1970] (see also RĂDULEŢ, TIMOTIN, ŢUGULEA and NICA [1978]) for the description of the skin effect and the relaxation properties of the electric lines. The condition that at a steady state the magnetic flux and the electric charge obey the usual linear laws reduces the number of the unknown constants of the model. The remaining constants have to be determined experimentally in the transient processes.

In Section 4 we construct a semilinear model in order to describe the corona effect by starting from the experimental results of MARUVADA, MENEMENLIS and MALEWSKI [1977]. This model was also checked on the experimental data of GARY, DRĂGAN and CRISTESCU [1976].

The last section concerns wave propagation in the presence of the corona effect by using a "rate independent" model. Such models are common to studies of corona effects on electric lines as well as in one-dimensional dynamic plasticity. Thus, one can directly use for studying electric lines the methods of numerical programming given in Chapter IV of this book. It was observed that such models cannot yield simultaneously very good agreements with the experimental data of WAGNER and LLYOD [1955] and those of GARY, DRĂGAN and CRISTESCU [1978]. This discrepancy is mainly due to the fact that relaxation phenomena are neglected. It is believed that a model of the type discussed in Section 4 can lead to a much better agreement with the experimental data. At the same time the obtained system of partial differential equations is a semilinear one and therefore, from the mathematical point of view, a much simpler one and, from the physical point of view, the model is much sounder since all perturbations propagate with a given constant speed, a characteristic of the line.

2. QUASILINEAR RATE-TYPE CONSTITUTIVE EQUATIONS, REAL AND SYMMETRIC WAVES FOR ELECTRIC LINES

2a. The constitutive equations discussed in this section are similar to the constitutive equations of Chapter II, Section 2. Since one needs here two constitutive equations, the discussion will be repeated on some details.

We assume that there are defined on $\mathscr{D}$ six real-valued continuous functions:

$$\hat{\mathscr{C}}_u, \hat{\mathscr{L}}, \hat{\mathscr{R}}, \hat{\mathscr{C}}, \hat{\mathscr{L}}_i, \hat{\mathscr{G}} : \mathscr{D} \to R,$$

$$\begin{aligned} \mathscr{C}_u &= \hat{\mathscr{C}}_u(u, i, \varphi, q), & \mathscr{L} &= \hat{\mathscr{L}}(u, i, \varphi, q), \\ \mathscr{R} &= \hat{\mathscr{R}}(u, i, \varphi, q), & \mathscr{C} &= \hat{\mathscr{C}}(u, i, \varphi, q), \\ \mathscr{L}_i &= \hat{\mathscr{L}}_i(u, i, \varphi, q), & \mathscr{G} &= \hat{\mathscr{G}}(u, i, \varphi, q). \end{aligned} \tag{10.2.1}$$

Here, any hatted letter denotes a function while the same letter without hat denotes the value of this function at a given state in $\mathscr{D}$.

We shall say that an electric line has a *rate-type behaviour* in $\mathscr{D}$ if for any state $(u, i, \varphi, q) \in \mathscr{D}$ and any given rates $\frac{\partial u}{\partial t}, \frac{\partial i}{\partial t}$ at this state (for a fixed section X) the rates of magnetic flux and electric charge are determined by the relations (see SULICIU [1978] and MIHĂILESCU-SULICIU and SULICIU [1979] and also SULICIU [1979]).

$$\begin{aligned} \frac{\partial \varphi}{\partial t} &= \mathscr{C}_u \frac{\partial u}{\partial t} + \mathscr{L} \frac{\partial i}{\partial t} + \mathscr{R}, \\ \frac{\partial q}{\partial t} &= \mathscr{C} \frac{\partial u}{\partial t} + \mathscr{L}_i \frac{\partial i}{\partial t} + \mathscr{G}, \end{aligned} \tag{10.2.2}$$

where $\mathscr{L} = \hat{\mathscr{L}}(u, i, \varphi, q)$, etc. In other terms, the constitutive equations (10.2.2) are stating that at any fixed moment t, the variation of the magnetic flux (respectively of the electric charge) is determined by the value of (u, i, φ, q) at that moment, by the variations Δu, Δi as well as by the time variation Δt, according to the relation

$$\Delta \varphi = \mathscr{C}_u \Delta u + \mathscr{L} \Delta i + \mathscr{R} \Delta t.$$

Therefore, when $u(t)$, $i(t)$, $t \in [0, T)$ are given, the constitutive equations (10.2.2) determine $\varphi(t)$ and $q(t)$ if the initial values

$$\varphi(0) = \varphi_0, \qquad q(0) = q_0 \tag{10.2.3}$$

are known.

We shall assume in the following that the functions in (10.2.1) are given so that for any piecewise smooth functions $u(t)$, $i(t)$. $t \in [0, T)$, $T > 0$, and any initial values (10.2.3) with $(u(0,)\, i(0), \varphi(0), q(0)) \in \mathscr{D}$, the solution $(u(t), i(t), \varphi(t), q(t))$, $t \in [0, T_1)$, $0 < T_1 \leqslant T$, of the initial value problem (10.2.2) — (10.2.3) is unique. (For sufficient conditions to ensure uniqueness as well as for a more detailed discussion on such problems see SULICIU [1973, 1974c] and also Chapter V, Section 3 of this book).

Let u_0 and i_0 be two real constants and let

$$u(t) = u_0, \quad i(t) = i_0, \qquad t \in [0, T), \quad T > 0. \tag{10.2.4}$$

Then the magnetic flux φ and the electric charge q will remain constant and equal to their initial values if and only if

$$\hat{\mathscr{R}}(u_0, i_0, \varphi_0, q_0) = 0, \qquad \hat{\mathscr{G}}(u_0, i_0, \varphi_0, q_0) = 0. \tag{10.2.5}$$

Any state $(u_0, i_0, \varphi_0, q_0) \in \mathscr{D}$ that satisfies (10.2.5) is called an *equilibrium state*. The state $u_0 = 0$, $i_0 = 0$, $\varphi_0 = 0$, $q_0 = 0$ may be called the *natural equilibrium state* (if it satisfies (10.2.5)).

An electric line is said to be *without losses* if

$$\left.\begin{aligned} \mathscr{R} &= \hat{\mathscr{R}}(u, i, \varphi, q) = 0, \\ \mathscr{G} &= \hat{\mathscr{G}}(u, i, \varphi, q) = 0 \end{aligned}\right\} \quad \text{in } \mathscr{D} \tag{10.2.6}$$

Thus, $\tilde{u} = 0$, $\tilde{i} = 0$ in (10.1.1) does not generally imply that the electric line is without losses. These may be present due to the constitutive equations for the magnetic flux and the electric charge.

2b. The system of partial differential equations (10.1.1) and (10.2.2) (with $\tilde{u} = 0$ and $\tilde{i} = 0$) is a complete system of four equations with four unknown functions. One may formulate initial and boundary value problems for this system and one may expect to get unique solutions if the functions in (10.2.1) verify certain conditions.

For a given point $(X, t) \in D$, at a given state $(u = u(X, t),\ i = i(X, t),\ \varphi = \varphi(X, t),\ q = q(X, t)) \in \mathscr{D}$, the propagation speeds of any continuous perturbation in u, i, φ or q are given by the slopes of the characteristic curves of the system (10.1.1) + (10.2.2). These slopes are the solutions of the algebraic equation

$$s^2[(\mathscr{C}_u\mathscr{L}_i - \mathscr{L}\mathscr{C})\, s^2 - (\mathscr{C}_u + \mathscr{L}_i)\, s + 1] = 0, \tag{10.2.7}$$

where

$$s = \frac{\mathrm{d}X}{\mathrm{d}t} \tag{10.2.8}$$

denotes the required slope. The equation (10.2.7) is obtained by standard methods (see Chapter II, Section 4 and the references indicated there).

The zero slope characteristic curves (stationary waves),

$$s = 0, \tag{10.2.9}$$

are specific for rate-type constitutive equations and they account for the possible losses in the quantities q and φ. The other two propagation speeds are the solutions of the equation

$$(\mathscr{C}_u\mathscr{L}_i - \mathscr{L}\mathscr{C})\, s^2 - (\mathscr{C}_u + \mathscr{L}_i)\, s + 1 = 0. \tag{10.2.10}$$

Since wave symmetry requires that at a given state $(u, i, \varphi, q) \in \mathscr{D}$, a certain perturbation propagates with the same speed in both positive and negative directions of the OX-axis, we must necessarily have (Mihăilescu-Suliciu and Suliciu [1979])

$$\mathscr{C}_u + \mathscr{L}_i = 0. \tag{10.2.11}$$

On the other hand, real waves exist if and only if

$$\mathscr{L}_i^2 + \mathscr{L}\mathscr{C} > 0, \tag{10.2.12}$$

and therefore the propagation speeds are given by

$$s^2 = \frac{1}{\mathscr{L}_i^2 + \mathscr{L}\mathscr{C}}. \tag{10.2.13}$$

3. LINEAR RATE-TYPE CONSTITUTIVE EQUATIONS

3a. Suppose there is a subdomain $\mathscr{D}_1 \subset \mathscr{D}$ (containing the natural equilibrium state $u = 0$, $i = 0$, $\varphi = 0$, $q = 0$) where the functions in (10.2.1) have the form

$$\begin{aligned} &\hat{\mathscr{C}}_u(u, i, \varphi, q) = C_u, \qquad \hat{\mathscr{L}}_i(u, i, \varphi, q) = L_i, \\ &\hat{\mathscr{L}}(u, i, \varphi, q) = L, \qquad \hat{\mathscr{C}}(u, i, \varphi, q) = C \end{aligned} \tag{10.3.1}$$

and

$$\begin{aligned} \hat{\mathscr{R}}(u, i, \varphi, q) &= R_u u + R_i i + R_\varphi \varphi + R_q q, \\ \hat{\mathscr{G}}(u, i, \varphi, q) &= G_u u + G_i i + G_\varphi \varphi + G_q q. \end{aligned} \tag{10.3.2}$$

The quantities C_u, L_i, L, C, R_u, R_i, R_φ, R_q, G_u, G_i, G_φ, G_q are real constants which will specify the given electric line (at any given section X) when working in a linear regime.

Due to (10.2.11) and (10.2.12), the constants C_u, L_i, L and C must satisfy

$$C_u + L_i = 0 \tag{10.3.3}$$

and

$$LC + L_i^2 > 0. \tag{10.3.4}$$

However, one cannot conclude that $L_i = 0$ and thus $C_u = 0$.

Formally, from (10.3.1), (10.3.2) and (10.2.2) one may obtain the telegrapher's equations by particularizing the constants in (10.3.1) and (10.3.2) (see SULICIU [1978]). However, if we do this, the magnetic flux φ and the electric charge q will loose their usual meaning.

By substituting (10.3.1) and (10.3.2) into (10.2.2) we find

$$\begin{aligned} \frac{\partial q}{\partial t} &= C\frac{\partial u}{\partial t} + L_i\frac{\partial i}{\partial t} + G_u u + G_i i + G_\varphi \varphi + G_q q, \\ \frac{\partial \varphi}{\partial t} &= C_u\frac{\partial u}{\partial t} + L\frac{\partial i}{\partial t} + R_u u + R_i i + R_\varphi \varphi + R_q q. \end{aligned} \tag{10.3.5}$$

A constitutive equation of the same type as (10.3.5) has been proposed for the first time by MAXWELL in 1867, in order to describe stress relaxation at constant strain. A mathematical theory of the mechanical behaviour of the bodies, based on such constitutive equations can be called a MAXWELL type viscoelasticity.

It is obvious that the constitutive equations (10.3.5) describe the relaxation properties of both electric charge q and magnetic flux φ.

3b. In the following it will be shown (see SULICIU [1978]) that (10.3.5) can be written in the same form as the constitutive equations with memory which were proposed by RĂDULEŢ, TIMOTIN and ŢUGULEA [1970] for the purpose of describing the skin effect (see RĂDULEŢ, TIMOTIN, ŢUGULEA and NICA [1978] where besides the relaxation effect also a numerical analysis is discussed); these equations are similar to those used in a viscoelasticity theory of VOLTERRA type.

Suppose that, for a fixed X, $i(t)$ und $u(t)$ are given functions of class C^1 on $[0, T)$, $T > 0$, and consider the initial value problem

$$\begin{aligned} \frac{\partial \varphi}{\partial t} &= C_u\frac{\partial u}{\partial t} + L\frac{\partial i}{\partial t}, & \varphi(0) &= \overline{\varphi} + C_u u(0) + Li(0), \\ \frac{\partial q}{\partial t} &= C\frac{\partial u}{\partial t} + L_i\frac{\partial i}{\partial t}, & q(0) &= \overline{q} + Cu(0) + L_i i(0). \end{aligned} \tag{10.3.6}$$

This problem has the following solution:

$$\begin{aligned} \varphi(t) &= C_u u(t) + Li(t) + \bar{\varphi}, \\ & \qquad\qquad\qquad\qquad t \in [0, T). \\ q(t) &= Cu(t) + L_i i(t) + \bar{q}, \end{aligned} \tag{10.3.7}$$

The quantities $\bar{\varphi}$ and $\bar{q}$ are now determined as functions of $t \in [0, T)$ by the LAGRANGE method of the variation of parameters so that (10.3.7) is a solution of (10.3.5), with the initial conditions

$$\bar{\varphi}(0) = 0, \qquad \bar{q}(0) = 0, \tag{10.3.8}$$

for any $u(t)$ and $i(t)$ of class C^1 on $[0, T)$.

Substituting (10.3.7) into (10.3.5) one gets

$$\begin{aligned} \frac{\partial \bar{\varphi}}{\partial t} &= R_\varphi \bar{\varphi} + R_q \bar{q} + \varphi_1, \qquad \bar{\varphi}(0) = 0, \\ \frac{\partial \bar{q}}{\partial t} &= G_\varphi \bar{\varphi} + G_q \bar{q} + q_1, \qquad \bar{q}(0) = 0, \end{aligned} \tag{10.3.9}$$

where

$$\begin{aligned} &\varphi_1 = A_u u + A_i i, \quad q_1 = B_u u + B_i i, \\ &A_u = R_u + R_\varphi C_u + R_q C, \quad A_i = R_i + R_\varphi L + R_q L_i, \\ &B_u = G_u + G_\varphi C_u + G_q C, \quad B_i = G_i + G_\varphi L + G_q L_i. \end{aligned} \tag{10.3.10}$$

The initial value problem (10.3.9) has a unique solution, determined as follows. Take $\bar{\varphi}$ and $\bar{q}$ of the particular form

$$\bar{\varphi}(t) = e^{rt}, \quad \bar{q}(t) = e^{rt}. \tag{10.3.11}$$

Let r be the real or complex number which is the solution of the algebraic equation

$$\begin{vmatrix} R_\varphi - r & R_q \\ G_\varphi & G_q - r \end{vmatrix} = (R_\varphi - r)(G_q - r) - G_\varphi R_q = 0. \tag{10.3.12}$$

First, we consider the case when the roots of (10.3.12) are real and distinct, i.e.

$$r_1 \neq r_2, \quad r_1, r_2 \text{ real and } G_\varphi \neq 0, R_q \neq 0. \tag{10.3.13}$$

Then the general solution of the homogeneous system

$$\begin{aligned} \frac{\partial \bar{\varphi}}{\partial t} &= R_\varphi \bar{\varphi} + R_q \bar{q}, \\ \frac{\partial \bar{q}}{\partial t} &= G_\varphi \bar{\varphi} + G_q \bar{q} \end{aligned} \tag{10.3.14}$$

can be written as

$$\bar{\varphi} = \alpha_1 e^{r_1 t} + \alpha_2 e^{r_2 t},$$
$$\bar{q} = \beta_1 e^{r_1 t} + \beta_2 e^{r_2 t}, \tag{10.3.15}$$

where α_1, α_2, β_1, β_2 must satisfy the following relations:

$$(R_\varphi - r_1)\,\alpha_1 + R_q\beta_1 = 0, \quad (G_q - r_1)\,\beta_1 + G_\varphi\alpha_1 = 0,$$
$$(R_\varphi - r_2)\,\alpha_2 + R_q\beta_2 = 0, \quad (G_q - r_2)\,\beta_2 + G_\varphi\alpha_2 = 0. \tag{10.3.16}$$

According to (10.3.12), the system (10.3.16) has only two independent equations and if one chooses α_1 and α_2 to be the two independent quantities in the solution, then

$$\beta_1 = \lambda_1\alpha_1, \quad \alpha_2 = \lambda_2\beta_2, \tag{10.3.17}$$

where

$$\lambda_1 = -\frac{R_\varphi - r_1}{R_q} = -\frac{G_\varphi}{G_q - r_1}, \quad \lambda_2 = -\frac{R_q}{R_\varphi - r_2} = -\frac{G_q - r_2}{G_\varphi},$$
$$\lambda_1\lambda_2 = \frac{R_\varphi - r_1}{R_\varphi - r_2} = \frac{G_q - r_2}{G_q - r_1}. \tag{10.3.18}$$

Applying the LAGRANGE method again, we determine α_1 and β_2 as functions of time such that the solution (10.3.15) of the homogeneous system (10.3.14) is a solution of the nonhomogeneous system (10.3.9) with zero initial conditions. Thus, from (10.3.15) and (10.3.9), by taking into account (10.3.17), we get

$$e^{r_1 t}\frac{\partial\alpha_1}{\partial t} + \lambda_2 e^{r_2 t}\frac{\partial\beta_2}{\partial t} = \varphi_1(t), \quad \alpha_1(0) = 0,$$
$$\lambda_2 e^{r_1 t}\frac{\partial\alpha_1}{\partial t} + e^{r_2 t}\frac{\partial\beta_2}{\partial t} = q_1(t), \quad \beta_2(0) = 0. \tag{10.3.19}$$

From (10.3.19) we obtain $\partial\alpha_1/\partial t$ and $\partial\beta_2/\partial t$ and, by integration,

$$\alpha_1(t) = \frac{1}{1 - \lambda_1\lambda_2}\int_0^t [\varphi_1(\tau) - \lambda_2 q_1(\tau)]\, e^{-r_1 t}\mathrm{d}\tau,$$
$$\beta_2(t) = \frac{1}{1 - \lambda_1\lambda_2}\int_0^t [q_1(\tau) - \lambda_1\varphi_1(\tau)]\, e^{-r_2 t}\mathrm{d}\tau. \tag{10.3.20}$$

Now, combining (10.3.20) and (10.3.14), we finally find the desired solution

$$\bar{\varphi}(t) = \frac{1}{1-\lambda_1\lambda_2}\int_0^t \{[A_u(e^{r_1(t-\tau)} - \lambda_1\lambda_2 e^{r_2(t-\tau)}) + \\ + \lambda_2 B_u(e^{r_1(t-\tau)} - e^{r_2(t-\tau)})]\, u(\tau) + [A_i(e^{r_1(t-\tau)} - \\ - \lambda_1\lambda_2 e^{r_2(t-\tau)}) + \lambda_2 B_i(e^{r_1(t-\tau)} - e^{r_2(t-\tau)})]\, i(\tau)\}\, d\tau,$$

$$\bar{q}(t) = \frac{1}{1-\lambda_1\lambda_2}\int_0^t \{[\lambda_1 A_u(e^{r_1(t-\tau)} - e^{r_2(t-\tau)}) + \\ + B_u(e^{r_2(t-\tau)} - \lambda_1\lambda_2 e^{r_1(t-\tau)})]\, u(\tau) + [\lambda_1 A_i(e^{r_1(t-\tau)} - \\ - e^{r_2(t-\tau)}) + B_i(e^{r_2(t-\tau)} - \lambda_1\lambda_2 e^{r_1(t-\tau)})]\, i(\tau)\}\, d\tau. \tag{10.3.21}$$

Therefore (10.3.7), with $\bar{\varphi}$ and $\bar{q}$ determined by (10.3.21), gives the solution of the system (10.3.5) with the initial conditions $q(0) = Cu(0) + L_i i(0)$ and $\varphi(0) = C_u u(0) + Li(0)$. In other words, the solution $\varphi(t)$, $q(t)$ of (10.3.5) with $\varphi(0) = C_u u(0) + Li(0)$, $q(0) = Cu(0) + L_i i(0)$ is thus completely determined by the formulas (10.3.7) and (10.3.21), for the case (10.3.13).

Suppose now that the roots of (10.3.12) are complex numbers,

$$r_1 = \gamma + j\omega, \quad r_2 = \gamma - j\omega, \quad j = \sqrt{-1}, \quad \omega \neq 0, \tag{10.3.22}$$

with γ and ω real numbers. Then

$$R_q \neq 0 \text{ and } G_\varphi = 0$$

and

$$\bar{\varphi} = e^{\gamma t}(\alpha_1 \cos \omega t + \alpha_2 \sin \omega t), \\ \bar{q} = e^{\gamma t}(\beta_1 \cos \omega t + \beta_2 \sin \omega t). \tag{10.3.23}$$

α_1 and α_2 can be taken as independent constants while β_1 and β_2 can be computed; thus $(10.3.23)_2$ becomes

$$\bar{q} = \frac{1}{R_q} e^{\gamma t}\{-[(R_\varphi - \gamma)\cos \omega t + \omega \sin \omega t]\,\alpha_1 + \\ + [\omega \cos \omega t - (R_\varphi - \gamma)\sin \omega t]\,\alpha_2\}. \tag{10.3.24}$$

By following the same method as above, one can determine α_1 and α_2 so that the functions $\bar{\varphi}$ and $\bar{q}$ given by $(10.3.23)_1$ and (10.3.24) become a solution of the nonhomogeneous initial value problem (10.3.9). One gets α_1 and α_2 as the following functions of t:

$$\alpha_1(t) = \frac{1}{\omega}\int_0^t e^{-\gamma\tau}\{\varphi_1(\tau)[\omega \cos \omega\tau - (R_\varphi - \gamma) \sin \omega\tau] - - R_q q_1(\tau) \sin \omega\tau\} \, d\tau,$$

$$\alpha_2(t) = \frac{1}{\omega}\int_0^t e^{-\gamma\tau}\{\varphi_1(\tau)\,[(R_\varphi - \gamma) \cos \omega\tau + \omega \sin \omega\tau] + + R_q q_1(\tau) \cos \omega\tau\} \, d\tau. \tag{10.3.25}$$

Thus, substituting (10.3.25) into $(10.3.23)_1$ and (10.3.24) and the obtained result into (10.3.7), one gets the desired solution in the case when the roots of (10.3.12) are complex numbers.

The case $r_1 = r_2$ can be treated similarly.

Consider now the case when in (10.3.12) one has

$$r_1 \neq r_2 \text{ and } R_q = 0 \text{ (or } G_\varphi = 0). \tag{10.3.26}$$

Then

$$r_1 = R_\varphi, \qquad r_2 = G_q \tag{10.3.27}$$

and from $(10.3.16)_3$, (10.3.17) and $(10.3.16)_2$,

$$\alpha_2 = 0, \qquad \lambda_2 = 0, \qquad \beta_1 = \lambda_1\alpha_1, \qquad \lambda_1 = \frac{G_\varphi}{R_\varphi - G_q}. \tag{10.3.28}$$

Therefore, by substituting (10.3.27) and (10.3.28) into (10.3.21), one obtains

$$\bar{\varphi}(t) = \int_0^t e^{R_\varphi(t-\tau)}(A_u u(\tau) + A_i i(\tau)) \, d\tau,$$

$$\bar{q}(t) = \int_0^t \left\{\left[\frac{G_\varphi}{R_\varphi - G_q} A_u(e^{R_\varphi(t-\tau)} - e^{G_q(t-\tau)}) + B_u e^{G_q(t-\tau)}\right] u(\tau) + + \left[\frac{G_\varphi}{R_\varphi - G_q} A_i(e^{R_\varphi(t-\tau)} - e^{G_q(t-\tau)}) + B_i e^{G_q(t-\tau)}\right] i(\tau)\right\} d\tau. \tag{10.3.29}$$

If in (10.3.29)

$$G_\varphi = 0, \quad G_i = 0, \quad C_u = 0, \quad L_i = 0, \quad R_u = 0, \tag{10.3.30}$$

i.e. if the constitutive equations (10.3.5) are uncoupled (which means that the constitutive equation $(10.3.5)_1$ contains the charge and the voltage only, while the constitutive equation $(10.3.5)_2$ contains only the magnetic flux and electric current), then $\varphi(t)$ is determined by the values of $i(\tau)$ on $[0, t)$ only and $q(t)$ is determined by the values of $u(\tau)$ on $[0, t)$ only.

In this case the constitutive equations (10.3.5) are a particular case of those considered by RĂDULEŢ, TIMOTIN and ŢUGULEA [1970] (see formula (3.11) of that paper, where r_{dc} and g_{dc} are equal to zero).

3c. The results presented above can be used to obtain information concerning the sign and the value of the constants involved in the general linear model. Thus, if for instance

$$\begin{aligned} i(t) &= i_0 = \text{const.}, \\ & \qquad\qquad t \in [0, \infty), \\ u(t) &= u_0 = \text{const.}, \end{aligned} \tag{10.3.31}$$

$\varphi(t)$ and $q(t)$ can be directly computed by means of the above formulas. In order that the obtained magnetic flux and electric charge have finite limits when $t \to \infty$, i.e.

$$\lim_{t\to\infty} \varphi(t) = \varphi_0 < \infty, \qquad \lim_{t\to\infty} q(t) = q_0 < \infty, \tag{10.3.32}$$

the coefficients in the model (10.3.5) must satisfy certain inequalities. If the roots of (10.3.12) are real, they have to be negative, and if these roots are complex, then the real part γ must be negative.

It is generally accepted that at a steady state the magnetic flux and the electric charge are proportional to the current and voltage respectively, i.e.

$$\varphi_0 = \tilde{L} i_0, \qquad q_0 = \tilde{C} u_0, \tag{10.3.33}$$

where $\tilde{L} = \text{const.} > 0$, $\tilde{C} = \text{const.} > 0$, while (i_0, u_0) are given by (10.3.31), and (φ_0, q_0) by (10.3.32). If the constants G_φ, G_q, R_φ and R_q are such that (10.3.32) holds, then taking into account (10.3.31) and (10.3.33) we obtain directly from (10.3.5) that

$$G_u u_0 + G_i i_0 + G_\varphi \tilde{L} i_0 + G_q \tilde{C} u_0 = 0,$$

$$R_u u_0 + R_i i_0 + R_\varphi \tilde{L} i_0 + R_q \tilde{C} u_0 = 0.$$

These relations must hold for any "small enough" u_0 and i_0; this yields

$$\begin{aligned} \tilde{L} &= -\frac{G_i}{G_\varphi} = -\frac{R_i}{R_\varphi} > 0, \\ \tilde{C} &= -\frac{R_u}{R_q} = -\frac{G_u}{G_q} > 0. \end{aligned} \tag{10.3.34}$$

Using (10.3.34), we can write the constitutive equations (10.3.5) as

$$\frac{\partial q}{\partial t} = C\frac{\partial u}{\partial t} + L_i\frac{\partial i}{\partial t} + G_\varphi(\varphi - \tilde{L}i) + G_q(q - \tilde{C}u),$$

$$\frac{\partial \varphi}{\partial t} = C_u\frac{\partial u}{\partial t} + L\frac{\partial i}{\partial t} + R_\varphi(\varphi - \tilde{L}i) + R_q(q - \tilde{C}u). \tag{10.3.35}$$

The constants G_φ and R_q may be called *delayed* coupling coefficients. If the effect of the "overcharge" $q - \tilde{C}u$ on the variation of the magnetic flux may be neglected, then one may set $R_q = 0$, and if the effect of the "overflux" $\varphi - \tilde{L}i$ on the variation of the electric charge may be neglected, then one may set $G_\varphi = 0$. Then the limits (10.3.32) exist if

$$G_q \leqslant 0 \quad \text{and} \quad R_\varphi \leqslant 0. \tag{10.3.36}$$

The coefficients G_q and R_φ may be called the *relaxation coefficients.* The coefficient $L_i = - C_u$ takes into account the instantaneous influence of the electric current on the electric charge and the instantaneous influence of the voltage on the magnetic flux. One would generally expect that $C \neq \tilde{C}$, $L \neq \tilde{L}$ and probably $L_i \neq 0$, unless the relaxation effects are negligible.

4. SEMILINEAR RATE-TYPE CONSTITUTIVE EQUATIONS FOR MODELLING THE CORONA EFFECT

4a. **General considerations.** In order to describe the corona effect one generally accepts the following constitutive equation for the magnetic flux φ (Wagner and Llyod [1955], Razewig [1959]):

$$\varphi = Li, \qquad L > 0. \tag{10.4.1}$$

Here L is the line inductance which is assumed to be constant. Then (10.2.2) and (10.2.11) imply

$$\mathscr{C}_u = 0, \quad \mathscr{R} = 0 \quad \text{and} \quad \mathscr{L}_i = 0. \tag{10.4.2}$$

Under the hypothesis (10.4.1) the system of four partial differential equations (10.1.1) and (10.2.2), reduces to the system of three equations

$$L\frac{\partial i}{\partial t}+\frac{\partial u}{\partial X}=0,$$
$$\frac{\partial q}{\partial t}+\frac{\partial i}{\partial X}=0, \tag{10.4.3}$$
$$\frac{\partial q}{\partial t}-\mathscr{C}\frac{\partial u}{\partial t}-\mathscr{G}=0,$$

where the functions $\mathscr{C}=\hat{\mathscr{C}}(u, i, Li, q)$, $\mathscr{G}=\hat{\mathscr{G}}(u, i, Li, q)$ may depend on u, i and q, while $\tilde{u}$ and $\tilde{i}$ are assumed to be zero (due to the experimental arrangement in studying the corona effect).

From $(10.4.1)_2$, $(10.4.2)_3$ and (10.2.12) we get

$$\mathscr{C}>0. \tag{10.4.4}$$

For a given transmission line one generally supposes that all perturbations propagate with a constant velocity. This assumption, used together with (10.4.1) and (10.2.13) lead to

$$\mathscr{C}=C=\text{const.}, \tag{10.4.5}$$

i.e. the function $\mathscr{C}$ in $(10.4.3)_3$ is a constant for a given line, and equal to the line geometric capacitance C.

According to $(10.4.2)_3$ and (10.4.5), the propagation velocity (10.2.13) of any perturbation along an electric line is the constant

$$s^2=\frac{1}{LC}, \tag{10.4.6}$$

and the constitutive equation $(10.4.3)_3$ takes the form

$$\frac{\partial q}{\partial t}=C\frac{\partial u}{\partial t}+\mathscr{G}(u, i, q). \tag{10.4.7}$$

Let $(u_0, i_0, Li_0, q_0)\in\mathscr{D}$. If, for $t\geqslant 0$, $u(t)$ and $i(t)$ are given such that $u(0)=u_0$, $i(0)=i_0$, then we can find a $T>0$ and a function $q(t)$ which is the solution of the equation (10.4.7) for $t\in[0, T)$, with the initial condition $q(0)=q_0$. If $(u(t), q(t))$, $t\in[0, T)$ is to be represented as a curve in the u, q plane, then

$$\left.\frac{dq}{du}\right|_{(u_0,q_0)}=\left.\frac{\dot{q}}{\dot{u}}\right|_{t=0}=C+\frac{\mathscr{G}(u_0, i_0, q_0)}{\dot{u}(0)}, \tag{10.4.8}$$

whenever $\dot{u}(0)\neq 0$. Thus we have the following property: At any equilibrium state (see (10.2.5)) the slope of the curve $q=q(u)$ is equal to the geometric capacitance C, that is

$$\left.\frac{dq}{du}\right|_{(u_0,i_0,q_0)}=C. \tag{10.4.9}$$

Constitutive equations of the same type as (10.4.7) are well known in mechanics (see Chapter II, Section 2 of this book). In the study of non-linear problems of viscoplastic strains they have been used for the first time by SOKOLOVSKIĬ [1948 *a*, *b*] and MALVERN [1951 *a*, *b*]. The analog of the case when C is a function of the state variables has been analysed by CRISTESCU [1963], but all these models are in fact generalizations of a linear constitutive equation proposed by MAXWELL in 1867.

The aim of the following considerations is to determine effectively the form of the function G in (10.4.7), on the basis of the experimental results of MARUVADA, MENEMENLIS and MALEWSKI [1977] (see also GARY, DRĂGAN and CRISTESCU [1976] and WAGNER and LLOYD [1955]). We shall show that (see MIHĂILESCU-SULICIU and SULICIU [1979] and also SULICIU [1979]), although these experimental data are incomplete, as neither u nor q are known as functions of time, we can nevertheless conclude that when G depends on u and q in a rather simple way, then the constitutive equation (10.4.7) allows a fair description of the corona effect due to some impulse voltages whose wavefront lengths may vary from the switching to the lightning ones.

4b. **A short description of the experimental results.** The experimental measurements of MARUVADA, MENEMENLIS and MALEWSKI [1977] were performed on single conductors with diameters of 30.48 mm (1.2″) and 46.30 mm (1.823″) as well as on bundle conductors of 4 × 30.48 mm (4 × 1.2″) and 6 × 46.30 mm (6 × 1.823″) and a length of 61 m. Each conductor was tested with four impulse forms covering the range between switching and lightning overvoltages (260/2300 μs, 75/2300 μs, 15/725 μs, 2.5/60 μs, where in the notation t_M/t_0, t_M represents the time required for the voltage to increase to its maximum value and t_0 is the time required for the voltage to decrease to half of its maximum value). The maximum applied voltage varied between the corona onset value and twice this value. For the description of the experimental set-up and measuring techniques the reader is referred to the above quoted paper.

There follow now a few general observations resulting from a global examination of the experimental results discussed in the above quoted paper (see the full lines in the figs. 10.4.2—10.4.5 reproduced after MARUDAVA, MENEMENLIS and MALEWSKI [1977]).

1. There does not exist a unique function g: $[0, u_M] \to R$ such that

$$q = g(u)$$

on the rising part of the function $u = u(t)$, for all four regimes that are experimentally studied.

2. The conductor corona onset voltage u_c increases with the voltage rate $\dot{u} = \frac{du}{dt}$ (on the rising part of $u = u(t)$). This fact is clearly noticed for impulses with small t_M ($t_M = 2.5\ \mu s$).

3. On the decreasing part of $u = u(t)$ one may take $q = C(u - u_N) + q_N$ (where C is the geometric capacitance, q_N is the maximum value of q and u_N is the voltage for which q reaches q_N) only for relatively small t_0 ($t_0 = 60\ \mu s$). For the other cases that were experimentally investigated, the measured values of q are, in general, much larger than those obtained by means of the above relation, especially for quite large time values as compared to t_M.

4. For impulses $u = u(t)$ with small t_M, starting from the moment t_c of the corona onset, one notices that q increases more rapidly than for larger values of t_M.

5. On the rising part of $u = u(t)$, the u, q curves get closer to the straight line $q = Cu$ as t_M decreases, i.e. as $\dot{u} = \frac{du}{dt}$ increases.

Remarks 1, 4 and 5 as well as the first part of Remark 3 are also confirmed by the experiments of GARY, DRĂGAN and CRISTESCU [1976], where impulse voltages with $t_M = 1.2\ \mu s$ and $t_M = 10\ \mu s$ have been used.

Let us note that all the above observations are familiar to those who work in dynamic experimental research on plastic deformation of metallic bars, provided the voltage is read as stress and the electric charge as strain.

In the following, we present a very simple function $G(u, q)$ to be placed in (10.4.7) with properties that are compatible with Remarks 1—5.

4c. **Constitutive equations used for numerical examples.** In the description of the corona effect one generally admits that the relation between the voltage u and the charge q is independent of the current i. This assumption will be accepted also here and therefore $\hat{\mathscr{G}}$ in (10.4.7) is considered as independent of i.

Let $\hat{\mathscr{G}}$ in (10.4.7) be a continuous and piecewise smooth function of u and q. We shall assume it is defined for $u \geqslant 0$ and $q \geqslant 0$ and, for $q \geqslant Cu$, it has the following form

$$\hat{G}(u, q) = \begin{cases} 0 & \text{if } q \geqslant q_0 + C_2(u - u_0), \\ k_2[q_0 + C_2(u - u_0) - q] & \text{if } \tilde{q}_0 + C_1(u - \tilde{u}_0) < q \leqslant q_0 + C_2(u - u_0), \\ k_2[q_0 + C_2(u - u_0) - q] + k_1[\tilde{q}_0 + C_1(u - \tilde{u}_0) - q] & \text{if } q \leqslant \tilde{q}_0 + C_1(u - \tilde{u}_0), \end{cases} \tag{10.4.10}$$

where

$$q_0 = Cu_0, \quad \tilde{q}_0 = C\tilde{u}_0, \quad \tilde{u}_0 \geqslant u_0 \geqslant 0, \tag{10.4.11}$$

$$0 < C < C_1 < C_2,$$

$$k_1 > 0, \quad k_2 > 0$$

are constants for a given electric line, independent of the form of the impulse $u = u(t)$; u_0 and $\tilde{u}_0$ are critical voltages, C is the geometric capacitance, C_1 and C_2 are "dynamic" capacitances while k_1 and k_2 can be interpreted as short-range and long-range electrical viscosity constants respectively.

If we introduce the corona overcharge q_c as

$$q_c = q - Cu, \tag{10.4.12}$$

the constitutive equation (10.4.7) with $\hat{\mathscr{G}}$ given by (10.4.10) can be written for $q_c \geqslant 0$ as

$$\dot{q}_c = \begin{cases} 0 & \text{if } q_c \geqslant (C_2 - C)(u - u_0), \\ k_2[(C_2 - C)(u - u_0) - q_c] & \text{if } (C_1 - C)(u - \tilde{u}_0) < q_c \leqslant (C_2 - C)(u - u_0), \\ k_2[(C_2 - C)(u - u_0) - q_c] + k_1[(C_1 - C)(u - \tilde{u}_0) - q_c] & \text{if } 0 \leqslant q_c \leqslant (C_1 - C)(u - \tilde{u}_0). \end{cases} \tag{10.4.13}$$

Starting from any state $(\bar{u}, \bar{q})$, with $\bar{u} \geqslant 0$, $\bar{q} \geqslant 0$ and $\bar{q} - C\bar{u} \geqslant 0$, and a given function $u(t) \geqslant 0$, $t \in [0, T)$, $T > 0$ with $u(0) = \bar{u}$, we can determine $q_c(t)$ as the solution of the equation (10.4.13) with the initial condition

$$q_c(0) = \bar{q}_c = \bar{q} - C\bar{u}, \tag{10.4.14}$$

and then $q(t) = q_c(t) + Cu(t)$.

For better understanding of the properties of the constitutive equation (10.4.7), we shall consider in the following the case when $u_0 = \tilde{u}_0$, $\bar{q} = 0$, $\bar{u} = 0$ and

$$u(t) = \begin{cases} at & \text{if } 0 \leqslant t < t_M = \dfrac{u_M}{a}, \\ u_M - b(t - t_M) & \text{if } t_M \leqslant t < t_M\left(\dfrac{1}{a} + \dfrac{1}{b}\right), \end{cases} \tag{10.4.15}$$

where $u_M > u_0$, $a > 0$, $b > 0$ are given constants.

We determine first $q_c(t)$ as the solution of equation (10.4.13) with $u(t)$ given by (10.4.15), and the initial condition $q_c(0) = 0$. Then we can draw the curves $(u(t), q(t) = q_c(t) + Cu(t))$ in the u, q-plane. For a given line, that is for fixed values of the constants in (10.4.11), we can fix the value u_M in (10.4.15), and therefore easily find the influence of the constants a and b (i.e. the voltage rate on the front and back of the impulse $u = u(t)$) on the form of the obtained curves $(u(t), q(t))$ (see fig. 10.4.1 and below). On the other hand, if the impulse (10.4.6) has a fixed form, i.e. if u_M, a and b are fixed, we can easily examine the influence of the constants in (10.4.2) on the form of the u, q curves.

We shall present now the solution $q_c(t)$ mentioned above.

For a time interval $[0, \widetilde{T}_1)$, $\widetilde{T}_1 > t_0 = u_0/a$, we can write

$$q_c(t) = \begin{cases} 0 & \text{if} \quad 0 \leqslant t < t_0 \\ \dfrac{a[k_1(C_1 - C) + k_2(C_2 - C)]}{k_1 + k_2} \left\{ t - t_0 - \dfrac{1}{k_1 + k_2} [1 - e^{-(k_1+k_2)(t-t_0)}] \right\} & \text{if} \quad t_0 \leqslant t < \widetilde{T}_1. \end{cases} \tag{10.4.16}$$

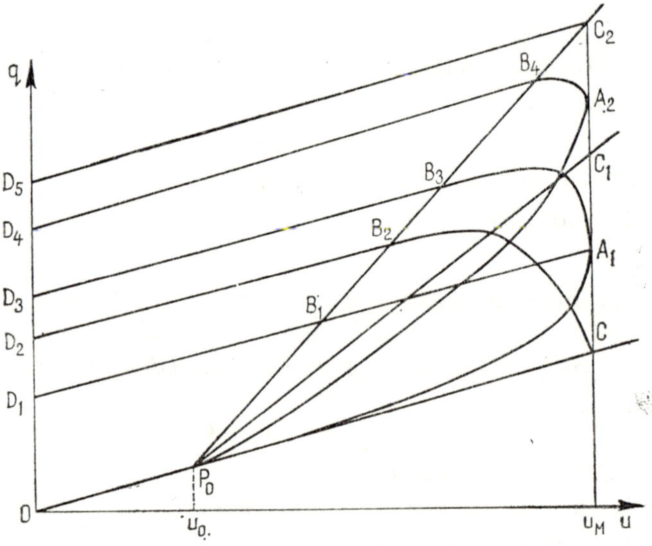

Fig. 10.4.1 Dependence of u, q curves on the constants a and b of (10.4.15) for fixed u_M, fixed constants (10.4.11) and $u_0 = \tilde{u}_0$

OP_0CP_0O	$(a \to \infty,\ b \to \infty)$
OP_0CC_2	$(a \to \infty,\ b \to 0)$
$OP_0CB_2D_2$	$(a \to \infty,\ b > 0)$
$OP_0A_1C_2$	$(a_1 > 0,\ b \to 0)$
$OA_1B_1D_1$	$(a_1 > 0,\ b \to \infty)$
$OA_1B_3D_3$	$(a_1 > 0,\ b > 0)$
$OP_0A_2B_4D_4$	$(a_2 > 0,\ b > 0),\ a_2 < a_1$
$OP_0B_1B_4C_2D_5$	$(a \to 0,\ b > 0)$

As we already noticed in discussing formula (10.4.9), the curve $(u(t), q(t) = q_c(t) + Cu(t))$ is tangent to the line OC (see fig. 10.4.1) at point $(u_0, q_0 = Cu_0)$, hence there exists a $\tilde{T}_1 > t_0$ such that (10.4.16) is the solution of equation (10.3.13) for $t \in [0, \tilde{T}_1)$. If $t_M \to \infty$, the function (10.4.16) remains a solution as long as the curve $(u(t), q(t))$ lies in the angle CP_0C_1. It is easily seen there exists a $T_1 > t_0$ such that

$$q_c(T_1) = (C_1 - C)\, a(T_1 - t_0), \tag{10.4.17}$$

where

$$T_1 = t_0 + y_1 = \frac{u_0}{a} + y_1 \tag{10.4.18}$$

and y_1 is the nontrivial solution of the equation

$$k_1(C_2 - C_1)\, y = \frac{k_1(C_1 - C) + k_2(C_2 - C)}{k_1 + k_2}\,[1 - e^{-(k_1+k_2)y}]. \tag{10.4.19}$$

y_1 is therefore independent of $u = u(t)$ given by (10.4.15). Thus, for any $u(t)$ given by (10.4.15) and any given constants (10.4.11), T_1 can be computed before the solution $q_c = q_c(t)$ is known. This solution has two forms depending on the position of t_M with respect to T_1.

If $T_1 \leqslant T_M$,

$$q_c(t) = \begin{cases} 0 & \text{if } 0 \leqslant t \leqslant \dfrac{u_0}{a} = t_0 \\[2ex] \dfrac{a[k_1(C_1 - C) + k_2(C_2 - C)]}{k_1 + k_2}\left\{t - t_0 - \dfrac{1}{k_1 + k_2}[1 - e^{-(k_1+k_2)(t-t_0)}]\right\} & \\ \qquad \text{if } t_0 \leqslant t \leqslant T_1 \leqslant t_M, & \\[2ex] q_c(T_1)\, e^{-k_2(t-T_1)} + (C_2 - C)\, a\left[t - t_0 - \dfrac{1}{k_2} + \left(\dfrac{1}{k_2} - T_1 + t_0\right) e^{-k_2(t-T_1)}\right] & \\ \qquad \text{if } T_1 \leqslant t \leqslant t_M, & \\[2ex] q_c(t_M)\, e^{-k_2(t-t_M)} + (C_2 - C)\left[u_M - u_0 - b(t - t_M) + \dfrac{1}{k_2}\right] - & \\ \quad - (C_2 - C)\left[u_M - u_0 + \dfrac{1}{k_2}\right] e^{-k_2(t-t_M)} \quad \text{if } t_M < t \leqslant T_2, & \\[2ex] q_c(T_2) & \text{if } T_2 < t, \end{cases} \tag{10.4.20}$$

where T_2 is that time value for which $(u(t), q(t))$ intersects the straight line P_0C_2, i.e. T_2 is obtained from the equation

$$q_c(T_2) = (C_2 - C)\,[u_M - u_0 - b(T_2 - t_M)]. \tag{10.4.21}$$

If $t_M \leqslant T_1$,

$$q_c(t) =$$

$$= \begin{cases} 0 & \text{if} \quad 0 \leqslant t \leqslant \dfrac{u_0}{a} = t_0, \\[2ex] \dfrac{a[k_1(C_1 - C) + k_2(C_2 - C)]}{k_1 + k_2} \left\{ t - t_0 - \dfrac{1}{k_1 + k_2} [1 - e^{-(k_1+k_2)(t-t_0)}] \right\} & \\ & \text{if } t_0 \leqslant t \leqslant t_M, \\[2ex] q_c(t_M)\, e^{-(k_1+k_2)(t-t_M)} + \dfrac{k_1(C_1 - C) + k_2(C_2 - C)}{k_1 + k_2} \Big\{ u_M - u_0 - & \\ - b(t - t_M) + \dfrac{b}{k_1 + k_2} - \left(u_M - u_0 + \dfrac{b}{k_1 + k_2} \right) e^{-(k_1+k_2)(t-t_M)} \Big\} & \\ & \text{if } t_M < t \leqslant T_1', \\[2ex] q_c(T_1')\, e^{-k_2(t-T_1')} + (C_2 - C) \Big\{ u_M - u_0 - b(t - t_M) + \dfrac{b}{k_2} - & \\ - \left[u_M - u_0 - b(T_1' - t_M) + \dfrac{b}{k_2} \right] e^{-k_2(t-T_1')} \Big\} & \text{if } T_1' < t \leqslant T_2', \\[2ex] q_c(T_2') & \text{if} \quad T_2' < t, \end{cases} \tag{10.4.22}$$

where

$$q_c(T_1') = (C_1 - C)\,[u_M - u_0 - b(T_1' - t_M)], \tag{10.4.23}$$
$$q_c(T_2') = (C_2 - C)\,[u_M - u_0 - b(T_2' - t_M)].$$

Fig. 10.4.1 contains the curves $(u(t), q(t))$ obtained for different values of the constants a and b.

Now, if the impulse voltage (10.4.15) is fixed and $u_0 = \tilde{u}_0$, C, C_1 and C_2 have given values, one can study the variation of the u, q curves with the line parameters k_1 and k_2. The u, q curves get closer to the curve $(u, q = Cu)$, for $t \in [0, t_M)$, when $k_1 + k_2$ is decreasing. If $k_2 \to 0$ and $k_1 \to \infty$, the u, q curves approach the straight line P_0C_1 for $t \in [t_0, t_M)$ and, for $t \in \left(t_M, u_M, \left(\dfrac{1}{a} + \dfrac{1}{b} \right) \right]$, they will become a straight line parallel to OC through the point C_1 (fig. 10.4.1).

This analysis shows that a constitutive equation (10.4.13) leads to u, q curves that behave according to the Remarks 1—5 of the previous subsection.

Now, for a given line, it remains to determine a set of constants (10.4.11) such that the u, q curves obtained by means of equation (10.4.13) for known functions $u = u(t)$, will represent a sufficiently good approximation of the experimental u, q curves; the determined set (10.4.11) has to be independent of the impulse form $u = u(t)$. This we shall do subsequently.

4d. **Comparison between numerical and experimental results.** We have selected from the existing experimental data the u, q curves for bundle conductors of 4 × 30.48 mm (4 × 1.2″) of MARUVADA, MENEMENLIS and MALEWSKI [1977], because of their regularity properties.

The voltage function $u = u(t)$ has been chosen as

$$u(t) = \tilde{u}(e^{-\alpha t} - e^{-\beta t}), \qquad (10.4.24)$$

where $\tilde{u}$, α and β are determined from the voltage maximum value and the time values t_M, t_0 as, being the only available numerical data on the voltage function.

The geometric capacitance C corresponding to relatively small voltage values has been determined from the experimental data of figs. 10.4.2 and 10.4.3 (for 2.5/60 and 15/725 impulse voltages) and it is equal to

$$C = 0.017 \frac{\mu C}{\text{m} \times \text{kV}} = 17 \text{ pF/m}. \qquad (10.4.25)$$

According to the previous subsection, the capacitance C_1 and the critical voltage $\tilde{u}_0$ have been determined by drawing a straight line through the points (u, q) where q reaches its maximum value, for the 2.5/60 impulse voltages. The result was:

$$C_1 = 0.031 \frac{\mu C}{\text{m} \times \text{kV}} = 31 \text{ pF/m}, \quad \tilde{u}_0 = 463 \text{ kV}. \qquad (10.4.26)$$

The last table of MARUVADA, MENEMENLIS and MALEWSKI [1977] shows that the value of $\tilde{u}_0$ is close to the smallest values of the critical voltage obtained by using that table.

The value taken for C_2 is

$$C_2 = 0.066 \frac{\mu \text{C}}{\text{m} \times \text{kV}} = 66 \text{ pF/m}. \qquad (10.4.27)$$

The first attempt to determine k_1 and k_2 was based on the choice $u_0 = \tilde{u}_0$ and $u_0 = 434$ kV; but the obtained u, q curves for impulse voltages with relatively large t_M ($t_M = 75\ \mu$s or $t_M = 260\ \mu$s) did increase too fast in comparison with the experimental u, q curves. A better agreement between numerical and experimental data has been obtained by using

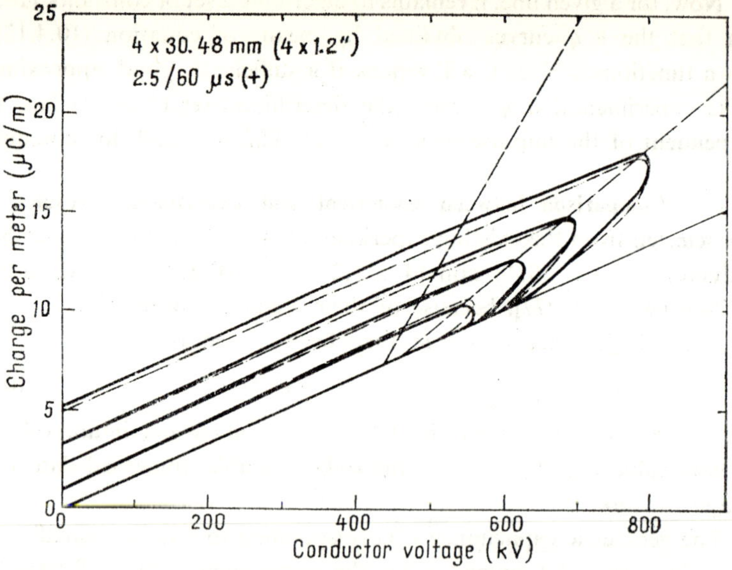

Fig. 10.4.2 The u, q curves. Full lines: experimental data; dotted lines: computed data

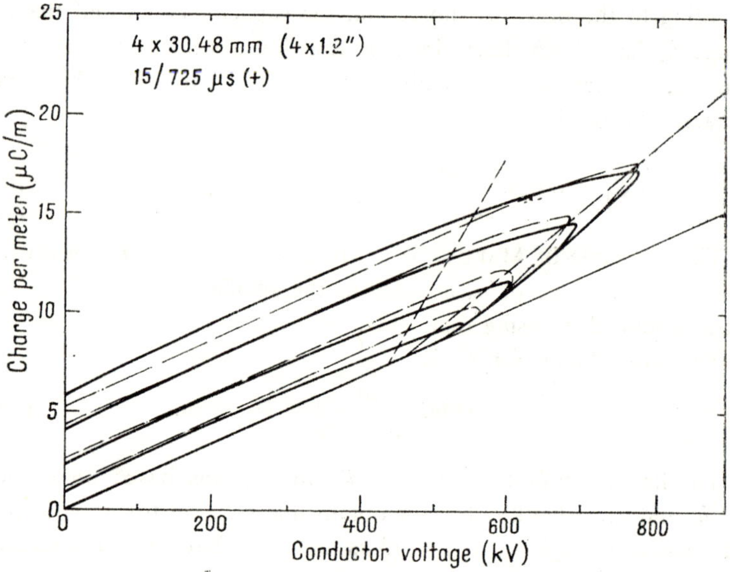

Fig. 10.4.3 The u, q curves. Full lines: experimental data; dotted lines: computed data

the constants (10.4.25)–(10.4.27) and

$$u_0 = 434 \text{ kV}, \; k_1 = 1 \frac{1}{\mu s}, \qquad k_2 = 8 \times 10^{-4} \frac{1}{\mu s} \qquad (10.4.28)$$

(see the dotted lines in figs. 10.4.2 to 10.4.5). However, the computed maximum corona overcharge q_c^M for the impulse voltages $u_M = 446$ kV, 75/230 and $u_M = 443$ kV, 260/2300 is too small. This discrepancy could

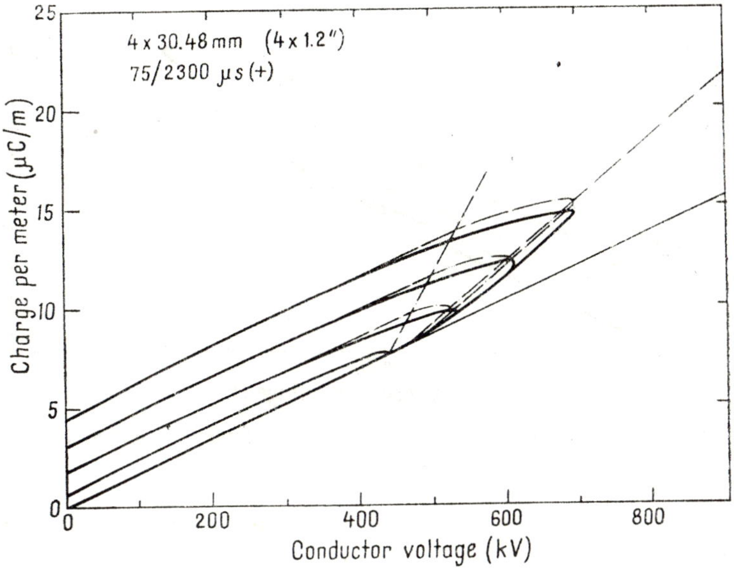

Fig. 10.4.4 The u, q curves. Full lines: experimental data; dotted lines: computed data

be removed by conveniently decreasing the values in one of the pairs of constants (u_0, k_2), (u_0, C_2) or (u_0, k_2, C_2) (or by reducing the values of (u_0, k_2) and increasing that of C_2); the other constants may remain unmodified as any change of these would affect only that part of the u, q curves which corresponds to the impulse voltage back.

The above discussed changes have not been made as the voltage function in (10.4.24) is merely postulated. As a matter of fact, the relation between u and q is strongly dependent on the form of the function $u = u(t)$ which follows clearly from the previous subsection. If, for instance, in fig. 10.4.2 we choose a function $u(t)$ which increases much slower than (10.4.24) on the first part of the interval [0, 2.5], but with the same maximum voltage value u_M, the u, q curves will remain at the beginning closer to the straight line $q = Cu$. If on the u, q curve corresponding to the largest u_M, the function $u(t)$ decreases at a much lower rate than (10.4.24) on an interval of several microseconds (say [2.5,5]), then the computed u, q curve will

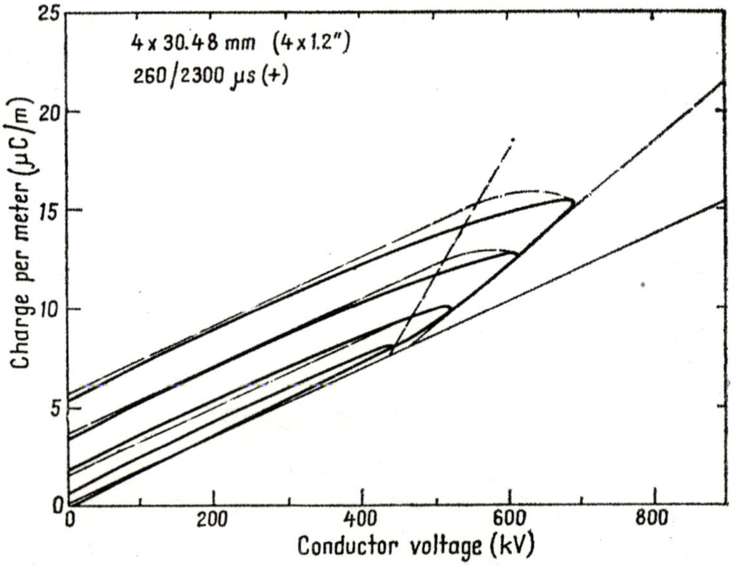

Fig. 10.4.5 The u, q curves. Full lines: experimental data; dotted lines: computed data

practically superpose over the experimental one. Similar remarks hold for the other figures too.

4e. **Remarks and conclusions concerning the discussed model**

1. The u, q curve always starts from a tangent point to the straight line $q = Cu$ (see formula (10.4.9)).

2. The propagation velocity of any perturbation is constant.

3. The observed increasing of the corona onset voltage with the voltage rate on the impulse front is incorporated by this model.

4. The present model does not lead to a unique u, q relation on the impulse front for every voltage rate, and this is in agreement with the experimental results.

5. For the impulse voltage back one obtains a larger q on the u, q curve, for larger values of t_0, where $2u(t_0) = u_{max}$.

6. The present model describes the behaviour of a transmission line over the whole range of the impulse voltages, from switching impulses

up to lightning ones, by means of the same constants (line parameters) which are independent of the impulse voltage form.

7. In wave propagation and decay problems, this model is simpler than the classical one, as all waves propagate with a constant speed and the characteristics of the partial differential system are straight lines. The maximum voltage decreases during propagation due to wave reflection as well as dissipation through the introduced electric viscosity with the coefficients k_1 and k_2.

8. For simplicity, the model has been chosen to contain two electric viscosity coefficients and two straight lines P_0C_1 and P_0C_2 (see fig. 10.4.1). It is obvious, however, that a better description of the line behaviour can be achieved by using in (10.4.7) a function $\hat{\mathscr{G}}(u, q)$ which is continuous and has continuous partial derivatives in the angle CP_0C_2 and, moreover, it has very large values for large voltage values above the straight line P_0C and close to it, and tends to zero when approaching P_0C_2 (where P_0C_2 can be replaced by a curve if necessary).

9. One can still improve the way constants are computed; a more precise determination of the line parameters can be performed, depending on the accuracy of the experimental results, if it is desired.

10. Finally, it seems that the relaxation coefficient k_1 (and perhaps k_2) depends on the dielectric, where the corona discharges take place (the air in this case), and it is independent of the conductor properties. This remark is confirmed, as far as k_1 is concerned, by describing in the same model the experimental data of both MARUVADA, MENEMENLIS and MALEWSKI [1977] and of GARY, DRĂGAN and CRISTESCU [1976]. The constants C, C_1, C_2 and u_0, $\tilde{u}_0$ are different in the two cases, since the conductors are different, but the same $k_1 = 1\ (\mu s)^{-1}$ can be used. One cannot draw definite conclusions about k_2 since the waveback in the paper of GARY, DRĂGAN and CRISTESCU [1976] is rather short so that the influence of k_2 (and of the entire angle $P_0C_1C_2$ — see fig. 10.4.1) is not too significant. The experimental data of MARUVADA, MENEMENLIS and MALEWSKI [1977] could be used to verify the validity of this remark, but a careful numerical analysis was not carried out.

In any case, the knowledge of the u, q curves only is not sufficient; one needs to know one of the following sets of quantities: a) u, q curves and $u = u(t)$, b) u, q curves and $q = q(t)$, c) $u = u(t)$ and $q = q(t)$.

5. RATE INDEPENDENT MODELS FOR THE CORONA EFFECT AND WAVE PROPAGATION

5a. We shall adopt also here the constitutive assumption of Section 4 *a* concerning the magnetic flux. The voltage drop on unit conductor length $\tilde{u}$ and the leakage current on unit length $\tilde{i}$ in the equations (10.1.1) will be also here neglected. Thus the balance laws (10.1.1) take the form $(10.4.3)_{1-2}$.

We may, formally, make the identifications

$$i = v, \; u = \sigma, \quad q = \varepsilon, \quad L = \rho_0, \tag{10.5.1}$$

where v is the particle velocity, σ is the stress (positive in compression), ε is the strain (positive in compression) and ρ_0 is the mass density. Then the equations $(10.4.3)_{1-2}$ and $(1.1.7)_{1-2}$ with $b = 0$ are identical.

The rate independent constitutive equations proposed in the literature (see WAGNER and LLOYD [1955], RAZEWIG [1959], Chapter III, Section 12, MARUVADA, MENEMENLIS and MALEWSKI [1977], HYLTEN-CAVALLIUS [1977]) for the description of the corona effect are then exactly of the form discussed in Chapter II, Section 1.

The numerical procedures for studying wave propagation on electric lines in the presence of corona discharges are the same as those discussed in Chapter IV (see also Appendix). This section deals with a comparison between numerical data (obtained, by using a rate independent constitutive equation, by CRISTESCU, CRISTESCU and SULICIU [1978]) and experimental data (of WAGNER and LLOYD [1955]).

5b. **Formulation of the problem.** The constitutive equation attached to the balance equations $(10.4.3)_{1-2}$ is of the form (2.1.4′), i.e.

$$\frac{\partial q}{\partial t} = \mathscr{F}\left(q, u, \operatorname{sgn}\frac{\partial u}{\partial t}\right)\frac{\partial u}{\partial t}, \tag{10.4.2}$$

where the function $\mathscr{F}$ is defined for $u \geqslant 0$ and $q \geqslant 0$ only (for negative impulses it can be defined in a similar way), that is

$$\mathscr{F}(q, u, \dot{u}) = \begin{cases} f'(u) & \text{if } q = f(u) \text{ and } \dot{u} \geqslant 0, \\ C_g & \text{if } \begin{cases} q > f(u), \\ \text{or} \\ q = f(u) \text{ and } \dot{u} < 0, \end{cases} \end{cases} \tag{10.5.2}$$

where $\dot{u} = \dfrac{\partial u}{\partial t}$ and

$$f(u) = \begin{cases} C_g u & \text{if } 0 \leqslant u \leqslant u_c, \\ q_0 + \left(\dfrac{u}{\beta}\right)^{1/\alpha} & \text{if } u_c < u, \end{cases} \tag{10.5.3}$$

where u_c is the corona onset (critical) voltage, C_g is the geometric capacitance, α and β are positive constants characterizing the given electric line, and $q_0 = C_g u_c$.

In order to use directly the computer program considered in Chapter IV, we may write the constitutive equation (10.5.2) in the form

$$\dot{q} = C_g \dot{u} + \Phi(q, u, \operatorname{sgn} \dot{u})\dot{u} = \dot{q}^R + \dot{q}^I, \tag{10.5.4}$$

where

$$\Phi(q, u, \operatorname{sgn} \dot{u}) = \begin{cases} \dfrac{1}{\alpha\beta}\left(\dfrac{u}{\beta}\right)^{\frac{1}{\alpha}-1} - C_g & \text{if } q = f(u) \text{ and } \dot{u} \geqslant 0, \\ 0 & \text{if} \begin{cases} q > f(u), \\ \text{of} \\ q = f(u) \text{ and } \dot{u} < 0; \end{cases} \end{cases} \tag{10.5.5}$$

$q^R = C_g u$ may be called the *reversible part of the charge* and it is the charge corresponding to the given voltage u in the absence of corona discharges, $q^I = q - q^R$ may be called the *irreversible part of the charge* and it may be interpreted as the overcharge spread out in the dielectric.

Let us denote by

$$s_0 = \frac{1}{\sqrt{LC_g}} \tag{10.5.6}$$

the speed of propagation (the characteristic slope) along the line of the small voltage impulses and by

$$s(u) = \frac{s_0}{\sqrt{1 + \Phi/C_g}} = \frac{1}{\sqrt{L}\sqrt{C_g + \Phi}}, \tag{10.5.7}$$

the variable propagation speed of the voltage, when the corona onset voltage is exceeded. Using the same procedure as in Chapter II, Section 4, we can easily show that the system of equations $(10.4.3)_{1-2}$ and (10.5.4) is of a hyperbolic type and it possesses three families of characteristic lines in the loading domain ($q = f(u)$ and $\dot{u} \geqslant 0$), namely

$$\mathrm{d}X = \pm\, s(u)\, \mathrm{d}t, \tag{10.5.8}$$

$$\mathrm{d}X = 0 \text{ (twice)}.$$

The differential relations satisfied along these characteristic lines are

$$
\begin{aligned}
\mathrm{d}u &= \mp Ls(u)\,\mathrm{d}i, \\
\mathrm{d}q^R &= C_g \mathrm{d}u, \qquad (10.5.9) \\
\mathrm{d}q^I &= \Phi\,\mathrm{d}u,
\end{aligned}
$$

respectively. In the unloading domain ($q > f(u)$ or $q = f(u)$ and $\dot{u} < 0$) $\Phi = 0$ holds and therefore $s = s_0$ and $\mathrm{d}q^I = 0$. Writing the constitutive equation in the form (10.5.4), we can easily pass from loading to unloading. The numerical method, which will be used below, is described in detail in Chapter IV and in the Appendix.

We assume that the electric line has length l; the initial conditions are taken in the form

$$
i(X, 0) = 0, \quad u(X, 0) = 0, \quad q(X, 0) = 0, \quad \text{for } 0 < X < l, \qquad (10.5.10)
$$

while the boundary conditions are

$$
u(0, t) = u_0(t) \quad \text{for } t \geqslant 0 \qquad (10.5.11)
$$

and

$$
u(l, t) = 0 \quad \text{for } t \geqslant 0. \qquad (10.5.12)
$$

For the boundary condition (10.5.11) we choose a function $u_0(t)$ which is in agreement with the experimental data of WAGNER and LLOYD [1955] presented in fig. 10.5.2, i.e.

$$
u_0(t) = \begin{cases} a t^{0.75} e^{-0.75t} & \text{for } \; 0 \leqslant t \leqslant 2\,\mu\text{s}, \\ u_0(2) - 1.053(t-2) & \text{for } \; 2\,\mu\text{s} < t \leqslant 3\,\mu\text{s}, \end{cases} \qquad (10.5.13)
$$

with $a = 3509.96$; t is given here in microseconds while u_0 is given in kilovolts. The function $u = u_0(t)$ in (10.5.13) is represented in fig. 10.5.2 by the dashed line which passes through the origin, while the corresponding experimental data (of WAGNER and LLOYD [1955]) are represented by full lines.

5c. **Numerical example.** (CRISTESCU, CRISTESCU and SULICIU [1978]). Let us start from the u, q curves given in fig. 11 of WAGNER and LLOYD [1955] and reproduced by full lines in fig. 10.5.1. We sought for several simple expressions of the form (10.5.3) in order to describe the behaviour of the electric line during the loading. For instance, the expression

$$
u = 195(q - 0.06)^{0.65}. \qquad (10.5.14)
$$

represented by a dashed line in fig. 10.5.1, reproduces rather accurately the experimental data during the loading.

Let us recall that if the constitutive equation $u \sim q$ for loading is of the form

$$q = f(u), \tag{10.5.15}$$

where $f(u)$ is a strictly increasing sm ooth function, then the propagation

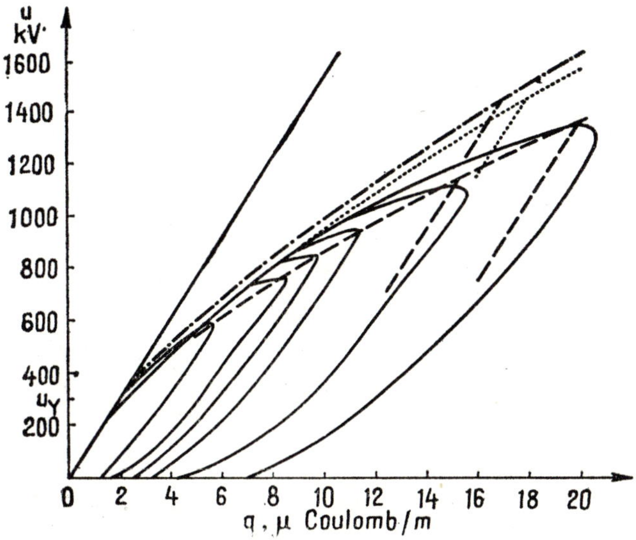

Fig. 10.5.1 The u, q curves. Full lines: experimental data of WAGNER and LLOYD [1955].Broken lines: $q = f(u)$ curves used in numerical analyses of propagation phenomena by CRISTESCU, CRISTESCU and SULICIU [1978]

speed of the loading waves is

$$s^2(u) = \left(\frac{\mathrm{d}X}{\mathrm{d}t}\right)^2 = \frac{1}{L}\frac{1}{\dfrac{\mathrm{d}f(u)}{\mathrm{d}u}}, \tag{10.5.16}$$

and it depends on the variable u alone. In other words, the propagation speed depends on the slope of the curve (10.5.15), and any fixed magnitude of u will propagate with the constant speed $s(u)$ in the loading domain. In the case of the initial and boundary value problem (10.5.10)—(10.5.12) with $u_0(t)$ given by (10.5.13) and for l large enough, the loading domain in the X, t plane where corona discharges take place contains the points (X, t), $0 \leqslant X \leqslant X_u(l)$ between the straight lines $X = s_0(t - t_c)$ and $X = s_0(t - t_M)$, where t_c and t_M are the time values where $u_0(t_c) = u_c$ and $u_0(t_M) = u_{max}$ (i.e. $\mathrm{d}u_0(t_M)/\mathrm{d}t = 0$); $X_u(l) > 0$ is here the point on the electric line where the first reflected (unloading) wave from the end $X = l$

appears. According to (10.5.16), we can compute for each position X the time t when the wave carrying a certain magnitude of u will reach that position. Thus, substituting (10.5.14) into (10.5.16) it is possible to draw the u, t curves in the loading domain at various positions X along the line. This is the way the dashed lines of fig. 10.5.2 have been drawn. The comparison with the experimental data (the full lines) shows that, especially

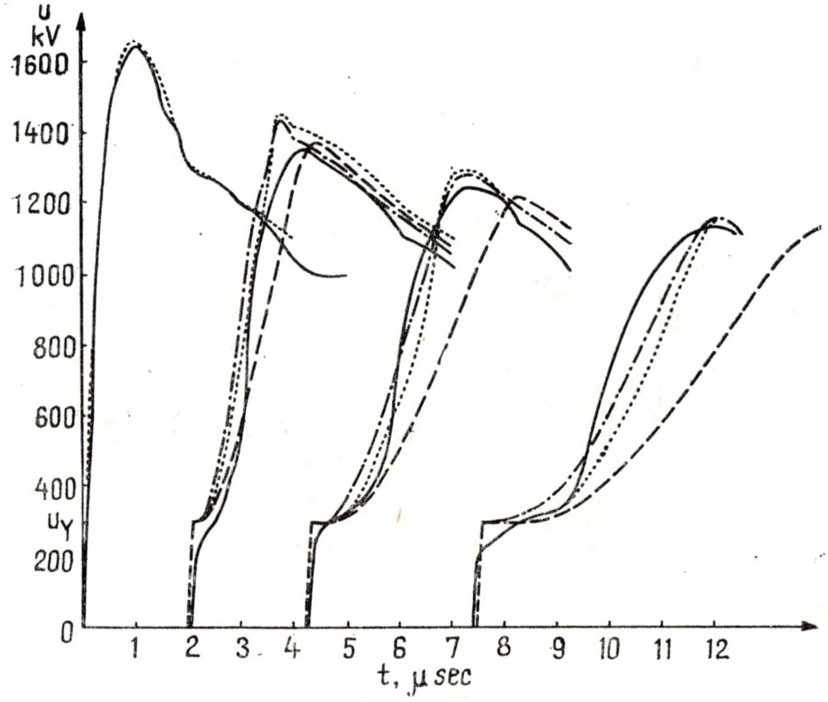

Fig. 10.5.2 The u, t curves. Full lines: experimental data of WAGNER and LLOYD [1955]. Broken lines: $u = u(t)$ computed curves using the corresponding $q = f(u)$ curves of Fig. 10.5.1 (CRISTESCU, CRISTESCU and SULICIU [1978])

for higher magnitudes of u, the constitutive equation (10.5.14) generally leads to too small propagation speeds, i.e. the arrival times of various given magnitudes of u are too big. In general, the maxima of u as obtained with (10.5.14) are in relatively good agreement with the experimental data.

Other considered relations between u and q are

$$u = 170.26(q + 0.15)^{0.76},$$
$$u = 187.48(q - 0.08)^{0.71}. \tag{10.5.17}$$

The first one is represented in fig. 10.5.1 by the dotted lines, and the second one by the border-lines. Both are experimental "quasistatic" curves. But, as the slopes of these two curves are for every u, higher than those obtained by means of (10.5.14), the propagation speed is higher too. Fig. 10.5.2 gives the u, t curves for various positions along the line, obtained according to (10.5.17) and (10.5.16); they are represented by the dotted lines and by the border lines. It must be mentioned here that the computations have been performed for various points along the conductor, but these points were not exactly those used in the experiments. Therefore the computed u, q curves have been plotted at the positions X closest to the ones for which the experimental data have been obtained. Comparison with the

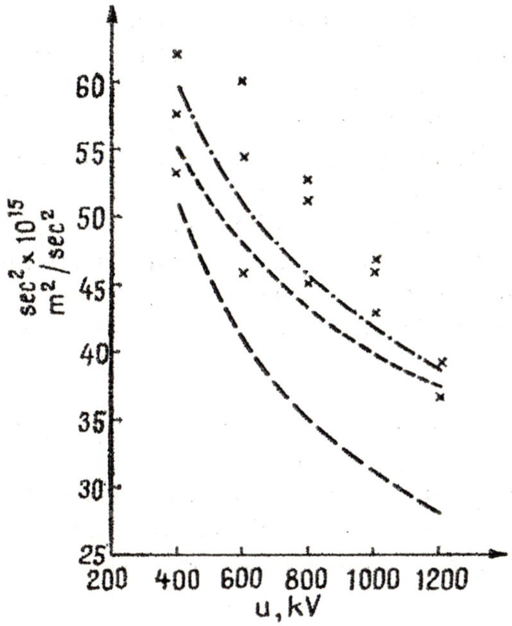

Fig. 10.5.3 The s^2, u curves. The dots are experimental data of WAGNER and LLOYD [1955]. Broken lines are $s^2 = s^2(u)$ curves computed using $q = f(u)$ curves of Fig. 10.5.1 (CRISTESCU, CRISTESCU and SULICIU [1978])

experiments shows that both relations (10.5.17) lead to maxima which are too high, but the arrival times are reasonably close to the experimental ones.

We can obtain relations of the type (10.5.17) by starting from the experimental data of fig. 10.5.2 and following a procedure that is well known in dynamic plasticity. By means of the experimental curves of fig. 10.5.2 we can determine the function $s(u)$ from the experimental data. Once $s(u)$ is known, and if a unique constitutive equation of the form (10.5.15) is assumed to exist in the loading domain, then from (10.5.16) we can obtain the slope $\mathrm{d}f(u)/\mathrm{d}u$ and, by integration, the function (10.5.15). To use this procedure, we plotted by crosses in fig. 10.5.3 the variation of s^2 with u resulting from the experimental data. One can see immediately that either due to a significant scatter of the experimental data or due to the fact that the corona effect cannot be described by a constitutive equation of type (10.5.15), one does not obtain a unique s^2, u curve. Accepting that the scatter of the crosses is only a scatter of the experimental data we have plotted several lines between these crosses, i.e. we have chosen several simple expressions for $s(u)$ which in virtue of the relation (10.5.16) lead, by integration, to the constitutive equation (10.5.15). Thus, starting from the dotted line and the border line of fig. 10.5.3, we obtained the relations (10.5.17). Returning to fig. 10.5.2, we observe that these two constitutive equations (determined by a procedure which could be termed "dynamic") lead to a much better agreement with the experimental data than those obtained by means of the constitutive equation (10.5.14) (determined by "quasistatic" data), since the arrival times of various magnitudes of the voltage are generally close to those experimentally obtained.

5d. **Concluding remarks.** From the previous analysis, taking into account the recent experimental results concerning the corona effect (see GARY, DRĂGAN and CRISTESCU [1976] and MARUVADA, MENEMENLIS and MALEWSKI [1977]) as well as the theory of propagation of elastic-plastic waves in bars, one can draw several conclusions.

The discussion related to fig. 10.5.3 emphasizes that there does not exist a unique relation between s and u in the loading domain; the uniqueness of such a relation would be equivalent to the existence of a unique constitutive equation (10.5.15) for loading. The experimental data of MARUVADA, MENEMENLIS and MALEWSKI [1977] do also support this conclusion. If a pulse with $t_M = 2.5\ \mu$s is produced at the end of a long enough electric line, then at a sufficiently large distance from this end t_M may increase up to $15\ \mu$s, say. But the above mentioned experimental data show that the u, q curves obtained with t_M equal to $2.5\ \mu$s and $15\ \mu$s are completely different.

The above computations were repeated by using other experimental data, namely those of GARY, DRĂGAN and CRISTESCU [1978], on wave propagations along a real electric line. The same scatter of the experimental data as mentioned in fig. 10.5.3 was obtained. All the other conclusions referring to the WAGNER and LLOYD data are valid in this case too.

It is believed that a rate-type constitutive equation of the type discussed in Section 4 of this chapter can lead to a better agreement between experimental and computed data. Such a constitutive equation could describe the behaviour of an electric line for a much wider range of forms of the given voltage impulse. Since this constitutive equation is continuous, the mathematical aspect of the problem of finding the solution becomes much simpler, once the line parameters are determined.

Appendix 1

NUMERICAL METHODS AND PROGRAMS USED IN THE STUDY OF EXPERIMENTAL DATA

1. INTRODUCTION

As is well known, the most difficult aspect in solving any problem in plasticity theory is the finding of the elastic/plastic boundary. In the elastic region the set of partial differential equations describing the deformation is distinct from the system of quasilinear partial differential equations valid in the plastic region. In dynamic problems things are even more involved due not only to the inertia forces, which now have to be considered, but also to the fact that in some cases one must use a rate-type constitutive equation.

In the previous chapters several numerical solutions have been given using either classical constitutive equations or rate-type constitutive equations. For both kinds of constitutive equations the interaction between the loading elastic-viscoplastic waves and elastic unloading waves have been studied. The boundary between the loading and unloading domains has been obtained using numerical methods for very general boundary and initial conditions. We shall give below a short exposition of these methods, used for both classical constitutive equations of the type (2.1.1) (which will be called *RI*, for rate independent) and rate-type constitutive equations of the form (2.3.1) (called *RD*, for rate dependent). We shall also give the flow chart and the programs used for the computer. We shall follow, in general, the paper by CRISTESCU and CRISTESCU [1973].

2. FORMULATION OF THE PROBLEM

The system of partial differential equations which has to be integrated is

$$\rho \frac{\partial v}{\partial t} = \frac{\partial \sigma}{\partial x}, \qquad -\frac{\partial v}{\partial x} = \frac{\partial \varepsilon}{\partial t},$$

$$\frac{\partial \varepsilon}{\partial t} = \frac{1}{E}\frac{\partial \sigma}{\partial t} + \Phi(\varepsilon, \sigma)\frac{\partial \sigma}{\partial t} + \Psi(\varepsilon, \sigma), \qquad \text{(A1.2.1)}$$

where the various forms of the coefficients are given in Chapter IV. The integration is carried out in the strip of the characteristic plane

$$a \leqslant x \leqslant b, \quad t \geqslant 0 \tag{A1.2.2}$$

with certain boundary conditions prescribed along $x = a$, $t > 0$ and $x = b$, $t > 0$ and initial conditions prescribed along $a \leqslant x \leqslant b$, $t = 0$.

We introduce the dimensionless variables

$$S = \frac{\sigma}{E}, \quad V = \frac{v}{c_0}, \quad T = Kt, \qquad X = \frac{K}{c_0}x. \tag{A1.2.3}$$

Here K is a constant of dimensions t^{-1}. Before integration in various domains of the strip (A1.2.2), the size of K has to be adjusted in order to modify the mesh size. With these dimensionless variables the characteristics (4.1.3) of the system (A1.2.1) are written as

$$\frac{\mathrm{d}X}{\mathrm{d}t} = \pm\frac{c}{c_0} = \pm\sqrt{\frac{1}{1+\Phi}}, \tag{A1.2.4}$$

$$\mathrm{d}X = 0 \text{ (twice)}.$$

Along them we have the differential relations

$$\begin{aligned} \mathrm{d}S &= \mp\frac{c}{c_0}\,\mathrm{d}V - \left(\frac{c}{c_0}\right)^2 \Psi\,\mathrm{d}T, \\ \mathrm{d}\varepsilon^P &= \Phi\,\mathrm{d}S + \Psi\,\mathrm{d}T, \\ \mathrm{d}\varepsilon^E &= \mathrm{d}S, \end{aligned} \tag{A1.2.5}$$

where the coefficients $\Psi = \dfrac{\psi}{K}$, $\Phi = E\phi$ are also dimensionless. The upper and lower signs in (A1.2.4) and (A1.2.5) correspond to each other.

The initial and boundary conditions were discussed in Chapter IV. Generally the "initial" conditions are such that for

$$\left.\begin{array}{c} t = 0 \\ a \leqslant x \leqslant b \end{array}\right\} : v(x, 0),\ \sigma(x, 0),\ \varepsilon(x, 0) \text{ are all prescribed,} \tag{A1.2.6}$$

while the "boundary" condition are such that for

$$\left.\begin{array}{c} x = a \\ t > 0 \end{array}\right\} : v(a, t) \text{ or } \sigma(a, t) \text{ are prescribed,}$$

and for

$$\left.\begin{array}{c} x = b \\ t > 0 \end{array}\right\} : v(b, t) \text{ or } \sigma(b, t) \text{ are prescribed.} \tag{A1.2.7}$$

We recall that we have discussed in Chapter IV the cases when, due to the specific boundary conditions, shock waves are also involved. In both cases, whether shock waves are involved or not, the solution is obtained with the aid of the computer.

3. PRINCIPLES OF THE ITERATION METHOD

Since it is very difficult to perform an integration with a computer using a nonregular grid of characteristic lines, and because two of the characteristic lines (A1.2.4) have variable slopes, a procedure which uses a regular grid has been employed, but at each vertex an iterative method was also necessary. The procedure is the following. First the maximum value taken by the characteristic slope in the whole domain of the characteristic plane under consideration must be found. This can be obtained easily from the constitutive equation and the initial and boundary conditions. For the present case this is just the "elastic slope" c_0. Next a basic net of characteristic lines of this (constant) slope is built. This net in turn will be used to find at each vertex the real slope of the characteristic line in that particular vertex and the first iteration of the solution at the same point.

To give a "physical" explanation of the method, the first iteration at each vertex is considered as being the "elastic" one. In other words, a regular grid of characteristic lines (A1.2.4) is built with $c = c_0$ (see fig. A1.1). This is obtained by taking $\mathcal{K} = 0$ (see A1.3.2)) or $\chi = 0$ (see A1.3.3)) in all equations. Therefore the grid consists of equally spaced lines,

$$\mathrm{d}X = \pm\, \mathrm{d}T$$

with the spacing $\Delta X = \Delta T$ chosen conveniently for each specific problem and which is sometimes changed even during the integration of a single problem.

The coordinates assigned to the vertices were chosen as follows: The vertical characteristics $\mathrm{d}X = 0$ have been designated by $0\Delta X$, $1\Delta X$, $2\Delta X, \ldots, M\Delta X, \ldots, M_F\Delta X = l$ (see fig. A1.1), while the succesive characteristics of positive slope $\mathrm{d}X = \mathrm{d}T$ have been designated by $0\Delta T$, $2\Delta T, \ldots$, $\ldots, 2N\Delta T, \ldots$

Let us now consider a loop of these characteristic lines (fig. Al. 2). The generic vertex where the solution is to be found is the point T(„top" or $(M, 2N + M)$). It is assumed that in the other three vertices of the loop, L ("left", or $(M-1, 2N+M-1)$), R("right", or $(M+1, 2N+M-1)$) and B („bottom", or $M, 2N + M - 2)$), the solution is already known from previous integration or from boundary and initial conditions. To find the solution at T, we first integrate equation (Al. 2.5)$_1$ along (Al.2.4) with $\chi = 0$. For this purpose, equation (Al.2.5)$_1$ is written in finite differences between L and T,

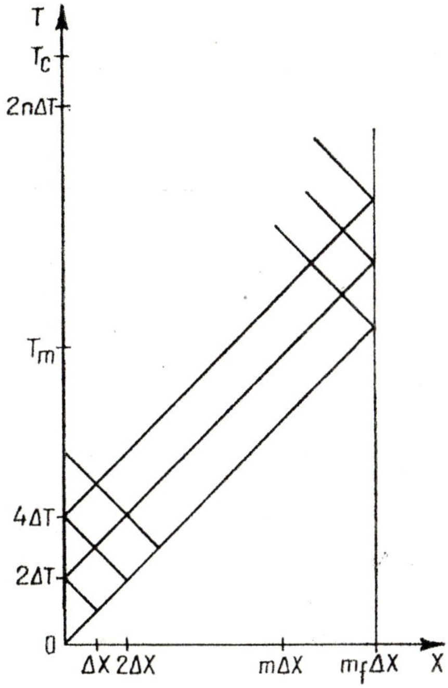

Fig. Al.1 Characteristic plane for numerical integration

and between R and T, respectively. Thus, S_T and V_T are obtained in the first ("elastic") approximation. Then (Al. 2.5)$_{2,3}$ are written in finite differences and these equations are integrated between B and T to obtain in a first approximation ε^E and ε^P. Always before computing ε^E, it is necessary to check if $S_T > S_B$. If this condition is not satisfied, then $\varepsilon_T^P = \varepsilon_B^P$ in the case *RI* when $\Psi = 0$, while in the case *RD* if $\Psi \neq 0$, (Al.2.5)$_2$ is written as

$$\varepsilon_T^P = \varepsilon_B^P + 2\Psi_B \Delta T .$$

These conditions express the physical requirement that always $\dot{\varepsilon}^P \geqslant 0$.

We recall now the inequalities involved in the definition of Ψ (see (4.1.13)),

$$\Psi(\varepsilon, \sigma) = \begin{cases} \dfrac{k(\varepsilon)}{E}[\sigma - f(\varepsilon)] & \text{if } \sigma > f(\varepsilon) \text{ and } \varepsilon \geqslant \dfrac{\sigma}{E}, \\ 0 & \text{if } \sigma \leqslant f(\varepsilon), \end{cases} \tag{Al.3.1}$$

and those involved in the definition of Φ, as used in the computer program. The coefficient function Φ is now multiplied by a factor χ, which in the *RI* case is defined (see (4.1.2)) as

$$\mathcal{X}(\sigma_m, \operatorname{sgn} \dot{\sigma}) = \begin{cases} 0 \text{ if } \sigma \leqslant \sigma_Y \text{ or } \sigma_Y \leqslant \sigma < \sigma_m(X), \\ \qquad \text{or } \sigma = \sigma_m \text{ and } \dot{\sigma} < 0, \\ 1 \text{ if } \sigma = \sigma_m \text{ and } \dot{\sigma} \geqslant 0, \end{cases} \qquad \text{(Al.3.2)}$$

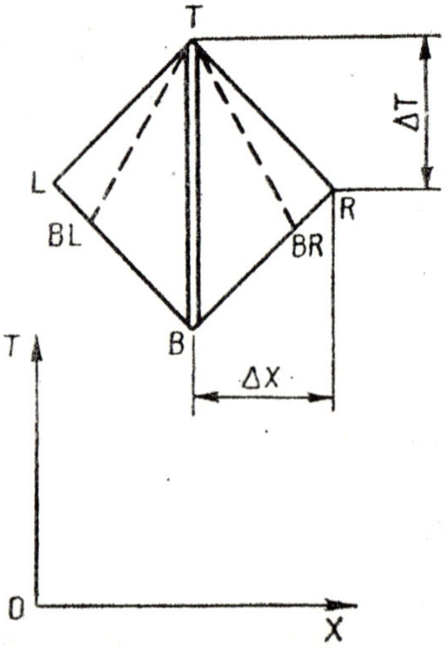

Fig. A.1.2 Generic loop of the characteristic mesh

while in the *RD* case it is given by

$$\chi(\sigma_m) = \begin{cases} 0 & \text{if} \quad \sigma \leqslant f(\varepsilon) \quad \text{or } \sigma < \sigma_m(X), \\ 1 & \text{if} \quad \sigma = \sigma_m(X). \end{cases} \qquad \text{(Al.3.3)}$$

According to the program, these inequalities are now tested. To do that, in the memory of the computer are kept the maximum values of S for any X including the value S_B just under consideration for a certain characteristic loop. If both $\Phi = 0$ and $\Psi = 0$, then the first approximation of the solution just obtained is "the solution" in that particular top of the loop and the results are printed. If $\Psi \neq 0$ or $\Phi \neq 0$, or both, then additional iterations are necessary.

If $\Phi \neq 0$ and $\Psi \neq 0$, then Φ and Ψ at T are computed with the first approximation just obtained for S and ε. Therefore, starting from T, the characteristics (Al.2.4) are drawn backwards (dotted lines in fig. Al.2) up to the intersection with the sides of the initial mesh at the points BL (bottom left) and BR (bottom right). After finding the coordinates of BL and BR by linear interpolation between B and L and between B and R, the values of all required unknown functions at the points BL and BR are found. We used the interpolation formula

$$S_{BL}^{(i)} = \frac{1}{\Delta T}\{(X_{BL}^{(i)} - (M-1)\Delta T)\, S_B + (M\Delta T - X_{BL}^{(i)})\, S_L\}, \quad \text{(AI.3.4)}$$

where the superscript (i) means the ith iteration and the subscripts show the point where the corresponding function is computed. Similar formulas have been used for ε^E, ε^P, V, Φ, and Ψ. Analogous formulas have been used for the point BR.

Now $(\text{AI.2.5})_1$ is integrated along $BL-T$ and $BR-T$, i.e. along (AI.2.4) to find a second and better approximation $S_T^{(i+1)}$ and $V_T^{(i+1)}$. Next $\varepsilon_T^{P(i+1)}$ and $\varepsilon_T^{E(i+1)}$ are obtained from $(\text{AI. 2.5})_{2,3}$ where the coefficient functions Φ and Ψ are computed for the $i+1$ iteration of ε_T^P, as $\frac{1}{2}(\Phi_B + \Phi_T^{(i)})$ and $\frac{1}{2}(\Psi_B + \Psi_T^{(i)})$. Then the values of the coefficient functions Φ and Ψ corresponding to this $i+1$ iteration are calculated from the specific formulas for Φ and Ψ, using the just computed values of $S_T^{(i+1)}$ and $\varepsilon_T^{(i+1)}$. This procedure is continued until the necessary precision is obtained. This requirement has been written in the program as

$$\left|\frac{S_T^{(i)}}{S_T^{(i+1)}} - 1\right| < 10^{-a},$$

where a is a positive integer (generally, for the examples given in Chapter IV, $a = 4$ was used). Practically, it was only seldom that more than 2 or 3 iterations were necessary. After each iteration, with the last obtained iteration at T, inequalities (AI. 3.1) and (AI.3.2) (or (AI. 3.3)) were checked. Therefore, at each iteration the slope of the characteristic lines (AI.2.4) was computed; the method assured the stability and convergence of iterations since always

$$\sqrt{\frac{f'_m}{\rho}} < \sqrt{\frac{f'(\varepsilon)}{\rho}} \leqslant \ldots \leqslant c^{(i)}(\varepsilon, \sigma) \leqslant c^{(i-1)}(\varepsilon, \sigma) \leqslant \ldots \leqslant c_0, \quad \text{(AI.3.5)}$$

where σ and ε stand for the values of these functions at the "top" of the considered characteristic loop, while $f'_m = \min_{\varepsilon} f'(\varepsilon)$ is a constant which can be found from the definition of f (it was assumed that $f''(\varepsilon) < 0$) and by estimating the maximum possible value of ε.

It is easy to show that the inequalities (Al.3.5) are equivalent to

$$\frac{1}{f'_m} - \frac{1}{E} > \frac{1}{f'(\varepsilon)} - \frac{1}{E} \geqslant \ldots \geqslant \Phi^{(i)}(\varepsilon, \sigma) \geqslant \Phi^{(i-1)}(\varepsilon, \sigma) \geqslant$$
$$\ldots \geqslant \Phi^{0}(\varepsilon, \sigma) \geqslant 0. \qquad \text{(A1.3.6)}$$

The extreme inequalities follow directly from the restriction (4.1.19) imposed on the function Φ by definition (see CRISTESCU [1972 *b*]), while the remaining ones have been checked for various particular expressions used for Φ.

In the case when $\Phi = 0$ and $\Psi \neq 0$ the iteration procedure is much simplified. In this case the points *BL* and *L* coincide and so do the points *BR* and *R* (that is, in (Al. 2.4) $c = c_0$). The iterations are still necessary for finding the correct value of ε^P as given by $(\text{Al. }2.5)_2$.

4. DESCRIPTION OF UNLOADING

Let us consider first the *RD* case, i.e. the case when the term $\Psi(\varepsilon, \sigma)$ is also present in (Al.2.1). We have to find by computations the loading-unloading boundary. It is possible that while computing the successive iterations at a point, and checking (Al.3.1), we may obtain for the first iteration that $\Psi \neq 0$, while for the second that $\Psi = 0$. This means that the computation takes place just on the boundary between the domains where the system is quasilinear and where it is linear. In this case the previously discussed program has been replaced by the so-called subroutine *X*. Since the successive iterations are situated above and under the curve $\sigma = f(\varepsilon)$ (see (Al.3.1)), according to subroutine *X* a point just on the curve $\sigma = f(\varepsilon)$ is chosen to get the solution. For this purpose a method called the chord method is used. The process of finding this point converges very fast.

For the case *RI*, when everywhere $\Psi(\varepsilon, \sigma) = 0$, the subroutine *X* is different. We mention here that during loading, in the *RI* case, the successive points (ε, σ) obtained in successive iterations, lie above the curve (2.1.1) and approach a point on this curve when the number of iterations is big enough. Now the boundary to be crossed is (2.1.2) which is a loading-unloading boundary and the inequalities to be used are (Al. 3.2). Therefore, if some of the successive iterations lie above and some under (2.1.1), subroutine *X* is used. However, due to the scatter of the successive iterations, the inequality $S_T < S_B$ is also used as an additional condition to be fulfilled to move to the program *X*. Concerning the program *X* itself, again a chord type method is used, or, if it is possible, the direct intersection of (2.1.1) and the straight line passing through the two successive iteration points is used. Both these subroutines were used.

5. BOUNDARY CONDITIONS

We recall that the examples discussed in Chapter IV refer to the symmetric impact of two bars. The boundary conditions considered in that chapter for a specimen initially at rest and undeformed are formulated for the two bar ends and thus two subroutines are necessary. At the free end

$$\left.\begin{array}{l} x = l \\ t \geqslant 0 \end{array}\right\} \sigma(l, t) = 0 \tag{AI.5.1}$$

and the corresponding subroutine is very simple. Further we shall consider in grater detail the boundary conditions for the impacted end.

The jumps of the data for v at $x = 0$, for $t = 0$ and $t > 0$ can be handled in two ways in two variants of the program. First, a shock wave can be introduced so that at the crossing of the front of this wave all the required functions involved in the problem may suffer a jump. This procedure was described in Chapter IV, Section 3. Second, the discontinuity for v can be smoothed out by assuming a smooth but very fast variation of v from zero to the maximal value v_{max}. In the latter case, at $x = 0$, three time intervals have been considered for which three kinds of subroutines have been prepared:

$$x = 0 \begin{cases} 0 \leqslant t \leqslant t_m\colon v(0,t) \text{ is prescribed and } 0 \leqslant v(0, t) \leqslant v_{max} \\ t_m \leqslant t \leqslant T_c\colon v_H = v_S \\ T_c \leqslant t\colon \sigma = 0. \end{cases} \tag{AI.5.2}$$

Here T_c is the so-called "time of contact" and is a computed quantity, while v_H is the hitter velocity and v_S the specimen velocity. We begin the discussion of the various cases of (AI. 5.2) by describing first the case *RD*.

Let us consider half of the characteristic loop of Fig. AI.2, as it is required by the subroutines at $x = 0$ (full lines in Fig. AI.3). For the case $(\text{AI.5.2})_1$ a smooth increase of v from zero to v_{max} is prescribed. Thus at T the value of v is always known from the boundary conditions. Then $S_T^{(0)}$ is found by integrating $(\text{AI.2.5})_1$, from B to T. Next, integrating $(\text{AI. 2.5})_{2,3}$ from B to T, $\varepsilon_T^{P(0)}$ and $\varepsilon_T^{E(0)}$ are obtained. The procedures follow then quite closely those of the main program (see Section 3).

Now let us explain the condition $(A1.5.2)_2$. This emerges from the physical aspect of the considered problem, the longitudinal collision of two identical bars. Let us assume that the bar from the left (called "hitter") is moving to the right with the initial velocity $v_{max}/2$, while the bar from

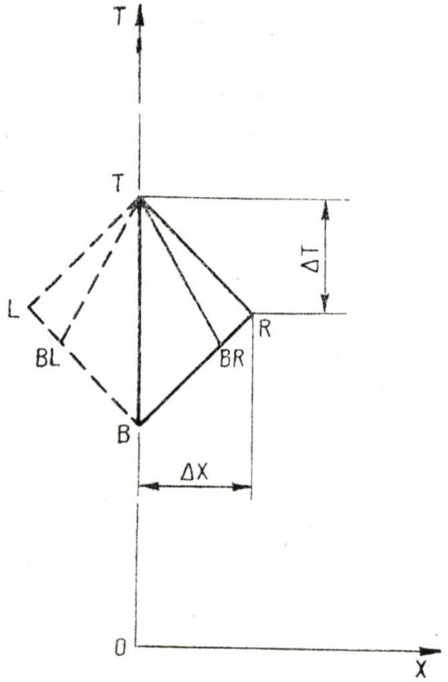

Fig. A1.3 A loop of characteristics at the end of bar

the right (called the "specimen") is moving to the left with the initial velocity $-v_{max}/2$. (see also Chapter III, Section 2). Thus, during impact, at the impacted ends $X = 0$, everything is symmetric, at B and T the values of all required functions are equal on the two sides of the interface $X = 0$. Besides this, by symmetry, it is assumed that $S_R = S_L$, $\varepsilon_R = \varepsilon_L$, etc., and $v_R = -v_L$, since $v_B = v_T = 0$. However, in the physical problem considered, the initial conditions for the specimen are in fact (A1.2.6), i.e., the specimen is at rest, while the hitter is initially moving to the right with the velocity v_{max}. Thus, besides the symmetry conditions assumed for S, ε, etc., the velocity condition used is

$$(\Delta V)_R = (-\Delta V)_L, \text{ i.e. } V_L = 2V_B - V_R. \qquad \text{(A1.5.3)}$$

Hence the half loop on fig. A1.3 is completed by the symmetric other half (dotted lines). The values of all required functions at L are obtained from those at R by the symmetry just described. From now on the procedure follows that of the main program described in Section 3.

The subroutine just described is used as long as $S > 0$. If at a certain time $S_B \leqslant 0$ we make $S_B = 0$ for physical reasons and the subroutine at $x = 0$ is again changed: we set $S_T = 0$, $\varepsilon_T^P = \varepsilon_B^P$, while V is obtained from (A1.2.5)$_2$ which for this case, when $\Psi = 0$ and $c = c_0$, becomes simply $V_T = V_R - S_R$. When $S = 0$, one has $\Phi = \psi = 0$, and no iterations are necessary. The moment when for the first time $S = 0$

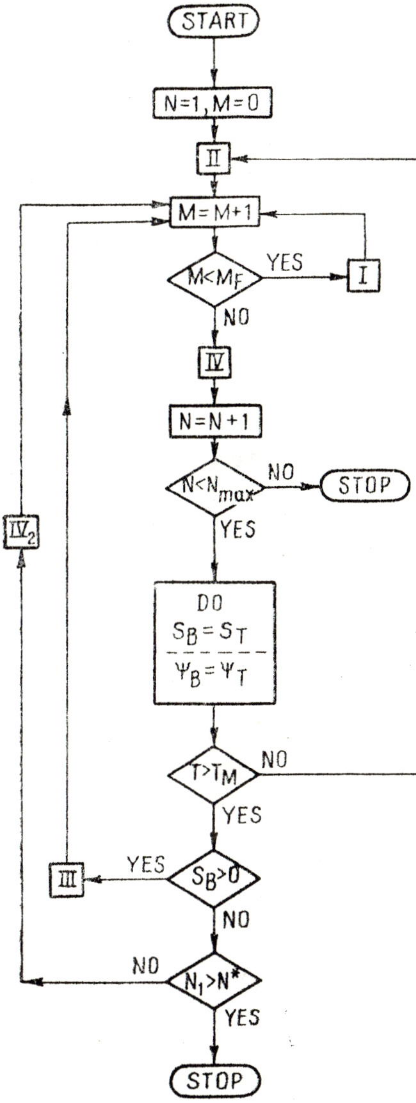

Fig. A1.4 Flowchart for the *RD* case

is called *time of contact* and denoted by T_c. The computations for $t \geqslant T_c$ are continued as long as it is considered necessary.

Thus, all three conditions (AI.5.2) have been written in the form of three subroutines. The passing from $(AI.5.2)_1$ to $(AI.5.2)_2$ is arbitrary and depends on the choice of t_m when the particle velocity at $X = 0$ reaches its maximum, while the passing from $(AI.5.2)_2$ to $(AI.5.2)_3$ is done automatically by the computer when $S = 0$.

The subroutine for the boundary $x = l$ corresponding to the boundary conditions (AI.5.1) is very similar to the subroutine for $x = 0$, when $S = 0$. Now the procedure is very simple, since at $x = l$ we have $S = \varepsilon^E = \varepsilon^P = 0$, and the only quantity to be computed is V. This is done with a formula of type $(AI.2.5)_1$.

The boundary conditions (AI.5.2) just described have been chosen since for this case a great deal of experimental data discussed in Chapter IV were available. It is not difficult, however, to contemplate writing some other subroutines which may correspond to some other kind of boundary conditions, if necessary.

6. FLOWCHART

A scheme of the flowchart for the *RD* case is given below (fig. AI.5). Though the flowchart is self-explanatory, some additional details will be given.

The computation starts with the vertex $N = 1$, $M = 0$. The subroutine used in connection with the boundary condition $(AI. 5.2)_1$ is denoted by II.

The computation is further continued at successive vertices lying on the same characteristic $N = 1$, as long as $M < M_F$. Now the main body of the program is used, which is denoted by I, and was described schematically in Section 3 above.

When $M = M_F$, that is, the boundary $x = l$ is reached, where the boundary conditions are (AI.5.1), the subroutine IV is used. This subroutine is quite similar to II and therefore will not be discussed here further.

At this stage the computation on the next characteristic line ($N = N + 1$) is started, and so on.

Now, the role of the characteristic N is replaced in the memory of the computer by $N + 1$, i.e. the operation is

$$\text{DO } S_B = S_T,\ V_B = V_T, \ldots, \Psi_B = \Psi_T$$

for all required unknown functions. Thus only the data on two successive characteristics are kept in the memory of the computer.

At this stage the condition $T > T_M$ is checked, where T_M corresponds to t_m in (Al.5.2)$_1$ which, as already mentioned, is arbitrarily chosen in con-

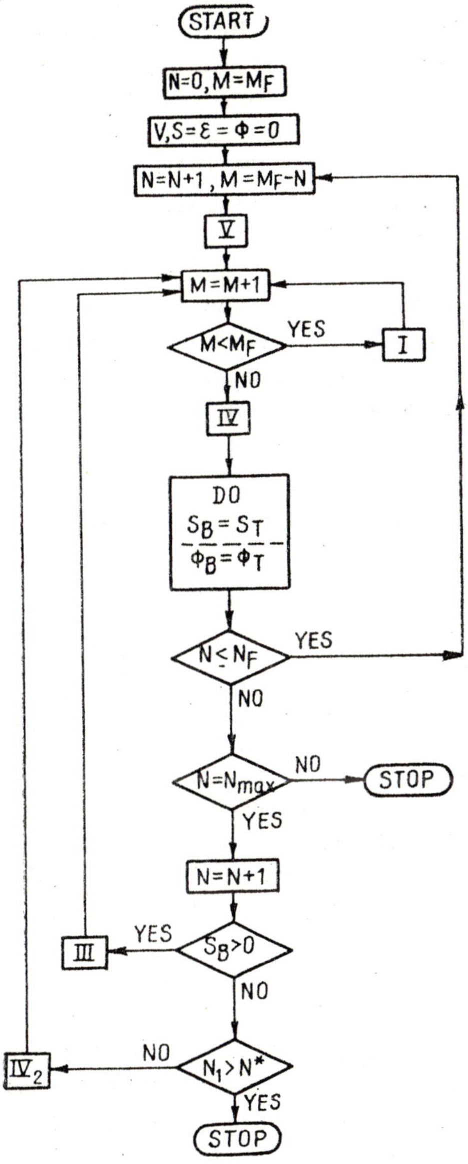

Fig. A1.5 Flowchart for the *RI* case

nection with a certain particular boundary condition. If $T < T_M$, the subroutine II is continued. If however $T > T_M$, one checks to see if $S_B > 0$. If $S_B > 0$, then subroutine III is applied corresponding to $(AI.5.2)_2$ which was explained in Section 5.

When for the first time at $X = 0$ one obtains $S_T = 0$ (if already $S_T < 0$ though $S_B > 0$, one makes $S_T = 0$), the corresponding time is denoted by T_c and another subroutine IV_2 corresponding to $(AI.5.2)_3$ is used. Starting from this moment the computation is carried out on a certain additional number N_1 of positive characteristic lines after which the computations are stopped when $N_1 > N^*$. The number N^* is chosen according to physical reasons. It should be observed that the problem may be considered solved when $T = T_c$ at $X=0$, and the computations are continued just to get the variation of various required functions for $T > T_c$.

The mesh size generally chosen was $\Delta X = 5 \times 10^{-4}$. However, if the boundary conditions $(AI.5.2)_1$ prescribe a very fast variation of $V(0, t)$, then in the neighborhood of the point $X = 0$, $T = 0$, the mesh size was gradually reduced down to $\Delta X = 10^{-5}$.

7. REMARKS CONCERNING THE *RI* CASE

In tackling the *RI* case (i.e. when $\Psi = 0$ everywhere) it was possible, due to the peculiarity of the problem, to economize on computing time by finding the solution by analytic formulas in the lower part of the strip of Fig. AI.1, under the straight line (see Chapter IV)

$$X = -(T - 2M_F \Delta T). \qquad \text{(A1.7.1)}$$

In other words, for any boundary conditions prescribed at $X = 0$, the solution along (AI.7.1) is found. These data are then used as "initial conditions" for the computation to be done above this line (see CRISTESCU [1970]). Some details concerning the main algorithm and the flowchart will be given below.

The main subroutine is in some respect close to the one described in connection with Fig. AI.2. The first "elastic" approximation is computed as before. Then the inequalities (AI.3.2) are checked. For this purpose for all cross-sections X considered, $S_{max}(X)$ is kept in the memory of the computer. If one of the inequalities $(AI.3.2)_2$ is satisfied, then the first iteration just obtained is the final solution for the particular vertex under consideration. If, however, $(AI.3.2)_1$ is satisfied, then there is a need for successive iterations, generally following those described above in connection with Fig. AI.2.

The flowchart is given above (Fig. AI.5). Now the computation has started from the vertex $N = 0$, $M = M_F$ (see Fig. AI.1). The computation is made on successive characteristics of positive slope, starting from the vertex lying on the straight line (AI.7.1). In the flowchart, V denotes the subroutine calculating the solution along (AI.7.1) by elementary analytic formulas. Subroutine IV is used for the boundary conditions (AI.5.1), subroutine IV_2 is used in conjunction with the boundary condition $(AI.5.2)_3$, and subroutine III corresponds to $(AI.5.2)_2$. Other details of the flowchart are self-explanatory.

8. OTHER VARIANTS OF THE METHOD OF CHARACTERISTICS

We gave above a short exposition of the variant of the method of characteristics which was used to obtain the examples given in Chapter IV. We presented mainly information concerning the practical aspect of the application of the method. There are certainly also other possible variants. Among all these only the variants using a regular distribution of the mesh points have a practical importance when using a computer. We recall also that the passing from loading to unloading and vice versa is the most difficult aspect of the program, whenever the constitutive equation used is of an elasto-viscoplastic type.

Another variant of the characteristic method starts by tracing equally spaced straight lines $T =$ const,, the interval ΔT between them being conveniently chosen. A typical mesh is shown on fig. AI.6. The position of the points L and R can be chosen in various ways, but always the stability conditions discussed in Section 3 must be fulfilled. Taking also into account that in our problem one has to pass from loading to unloading, it is useful to build the characteristic mesh of fig. AI.6 with $\Delta X = \Delta T$. Thus the iteration procedure described in Section 3 above can be adapted for such a mesh. Therefore, in principle, this variant uses also the straight line $dx = \pm c_0 dt$ for the basic net necessary in the numerical computation.

When writing the program some difficulties may arise if it is not possible to consider regular net points only. We mention here two such cases: a) the case when the boundary conditions are not prescribed on a fixed cross-section $X =$ const. but on a floating cross-section $X = h(T)$ varying in time, and b) the case when there is a shock wave propagating also in the field of acceleration waves.

From a mechanical point of view the case a) occurs when the longitudinal impact is transmitted to the bar not on a fixed cross-section but on a floating boundary, for instance through a ring which can slip along the bar

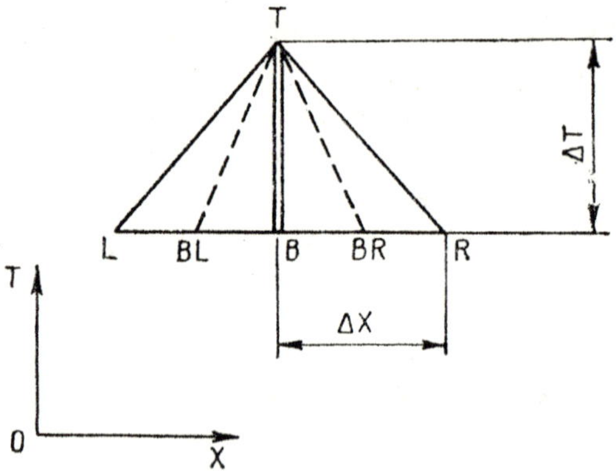

Fig. A1.6 Characteristic loop

during impact. Let $X = h(T)$ (fig. A1.7) be the equation of motion of the moving boundary. Sometimes this equation is known a priori, but in other cases this equation is to be determined step by step while obtaining the solution by numerical methods. Generally, this boundary is subjected only to the condition that it should be a "time-like" curve. We recall that a curve $X = h(T)$ is called "time-like" if tracing for increasing time the characteristics which are passing through an arbitrary point P on this curve we get some of these characteristics situated on one side of the curve and the others on the other side (fig. A1.8). We mention also that if $c_m =$

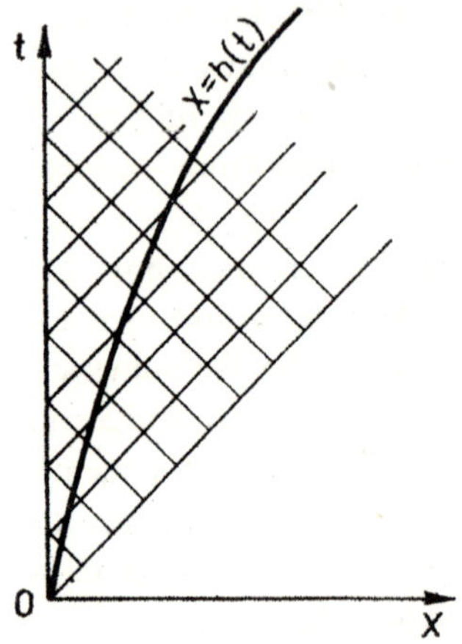

Fig. A1.7 Floating boundary

max $\underset{\varepsilon,\sigma}{} c(\varepsilon, \sigma)$, then the condition $h' < c_m$ must be fulfilled, i.e. the floating boundary must be "subsonic". In the cases considered in this book, generally $c_m \leqslant c_0$. The boundary conditions on a time-like floating boundary can be prescribed as discussed above (see (Al.2.7)) i.e. on

$$X = h(T) \text{ there is prescribed } v(h(T), T) \text{ or } \sigma(h(T), T). \qquad \text{(Al.8.1)}$$

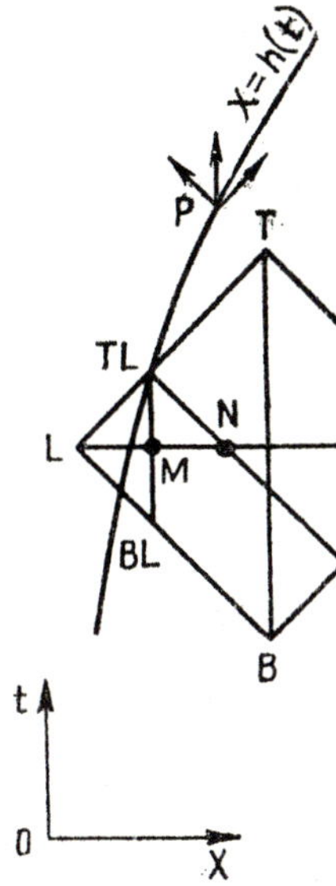

Fig. A1.8 Characteristic loop for the floating boundary case

Both cases are physically feasible.

The solution at points of the inner mesh can be obtained following the already mentioned procedure. Let us consider a mesh of characteristics drawn with the characteristics $dx = \pm c_0 dt$ intersecting the boundary $X = h(T)$ (fig. Al.8; other possible cases can be considered by similar procedures). The coordinates of the mesh point *TL* can be easily obtained, and further those of the mesh points *BR* and *BL*, by drawing the characteristics $dx = -c_0\, dt$ and $dx = 0$ which pass through *TL*. Let us assume that the solutions at the mesh points *B* and *R* are known from previous integration or from initial conditions. A solution at *BR* and *BL* can be obtained by interpolation or by extrapolation, starting from the neighbouring mesh points where the solution is already obtained. Then using (Al.8.1), we can integrate the equations (Al.2.5) along the characteristics *BR*—*TL* and *BL*—*TL* in order to get the first approximation for the solution at *TL*. For the next approximations the procedure is similar to the one already discussed. The solution at *TL* can certainly be obtained with other variants which use the mesh points M and N, where the solution is assumed to be already obtained by preliminary extrapolation. After finding the solution at TL one can proceed to find it at T; the procedure uses the mesh point TL instead of L and certainly the mesh points B and R, and follows a scheme quite close to that described in Section 3.

The case b) mentioned above can be treated somewhat similarly. In this case a shock wave is crossing the field of acceleration waves. The simplest case is the one in which $U = c_0$. This is the case which we discussed in Chapter IV, Section 3 and used in several numerical examples (see also CRISTESCU [1970]). For the case $U \neq c_0$ additional subroutines are necessary in the program. At the crossing of the shock wave the jump conditions discussed in Chapter I are used. In principle, the solution at a mesh point TL (fig. AI.8) is obtained computing the values of the variables on both sides of the shock wave in the field of acceleration waves, using the jump conditions.

9. REMARKS

We have described above very shortly one of the variants of the method of characteristics, used for the quasilinear hyperbolic systems, a method which is generally known as the MASSAU method (see MASSAU [1899]). The method was significantly developed and adapted to computers. The reader interested to get additional information concerning the numerical methods applicable to hyperbolic systems may consult various books (see for instance COURANT and FRIEDRICHS [1948], BEREZIN and JIDKOV [1960] Vol. II, FORSYTHE and WASOW [1960], JEFFREY and TANIUTI [1964], CRISTESCU [1967], GODUNOV and RYABENKII [1977], KUKUDJANOV [1976]).

REFERENCES

ACHENBACH, J. D., 1968, J. Mech. Phys. Solids, **16**, 73.

AHRENS, T. J., G. E. DUVALL, 1966, J. Geophys. Res., **71**, 4349–4360.

ALTER, B. E., C. W. CURTIS, 1956, J. Appl. Phys., **27**, 1079–1085.

BAKLASHOV, I. V., B. A. KARTOZYA, 1975, *Mechanics of rocks*, Nedra, Moscow (in Russian).

BALTOV, A., 1971, Teor. i. prikl. Mehanika, **3**, 41–51.

BANERJEE, A. K., L. E. MALVERN, 1974, Trans. ASME, Ser. E, J. Appl. Mechn., **41**, 3, 615–618.

BARONCEA, A., N, CRISTESCU, E. GLODEANU, 1977, St. Cerc. Mec. Appl., **36**, 1, 31–40.

BELL, J. F., 1951, *Propagation of plastic waves in pre-stressed bars*, U. S. Navy Tech. Rep. No. 5, Baltimore, The Johns Hopkins University.

BELL, J. F., 1960, J. Appl. Phys., **31**, 2188–2195.

BELL, J. F., 1961, J. Appl. Phys., **32**, 1982–1993.

BELL, J. F., 1968, *The Physics of large deformation of crystalline solids*, Springer Tracts in Natural Philosophy, Vol. 14, Springer Verlag, Berlin–Heidelberg–New York.

BELL, J. F., 1969, Int. J. Mech. Sciences, **11**, 633–657.

BELL, J. F., 1973, *The Experimental Foundations of Solid Mechanics*, Handbuch der Phys., Vol. VIa/1, Springer Verlag, Berlin–Heidelberg–New York.

BELL, J. F., A. STEIN, 1962, J. Mécan., **1**, 395–412.

BEREZIN, I. S., N. P. JIDKOV, 1960, *Computation methods*, Moscow, (in Russian).

BIANCHI, G., 1964, *Some experimental and theoretical studies on the propagation of longitudinal plastic waves in a strain-rate dependent material.* In: H. Kolsky and W. Prager, eds., Stress waves in anelastic solids, Springer Verlag, Berlin–Gottingen –Heidelberg, 101–117.

BINGHAM, E. C., 1922, *Fluidity and plasticity*, McGraw Hill, New York.

BOWEN, R. M., 1968, J. Chem. Phys., **49**, 1625–1637.

BOWEN, R. M., 1969, Arch. Rational Mech.. Anal., **33**, 169–180.

BOWEN, R. M., P. J. CHEN, 1972, Acta mech., **15**, 95.

CAMPBELL, J. D., 1973, Materials Science and Engineering, **12**, 3–21.

CAMPBELL, J. D., J. L. LEWIS, 1969, Report No. 1080, 69, Dept. of Eng. Science, University of Oxford.

CAMPBELL, J. D., M. C. C. TSAO, 1972, Q. J. of Mech. Appl. Mathematics, *XXV*, **2**, 173–184.

CARLSON, D. E., 1972, *Linear Thermoelasticity*, Handbuch der Physik, Vol. 6a/2, Springer Verlag, Berlin.

CATTANEO, C., 1948, Atti del Seminario Matem. Fiz., Univ. di Modena, **3**, 83.

CATTANEO, C., 1958, Compt. Rend. Sci., **247**, 431.

CHEN, P. J., 1973, *Growth and decay of waves in solids*, Handbuch der Physik, Vol. 6a/3, Springer Verlag, Berlin.

CHEN, P. J., 1976, *Selected topics in wave propagations*, Noordhoff, Leyden, The Netherlands.

CHEN, P. J., M. E. GURTIN, 1971, Phys. Fluids, **14**, 1091–1094.

CHEN, P. J., M. E. GURTIN, 1972, J. Appl. Math. Phys. (ZAMP), **23**, 69–79.

CLIFTON, R. J., 1966, *An Analysis of Combined Longitudinal and Torsional Plastic Waves in a thin-walled Tube*, in Proc. 5-th U. S. National Congr. of Applied Mechanics, ASME, New York, 465–480.

CLIFTON, R. J., 1971, *On the Analysis of Elastic/Visco-Plastic Waves of Finite Uniaxial Strain*, In "Shock waves and the mechanical properties of solids" Syracuse Univ. Press, Syracuse, N.Y., 73–116.

CLIFTON, R. J., 1972, *Plastic Waves. Theory and Experiment*. In Mechanics Yearbook, ed. Nemat-Nasser, Pergamon Press, 102–167.

CODDINGTON, E. A., N. LEVINSON, 1955, *Theory of Ordinary Differential Equations*, McGraw Hill, New York.

COLEMAN, B. D., M. E. GURTIN, 1965, Arch. Rational Mech. Anal., **19**, 266–298.

COLEMAN, B. D., M. E. GURTIN, 1967, J. Chem. Phys., **47**, 597–613.

COLEMAN, B. D., M. E. GURTIN, 1967 a, Phys. Fluids, **10**, 1454–1458.

COLEMAN, B. D., M. E. GURTIN, I. R. HERRERA, C. TRUESDELL, 1965. In: Wave Propagation in Dissipative Materials, Springer Verlag.

COLEMAN, B. D., W. NOLL, 1959, Arch. Rational Mech. Anal., **4**, 97.

COLEMAN, B. D., W. NOLL, 1963, Arch. Rational Mech. Anal., **13**, 167–178.

COURANT, R., 1962, *Methods of Mathematical Physics*, Vol. II, *Partial differential equations*, Interscience Publ., New York.

COURANT, R., K. O. FRIEDRICHS, 1948, *Supersonic flow and shock waves*, Interscience Publ., New York.

CRISTESCU, C., N. CRISTESCU, 1973, *A numerical method to describe unloading in dynamic plasticity*. In: Proc. of the second Intern. Conf. on Structural Mechanics in Reactor Technology, Berlin, Vol. 5, Part L.

CRISTESCU, C., N. CRISTESCU, I. SULICIU, 1978, Bull. Acad. Polon. Sci., Serie de Sci. techn., **24**, 417–425.

CRISTESCU, N., 1955, Prikl. Mat. Mekh., **19**, 4, 433–442.

CRISTESCU, N., 1956, Comunicările Acad. R. P. Române, **6**, 1, 19–28.

CRISTESCU, N., 1958, *Probleme Dinamice în Teoria Plasticității*, Ed. Tehnică, București.

CRISTESCU, N., 1959, Prikl. Mat. Mekh., **23**, 6, 1124–1128.

CRISTESCU, N., 1960, *Some observations on the propagation of plastic waves in plates*, In: Plasticity, Proc. Second Symp. on Naval Structural Mechanics, Pergamon Presss, London, 501–510.

CRISTESCU, N., 1963, Bull. Acad. Pol. Sci., **11**, 129–133.

CRISTESCU, N., 1964, *Some problems on the mehcanics of extensible strings*, in Stress Waves in Anelastic Solids, eds. H. Kolsky and W. Prager, Springer Verlag, 118–132.

CRISTESCU, N., 1965, Arch. Mech. Stos., **17**, 291–305.

CRISTESCU, N., 1967, *Dynamic Plasticity*, North Holland Publ., Amsterdam.

CRISTESCU, N., 1968, *Dynamic Plasticity Under Combined Stress*. In: Mechanical Behavior of Materials under Dynamic Loads, ed. U. S. Lindholm, Springer Verlag, New York, 329–342.

CRISTESCU, N., 1970, Int. J. Mech. Sciences, **12**, 723–738 (or Mekhanika, no. 5, 1971, 125–141).

CRISTESCU, N., 1971a, Rev. Roumaine Sci. Techn., Ser. Mec. Appl., **16**, 797–809.

CRISTESCU, N., 1971b, *Introduction to rate-dependent plasticity*, Lectures held at International Centre for Applied Mechanics, Udine, Italy.

CRISTESCU, N., 1972, Int. J. Solids Structures, **8**, 511–531.

CRISTESCU, N., 1974, *Rate-type constitutive equations in dynamic plasticity*. In: Problems of plasticity, Noordhoff Intern. Publ., Leyden, 287–310.

CRISTESCU, N., 1977, Rev. Roumaine de Mécanique Appliquée, **22**, 3, 391–399.

CRISTESCU, N., 1979, *A viscoplastic constitutive equation for rocks*, INCREST Preprint series in Mathematics No. 49/1979, Institutul de Matematică, București.

CRISTESCU, N., J. F. BELL, 1970, *On unloading in the symmetrical impact of two aluminium bars*. In: Inelastic Behaviour of Solids, eds. M. F. Kannienen, W. F. Adler, A. R. Rosenfield, R. I. Jaffee, McGraw Hill, New York, 397–421 (or Mekhanika, No. 1, 1973, 151–168).

CRISTESCU, N., I. SULICIU, 1976, *Vîscoplasticitate*, Ed. Tehnică, București.

DALLY, J. W., W. F. RILEY, 1965, *Experimental Stress Analysis*, Mc Graw Hill, New York.

DUFFY, J., J. D. CAMPBELL, R. H. HAWLEY, 1971, Trans. ASME, Series E, J. of Appl. Mechanics, **38**, 89–91.

Int J Sol S 20, 921

DUFFY, J., R. H. HAWLEY, R. A. FRANTZ, Jr., 1972, Trans. ASME, Series E, J. of Appl. Mechanics, **39**.

DIEUDONNÉ, J., 1968, *Elements d'Analyse*, Vol. I, Gauthier Villars, Paris.

EFRON, L., L. E. MALVERN, 1969, Experimental Mech., **9**, 6, 1–8.

EIRICH, F. R., editor, 1956, *Rheology. Theory and Applications*. Academic Press, New York.

ERJANOV, J. S., A. S. SAGYNOV, G. N. GUMENIUK, I. A. VEKSLER, G. A. NESTEROV, 1970. *Creep of sedimentary rocks*, Nauka, Moscow (in Russian).

FILBEY, G. L., 1961, *Intensive plastic waves*. Ph. D. dissertation, The Johns Hopkins University, Baltimore.

FOKUOKA, H., K. MASUI, 1970, *Experiment on incremental impact loading of plastically pre-stressed aluminium*. The Thirteenth Japan Congress on material research, Metallic materials, March 1970.

FORSYTHE, G. E., W. R. WASOW, 1960, *Finite-difference methods for partial differential equations*. John Wiley and Sons, New York.

FOWLES, G. R., 1973, *Experimental Technique and Instrumentation*. In: Dynamic Response of Materials to Intensive Impulsive Loading, eds. P. C. Chou, A. K. Hopkins, Air Force Materials Laboratory, 405–480.

FRANK, R., P. QUELLEC, 1974, Bull. Liaison Labo. P. et Ch., **73**, 45–51.

FRANTZ, R. A. JR., J. DUFFY, 1972, Trans. ASME, Series E, J. of Appl. Mechanics, **39**.

FREUDENTHAL, A. M., 1958, *The mathematical theories of the inelastic continuum*. Handbuch der Physik, Vol. 6, Springer Verlag, Berlin.

GARY, C., G. DRĂGAN, D. CRISTESCU, 1976, Rev. Roum. Sci. Techn. – Electrotechn. et Energ., **21**, 237–258.

GARY, C., G. DRĂGAN, D. CRISTESCU, 1978, Private communication.

GIBBS, J. W., 1928, *The Collected Works of J. W. Gibbs*, Vol. 1, Thermodynamics, Yale Univ. Press, New Haven, Conn.

GODUNOV, S. K., V. S. RYABENKY, 1977, *Difference methods*, Moscow, (in Russian).

GOEL, R. P., L. E. MALVERN, 1970, Trans. ASME, Series E, J. Appl. Mech., **37**, 4, 1100–1106.

GOEL, R. P., L. E. MALVERN, 1972, Trans. ASME, Series E, J. Appl. Mech., Paper No. 71–APM–10.

GOLDSMITH, W., 1960, *Impact*, Arnold, London.

GOLDSMITH, W., J. L. SACKMAN, 1973, *Wave transmission in rocks*, ASME, Detroit, Symposium on Rock Mechanics.

GREEN, A. E., R. S. RIVLIN, 1964a, Arch. Rational Mech. Anal., **16**, 113–147.

GREEN, A. E., R. S. RIVLIN, 1964b, Arch. Rational Mech. Anal., **17**, 325–353.

GREENBERG, J. M., 1968, Quart. Appl. Math., **26**, 27–34.

GURTIN, M. E., 1972, *The linear theory of elasticity*, Handbuch der Physik, Vol. 6a/2, Springer Verlag, Berlin.

GURTIN, M. E., A. C. PIPKIN, 1968, Arch. Rational Mech. Anal., **31**, 113.

GURTIN M. E., I. SULICIU, 1976, *Thermodynamics of rate-type constitutive equations*. In: Cristescu, N., and Suliciu, I., Vîscoplasticitate, Ed. Tehnică, București.

GURTIN, M. E., I. SULICIU, W. O. WILLIAMS, 1979, *On rate-type constitutive equations and the energy of viscoelastic and viscoplastic materials* (to appear in Int. J. Solids Structure).

GURTIN, M. E., W. O. WILLIAMS, 1966, Arch. Rational Mech. Anal., **23**, 163.

HADAMARD, J., 1903, *Leçons sur la Propagation des Ondes et les Equations de l'Hydrodynamique*, Hermann, Paris.

HALANAY, A., 1963, *Teoria Calitativă a Ecuațiilor Diferențiale*, Editura Acad. R. P. Române, București.

HANSON, D., M. A. WHEELER, 1931, J. Inst. Metals, **45**, 229–264.

HARTMAN, P., *Ordinary differential equations*, John Wiley and Sons, New York.

HAUSER, F. E., J. A. SIMMONS, J. E. DORN, 1960, *Strain Rate Effects in Plastic Wave Propagation*. In: Response of Metals to High Velocity Deformation, Intersci. Publ., New York–London.

HAYES, M., 1960, Arch. Rational Mech. Anal., **33**, 188–191.

HILL, R., 1950, *The Mathematical Theory of Plasticity*, Clarendon Press, Oxford.

HOHENEMSER, K., W. PRAGER, 1932, Z. Angew. Math. Mech., **12**, 215–226.

HYLTEN-CAVALLIUS, W., 1977, AIEE Trans., PAS, **96**, 109.

Ilyushin, A. A., V. S. Lenskiĭ, 1959, *Strength of materials*, Moscow, (in Russian).
Jacob, C., 1959, *Introduction mathématique à la mécanique des fluides*, Ed. Acad. R.P.R. Bucureşti and Gauthier Villar, Paris.
Jeffrey, A., 1976, *Quasilinear hypeborlic systems and waves*, Pitman Publ., London.
Jeffrey, A., T. Taniuti, 1964, *Non-linear wave propagation with applications to physics and magnetohydrodynamics*, Academic Press, New York—London.
Johnson, W., 1972, *Impact Strength of Materials*, Arnold, London.
Johnston, W. G., J. J. Gilman, 1959, J. Appl. Phys., **30**, 129.
Jubaev, N. J., 1979, *Onedimensional elastic-plastic waves for combined stresses* (in Russian), Nauka, Alma-Ata.
Jukov, A. M., 1960, Inj. Sbornik, **30**, 3—16.
Jumikis, A. R., 1979, *Rock Mechanics*, Trans. Tech. Publ., Aedermannsdorf, Switzerland.
Kármán, Th. von, P. Duwez, 1950, J. Appl. Phys., **21**, 987.
Kartashov, I. M., 1973, *Dynamic methods to determine rheological properties of rocks* (in Russian), Nedra, Moscow.
Kenig, M. J., 1967, Trans. ASME, Series E, J. of Appl. Mechanics, **34**, 493—495.
Kenig, M. J., O. W. Dillon, Jr., 1966, Trans. ASME, Series E, J. of Appl. Mechanics, **33**, 907—916.
Klepaczko, J., 1967, Arch. Mech. Stos., **19**, 2, 211—229.
Klepaczko, J., 1973, *Some experimental investigations of the elastic-plastic wave propagation in bars*. In: Foundations of Plasticity, ed. A. Sawczuk, Noordhoff Int. Publ., Leyden, 451—462.
Kolsky, H., 1963, *Stress Waves in Solids*, Dover Publ., New York.
Kosinski, W., 1975, Arch. Mech. Stos., **27**, 733—748.
Kosinski, W., P. Perzyna, 1972, Arch. Mech. Stos., **24**, 629.
Krafft, J. M., 1961, *Instrumentation for High-Speed Strain Measurement*. In: Response of metals to high velocity deformation, eds. P. G. Shewmon, V. F. Zackay, Interscience Publ., New York—London, 9—48.
Kratochvil, J., 1973, Acta Mech., **16**, 127—142.
Kratochvil, J., O. W. Dillon. 1969, J. Appl. Phys., **40**, 3207—3218.
Kratochvil, J., O. W. Dillon, 1970, J. Appl. Phys., **41**, 1470—1479.
Kröner, E., C. Teodosiu, 1974, *Lattice defect approach to plasticity and viscoplasticity*. In: Foundations of Plasticity, Noordhoff Int. Publ., 45—88.
Kukudjanov, V. N., 1967, *Propagation of elastic-plastic waves in rods taking into account the rate influence*, Computing Centr. Acad. Sci. USSR, Moscow (in Russian).
Kukudjanov, V. N., 1976, *Numerical solution of stress wave propagation in solid bodies*, Computing Centre Acad. Sci. USSR, No. 6, Moscow (in Russian).
Kukudjanov, V. N., 1977, *One dimensional problems of stress wave propagation in bars*, Computing Centre Acad. Sci. USSR, No. 7, Moscow (in Russian).
Kuntysh, M. F., R. I. Teder, 1970, *The influence of the loading rate on the yield limit in compression for rocks*. In: The study of physical-mechanical properties and fracture of rocks, Nauka, Moscow, 60—70, (in Russian).
Kuriyama, S., K. Kawata, 1973, J. Appl. Phys., **44**, 3445—3454.
Lama, R. D., V. S. Vutukuri, 1978, *Handbook on Mechanical Properties of Rocks*, Vol. II, Trans. Tech. Publ.
Lax, P. D., 1957, Commun. Pure Appl. Math., **10**, 537.
Lee, E. H., 1953, Quart. Appl. Math., **10**, 335—346.
Lee, E. H., 1956, *Wave propagation in inelastic materials*. In: IUTAM Colloq. Deformation and Flow of Solids, p. 129.
Lee, E. H., 1960, *The theory of wave propagation in anelastic materials*. In: Stress Wave Propagation in Materials, ed. N. Davids, Interscience Publ., 199—228.
Levykin, A. I., V. V. Vavakin, 1978, Izv. Akad. Nauk SSSR, Fiz. Zemli, no. 5, 42—51.
Lord, H. W., Y. Shulman, 1967, J. Mech. Phys. Solids, **15**, 299—309.
Lubliner, J., 1964, J. Mech. Phys. Solids, **12**, 59—65.
Ludwig, P., 1909, Phys. Z., **10**, 411—417.
Lykov, A. V., 1965, Inj. Fiz. J., **9**, 287—304.
Malvern, L. E., 1951a, Quart. Appl. Math., **8**, 405—411.
Malvern, L. E., 1951b, J. Appl. Mech., **18**, 203—208.

MALVERN, L. E., 1965, *Experimental studies of strain-rate effects and plastic wave propagation in annealed aluminium*. In: Behaviour of Materials under Dynamic Loading, ASME Colloq.

MALYSHEV, V. M., 1960, Izv. Akad. Nauk SSSR, OTN, Mekh. Mat., **2**, 120–124.

MALYSHEV, V. M., 1961, J. Prikl. Mekh. Techn. Fiz., **2**, 104–110.

MANDEL, J., 1972, *Plasticité classique et viscoplasticité*, Springer Verlag, Wien–New York.

MASSAU, J., 1899, *Mémoire sur l'intégration graphique des équations aux dérivées partielles*, F. Meyer-Van Loo, Gent.

MASSON, A. P., 1841, Annales de Chimie et de Physique, 3 série, 451–462.

MAZILU, P., 1978, Rev. Roum. Math. Pures et Appl., **23**, 419–435.

MCCARTHY, M., 1970, Int. J. Engng. Sci., **8**, 467–474.

MCCARTHY, M., 1972, Int. J. Engng. Sci., **10**, 593–602.

MIHĂILESCU, M., I. SULICIU, 1975a, J. Math. Anal. Appl., **52**, 10–24.

MIHĂILESCU, M., I. SULICIU, 1975b, Rev. Roum. Math. Pures et Appl., **20**, 551–559.

MIHĂILESCU, M., I. SULICIU, 1976, Int. J. Solid Struct., **12**, 559–575.

MIHĂILESCU-SULICIU, M., I. SULICIU, 1978, Analele Ştiinţifice ale Univ. ,,Al. I. Cuza", Iaşi, **24**, S. Ia, f. 2, 403–416.

MIHĂILESCU-SULICIU, M., I. SULICIU, 1979, *Energy for Hypoelastic Constitutive equations*. Arch. Rational Mech. Anal., **71**, 327–344.

MIHĂILESCU-SULICIU, M., I. SULICIU, 1979a, *A rate type constitutive equations for the description of corona effect*. INCREST Preprint series in Mathematics, No. 32/1979.

MISICU, M., 1967, *Mecanica mediilor deformabile*, Editura Acad. R. S. Romania, Bucureşti.

MILLER, I., 1971, *Entropy, Absolute temperature and Coldness in thermodynamics. Boundary conditions in Porous Materials*, Udine 1971, Springer Verlag, Wien, New York.

MROZ, Z., H. P. SHRIVASTAVA, R. N. DUBEY, 1976, Acta Mechanika, **25**, 51–61.

NEYFEH, A. H., S. NEMAT-NASSER, 1972, J. Appl. Math. Phys. (ZAMP), **23**, 50.

NICHOLAS, T., J. D. CAMPBELL, 1972, Experim. Mech., **12**, 441.

NICOLESCU, M., 1958, *Analiză Matematică*, Vol. II, Editura Tehnică, Bucureşti.

NOLL, W., 1958, Arch. Rational Mech. Anal., **2**, 197–226.

NOLL, W., 1972, Arch. Rational Mech. Anal., **48**, 1–50.

NOWACKI, W. K., 1978, *Problems of wave propagation in theory of plasticity*, (in Russian), Mir, Moscow.

NUNZIATO, J. W., E. K. WALSH, K. W. SCHULER, L. M. BARKER, 1974, *Wave propagation in nonlinear viscoelastic solids*. Handbuch der Physik, Vol. 6a/4, Springer Verlag, Berlin.

OBERT, L., W. I. DUVALL, 1967, *Rock Mechanics and the Design of Structures in Rock*, John Wiley and Sons, New York.

OGYBALOV, P. M., A. H. MIRZADJANDZADE, 1970, *Nonstationary motions of visco-plastic media*, Moscow, (in Russian).

OLSZAK, W., P. PERZYNA, A. SAWCZUK, 1970, *Teoria Plasticităţii*, Ed. Tehnică, Bucureşti.

ONAT, E. T., 1968, *The notion of state and its implications in thermodynamic of inelastic solids*, In: Proc. IUTAM Symp., Springer Verlag, Berlin, 292–314.

ONAT, E. T., F. FARDSHISHEH, 1972, *Representation of Creep of Metals*. Oak Ridge National Lab., ORNL-4783, Oak Ridge, Tenn.

OWEN, R. D., 1968, Arch. Rational Mech. Anal., **31**, 91–112.

PERKINS, R. D., S. J. GREEN, M. FRIEDMAN, 1970, Int. J. Rock Mech. Min. Sci., **7**, 527–535.

PERZYNA, P., 1963a, Quart. Appl. Math., **20**, 321–332.

PERZYNA, P., 1963b, Arch. Mech. Stos., **15**, 113–130.

PERZYNA, P., 1971a, Bull. Acad. Polon. Sci., Ser. Sci. Techn., **19**, 177–181, 182–188.

PERZYNA, P., 1971b, *Viscoplasticity. Theory and Applications*. Lectures held at International Centre for Applied Mechanics, Udine, Italy.

PERZYNA, P., 1973, Bull. Acad. Polon. Sci., Ser. Sci. Techn., **21**, 123–128, 129–139.

PERZYNA, P., W. WOJNO, 1968, Arch. Mech. Stos., **20**, 499–511.

PIAU, M., 1978, Int. J. Engng. Sci., **16**, 185–201.

PODSTRIGATCH, I. S., I. M. KOLYANO, 1976, *Generalized thermomechanics*, Naukova Dumka, Kiew, (in Russian).

POLUKHIN, P. I., G. I. GUN, A. M. GALKIN, 1976, *Resistance to plastic deformation of metals and alloys*, Metallurgya, Moscow, (in Russian).
PONOMAREV, S. D., V. L. BIDERMAN, K. K. LIKHAREV, V. M. MAKUSHIN, N. N. MALININ, V. I. FEODOSIEV, 1956, *Strength calculation in machine construction*, Mashgyz, Moscow (in Russian).
POPOV, E. B., 1967, Prikl. Mat. Mekh., **31**, 2, 328–333.
PORTEVIN, A., F. LE CHATELIER, 1923, C. R. Acad. Sci., Paris, **176**, 507–510.
PRANDTL, L., 1928, Z. Angew. Math. Mech., **8**, 85–106.
RAKHMATULIN, H. A., 1945, Prikl. Mat. Mekh., **9**, 91–100.
RAKHMATULIN, H. A., 1958, Prikl. Mat. Mekh., **22**, 6, 759–765.
RAKHMATULIN, H. A., I. A. DEMIANOV, 1961, *Strength under intensive momentary loads*, Moscow, (in Russian).
RAZEVIG, D. V., 1959, *Atmospherical overstresses in electro-transmission cables*, Moscow-Leningrad, (in Russian).
RĂDULEŢ, R., A. TIMOTIN, A. ŢUGULEA, 1970, Rev. Roum. Sci. Techn., Electrotechn. et Energ., **15**, 585–599.
RĂDULEŢ, R., A. TIMOTIN, A. ŢUGULEA, A. NICA, 1978, Rev. Roum. Sci. Techn., Electrotechn. et Energ., **23**, 3–19.
REINER, M., 1958, *Rheology*, Handbuch der Physik Vol. 6, Springer Verlag, Berlin.
REINER, M., 1960, *Deformation, Strain and Flow*. H. C. Lewis. Co., London.
ROZHDESTVENSKIĬ, B. L., N. N. YANENKO, 1968, *Systems of quasilinear equations*, Moscow, (in Russian).
RUBIN, R. J., 1954, J. Appl. Phys., **25**, 528–531.
RUSCH, H., 1960, *Researches toward a general flexural theory for structural concrete*. Proc. Amer. Conc. Inst., **57**, 1–28.
SARMA MARUVADA, P., E. MENEMENLIS, R. MALEWSKI, 1977, AIEE, Trans. PAS, 96, 102–115.
SAVART, F., 1837, Annales de Chimie et de Physique, 2 série, **65**, 337–402.
SHARPE, W. N. JR., 1966, J. Mech. Phys. Solids, **14**, 187–202.
SHARPE, W. N. JR.. 1970, Exper. Mech., **10**, 12, 89–92.
SOKOLOVSKIĬ, V. V., 1948a, Dokl. Akad. Nauk SSSR, **60**, 775–778.
SOKOLOVSKIĬ, V. V., 1948b, Prikl. Mat. Mekh., **12**, 261–280.
SOLOMON, L., 1969, *Elasticitate liniară*, Editura Acad. R. S. Romania, Bucureşti.
STEPHENS, D. R., E. M. LILLEY, H. LOUIS, 1970, Int. J. Rock Mech., Min. Sci., **7**, 257–296.
STERNGLASS, .E J., D. A. STUART, 1953, J. Appl. Mech., **20**, 427–434.
SULICIU, I., 1966, Analele Universităţii Bucureşti, Seria Matematică–Mecanică, **15**, 53–62.
SULICIU, I., 1973, St. Cerc. Mat., **25**, 53–170.
SULICIU, I., 1974a, Mech. Res. Comm., **1**, 101–106.
SULICIU, I., 1974b, Mech. Res. Comm., **1**, 167–172.
SULICIU, I., 1974c, Arch. Mech. Stos., **26**, 675–699.
SULICIU, I., 1975a, Arch. Mech. Stos., **27**, 665–667.
SULICIU, I., 1975b, Arch. Mech. Stos., **27**, 841–856.
SULICIU, I., 1975c, Mech. Res. Comm., **2**, 79–83.
SULICIU, I., 1978, Rev. Roum. Sci. Techn., Electrotechn. et Energ., **23**, 455–465.
SULICIU, I., 1979, *An analogy between the constitutive equations of electric lines and those of one dimensional viscoplasticity*, INCREST Preprint Series in Mathematics No. 32/1979, Institutul de Matematică, Bucureşti.
SULICIU, I., L. E. MALVERN, N. CRISTESCU, 1972, Arch. Mech. Stos., **24**, 999–1011.
SULICIU, I., L. E. MALVERN, N. CRISTESCU, 1973, Mekhanika, **6**, 140–149.
SULICIU, I., L. E. MALVERN, N. CRISTESCU, 1974, Int. J. Solids Structures, **10**, 21–33.
TAYLOR, G. I., 1964, J. Inst. Civil Engrs., **2**, 486–519.
TEODOSIU, C., F. SIDOROFF, 1976, Int. J. Engng. Sci., **14**, 713–723.
THOMSEN, E. G., C. T. YANG, S. KOBAYASHI, 1968, *Mechanics of plastic deformation in metal processing*, Macmillan Co., London.
THURSTON, R. N., 1974, *Waves in Solids*, Handbuch der Physik, Vol. 6a/4, Springer Verlag, Berlin.

Ting, T. C. T., 1968, *On the initial slope of elastic-plastic boundaries in combined longtudinal and torsional wave propagation*, Technical Report 188, Dept. Appl. Mech., Standford Univ.
Ting, T. C. T., P. S. Symonds, 1964, J. Appl. Mech., **31**, Ser. E, 199–207.
Ting, C. S., 1975, *Incremental plastic stress waves in aluminium and copper rods under axial quasistatic preloading*, Philos. Dissertation, Univ. of Florida.
Truesdell, C., 1961, Arch. Rational Mech. Anal., **8**, 263.
Truesdell, C., 1969, *Rational Thermodynamics*, A course of lectures on selected topics, McGraw Hill, New York.
Truesdell, C., 1975, *A first course in Rational Continuum Mechanics*, Mir, Moscow, (in Russian).
Truesdell, C., W. Noll, 1965, *The non-linear field theories of mechanics*, Handbuch der Physik, Vol. III/3, Springer Verlag, Berlin.
Truesdell, C., R. E. Toupin, 1960, *The Classical Field Theories*, Handbuch der Physik, Vol. III/1, Springer Verlag, Berlin.
Valanis, K. C., 1968, J. Math. Phys., **47**, 262–275.
Vasin, R. A., V. S. Lenskiĭ, E. V. Lenskiĭ, 1975, *Dynamic stress-strain relations.* In: Problems of dynamics of elastic-plastic bodies, ed. G. S. Shapiro, Mir, Moscow, 7–38 (in Russian).
Volarovitch, M. P., 1977, *Deformation and strength properties of rocks.* In: High presure and temperature studies of physical properties of rocks and mineral, Naukova Dumka, Kiew, 126–134 (in Russian).
Volarovitch, M. P., E. I. Bayuk, A. I. Levykin, I. C. Tomashevskaya, 1974, *Physical mechanical properties of rocks and mineral at high pressures and temperatures*, Nauka, Moscow, (in Russian).
Voloshenko-Klimovitskiĭ, Yu. Ya., 1965, *Dynamic yield stress*, Nauka, Moscow, (in Russian).
Vyalov, S. S., 1978, *Rheological foundations for soil mechanics*, Vysh. shkola, Moscow, (in Russian).
Wagner, C. F., B. L. Lloyd, 1955, Trans. AIEE, PAS, **74**, part III, 858–872.
Wang, C. C., 1969, Arch. Ration, Mech. Anal., **32**, 1.
Wang, C. C., 1970, Arch. Ration. Mech. Anal., **36**, 166, 198.
Wang, C. C., 1971, Arch. Ration. Mech. Anal., **43**, 392.
Wang, C. C., C. Truesdell, 1973, *Introduction to Rational Elasticity*, Noordhoff Int. Publ., Leyden.
Wood, E. R., A. Phillips, 1967, J. Mech. Phys. Solids, **15**, 241–254.
Zarka, J., 1970, *Sur la viscoplasticité des métaux*, Mém. l'Art. Franc., fasc. 2e.
Yew, C. H., H. A. Richardson, Jr., 1969, Exper. Mechanics, **9**, 366–373.

SUBJECT INDEX

AUTHOR INDEX